T0298682

Fabrication and Applications of Biomass-Derived Porous Carbon

This book systematically introduces the fundamentals, preparation technology, state-of-the-art applications, and future development of biomass-derived porous carbon materials.

The authors provide a theoretical foundation that demonstrates the microstructure and physicochemical properties of carbon materials. The fabrication methods, including physical activation methods, chemical activation methods, and advances in other new fabrication methods, are explicitly described. The book also identifies many potential applications of biomass (especially biomass-derived porous carbon materials), such as supercapacitors, removal of organic pollutants from water, CO_2 capture, photocatalytic application, and farmland restoration.

The book will be a valuable resource for researchers, scientists, and engineers working in the field of biomass-derived porous carbon materials, carbon resource development, and environmental protection.

Kai Yan is a full professor at Sun Yat-sen University. He earned his Ph.D. degree from the Max Planck Institute for Coal Research and RWTH Aachen University in 2011. His current interests are mainly focused on biomass utilization and environmental applications.

Yetao Tang is currently a professor at the School of Environmental Science and Engineering, Sun Yat-sen University. His research interests include remediation of water and soil pollution, ecological remediation of mine pollution, phytoremediation, and biogeochemistry of transition metal elements.

Rongliang Qiu is the current vice president of South China Agricultural University. His current interests include plant–chemical–microbial combined remediation of heavy metal–contaminated soil, chemical remediation, and bioremediation of organic pollution in water and soil environment.

Fabrication and Applications of Biomass-Derived Porous Carbon

Edited by
Kai Yan, Yetao Tang and Rongliang Qiu

CRC Press
Taylor & Francis Group
Boca Raton London New York

CRC Press is an imprint of the
Taylor & Francis Group, an **informa** business

This book is published with financial support from the National Key R&D Program of China (2023YFC3905804), the National Natural Science Foundation of China (41920104003, 22078374, 22378434, and 22309210), the Scientific and Technological Planning Project of Guangzhou (202206010145)

Designed cover image: © Prof. Rongliang Qiu's Team

First edition published 2025
by CRC Press
2385 NW Executive Center Drive, Suite 320, Boca Raton FL 33431

and by CRC Press
4 Park Square, Milton Park, Abingdon, Oxon, OX14 4RN

CRC Press is an imprint of Taylor & Francis Group, LLC

ISBN: 978-1-032-67198-7 (hbk)
ISBN: 978-1-032-86006-0 (pbk)
ISBN: 978-1-003-52056-6 (ebk)

DOI: 10.1201/9781003520566

Typeset in Minion
by Apex CoVantage, LLC

Contents

Contributors

Chapter 1 Fundamentals of Biomass-Derived Porous Carbon

Yuwen Chen
Guangdong Provincial Key Laboratory of Environmental Pollution Control and Remediation Technology
School of Environmental Science and Engineering, Sun Yat-sen University
Guangzhou, China

Ke Li
Guangdong Provincial Key Laboratory of Environmental Pollution Control and Remediation Technology
School of Environmental Science and Engineering, Sun Yat-sen University
Guangzhou, China

Yuchen Wang
Guangdong Provincial Key Laboratory of Environmental Pollution Control and Remediation Technology
School of Environmental Science and Engineering, Sun Yat-sen University
Guangzhou, China

Zhenhao Xu
Guangdong Provincial Key Laboratory of Environmental Pollution Control and Remediation Technology
School of Environmental Science and Engineering, Sun Yat-sen University
Guangzhou, China

Kai Yan
Guangdong Provincial Key Laboratory of Environmental Pollution Control and Remediation Technology
School of Environmental Science and Engineering, Sun Yat-sen University
Guangzhou, China

Yongjian Zeng
Guangdong Provincial Key Laboratory of Environmental Pollution Control and Remediation Technology
School of Environmental Science and Engineering, Sun Yat-sen University
Guangzhou, China

Yutong Zhang
School of Environmental Science and Engineering, Guangdong University of Technology
Guangzhou, China

Ke Zhu
Guangdong Provincial Key Laboratory of Environmental Pollution Control and Remediation Technology
School of Environmental Science and Engineering, Sun Yat-sen University
Guangzhou, China

Chapter 2 Fabrication of Biomass-Derived Porous Carbon

Wanrong Bu
Guangdong Provincial Key Laboratory of Environmental Pollution Control and Remediation Technology
School of Environmental Science and Engineering, Sun Yat-sen University
Guangzhou, China

Ke Li
Guangdong Provincial Key Laboratory of Environmental Pollution Control and Remediation Technology
School of Environmental Science and Engineering, Sun Yat-sen University
Guangzhou, China

Yuchen Wang
Guangdong Provincial Key Laboratory of Environmental Pollution Control and Remediation Technology
School of Environmental Science and Engineering, Sun Yat-sen University
Guangzhou, China

Zhenhao Xu
Guangdong Provincial Key Laboratory of Environmental Pollution Control and Remediation Technology
School of Environmental Science and Engineering, Sun Yat-sen University
Guangzhou, China

Yutong Zhang
School of Environmental Science and Engineering, Guangdong University of Technology
Guangzhou, China

Ke Zhu
Guangdong Provincial Key Laboratory of Environmental Pollution Control and Remediation Technology
School of Environmental Science and Engineering, Sun Yat-sen University
Guangzhou, China

Chapter 3 Biomass-Derived Porous Carbon for Supercapacitors

Prakash Chand
Department of Physics
National Institute of Technology
Kurukshetra, India

Aman Joshi
Department of Physics
J.C. Bose University of Science and Technology, YMCA
Faridabad, Haryana, India

Sunaina Saini
Department of Physics
National Institute of Technology
Kurukshetra, India

Chapter 4 Biomass-Derived Porous Carbon for Removal of Organic Pollutants from Water

Xiaoning Liu
College of Environmental Science and Engineering
Nankai University
Tianjin, China

Xinhua Qi
College of Environmental Science and Engineering
Nankai University
Tianjin, China

Xiaoping Wang
College of Environmental Science and
    Engineering
Nankai University
Tianjin, China

Haiqing Zhang
College of Environmental Science and
    Engineering
Nankai University
Tianjin, China

**Chapter 5 Biomass-Derived Porous
                Carbon for CO$_2$ Capture**
Shuai Deng
Key Laboratory of Efficient
    Utilization of Low and
    Medium Grade Energy
    (Tianjin University)
Ministry of Education
    of China
Tianjin, China

Wenhui Jia
Ministry of Education of Key Laboratory
    of Energy Thermal Conversion and
    Control
School of Energy and Environment
Southeast University
Nanjing, China

Shuangjun Li
Department of Chemical & Biological
    Engineering
Korea University
Seoul, Republic of Korea

Junyao Wang
School of Materials and Energy
Guangdong University of Technology
Guangzhou, China

Xiangzhou Yuan
Ministry of Education of Key
    Laboratory of Energy
    Thermal Conversion
    and Control
School of Energy and
    Environment
Southeast University
Nanjing, China

Huiyan Zhang
Ministry of Education of Key Laboratory
    of Energy Thermal Conversion and
    Control
School of Energy and Environment
Southeast University
Nanjing, China

**Chapter 6 Hydrothermal Carboni-
                zation of Biomasses
                for Photocatalytic
                Application**
Chengyu Duan
School of Environmental Science and
    Engineering
Guangdong Provincial Key Laboratory
    of Environmental Pollution
    Control and Remediation
    Technology
Sun Yat-sen University
Guangzhou, China

Zhuofeng Hu
School of Environmental Science and
    Engineering
Guangdong Provincial Key Laboratory
    of Environmental Pollution
    Control and Remediation
    Technology
Sun Yat-sen University
Guangzhou, China

Huimin Liu
School of Environmental Science and
    Engineering
Guangdong Provincial Key Laboratory of
    Environmental Pollution Control and
    Remediation Technology
Sun Yat-sen University
Guangzhou, China

Yinglong Lu
School of Environmental Science and
    Engineering
Guangdong Provincial Key Laboratory of
    Environmental Pollution Control and
    Remediation Technology
Sun Yat-sen University
Guangzhou, China

Guanghui Luo
School of Environmental Science and
    Engineering
Guangdong Provincial Key Laboratory of
    Environmental Pollution Control and
    Remediation Technology
Sun Yat-sen University
Guangzhou, China

Zheshun Ou
School of Environmental Science and
    Engineering
Guangdong Provincial Key Laboratory of
    Environmental Pollution Control and
    Remediation Technology
Sun Yat-sen University
Guangzhou, China

Mengdi Sun
School of Environmental Science and
    Engineering
Guangdong Provincial Key Laboratory of
    Environmental Pollution Control and
    Remediation Technology

Sun Yat-sen University
Guangzhou, China

Ruilin Wang
School of Environmental Science and
    Engineering
Guangdong Provincial Key Laboratory of
    Environmental Pollution Control and
    Remediation Technology
Sun Yat-sen University
Guangzhou, China

Quan Zhou
School of Environmental Science and
    Engineering
Guangdong Provincial Key Laboratory of
    Environmental Pollution Control and
    Remediation Technology
Sun Yat-sen University
Guangzhou, China

**Chapter 7 Biomass-Derived Porous
    Carbon for Farmland
    Restoration**

Nishu
Guangzhou Institute of Energy Conversion
Chinese Academy of Sciences
CAS Key Laboratory of Renewable Energy
Guangdong Provincial Key Laboratory of
    New and Renewable Energy Research
    and Development
Guangzhou, China

Gaixiu Yang
Guangzhou Institute of Energy Conversion
Chinese Academy of Sciences
CAS Key Laboratory of Renewable Energy
Guangdong Provincial Key Laboratory
    of New and Renewable
    Energy Research and
    Development
Guangzhou, China

Juntao Yang
Guangzhou Institute of Energy Conversion
Chinese Academy of Sciences
CAS Key Laboratory of Renewable Energy
Guangdong Provincial Key Laboratory of
  New and Renewable Energy Research
  and Development
Guangzhou, China

**Chapter 8 Other Potential Applications
  for Biomass-Derived Porous
  Carbon**

Song Yang
College of Chemical Engineering and
  Technology
Taiyuan University of Technology
Taiyuan, China

CHAPTER 1

Fundamentals of Biomass-Derived Porous Carbon

Ke Zhu, Ke Li, Yongjian Zeng, Yuchen Wang, Yuwen Chen, Yutong Zhang, Zhenhao Xu, and Kai Yan*

ABSTRACT

This chapter lays the groundwork for biomass-derived carbon materials, offering an overview that elucidates their significance, synthesis, and applications. It emphasizes the crucial role of precursor selection and pretreatment techniques in shaping the texture, porosity, and chemical composition of the resulting carbon. The chapter delineates the diverse origins of materials used for biomass-derived porous carbon. Additionally, it underscores the correlation between physicochemical properties and the behavior of porous carbon. The introduction of characterization methods aims to enhance understanding of the microstructure and physicochemical attributes of carbon materials. Overall, this chapter provides an introductory exploration of biochar, spanning from basic biomass to intricate and functional carbon structures, offering an engaging journey from biomass to engineered materials.

1.1 INTRODUCTION

Growing environmental concerns and the diminishing availability of non-renewable resources have prompted extensive research into the development of sustainable and innovative materials. A particularly intriguing field is the synthesis of functional carbon materials from biomass, which presents a renewable and abundant resource rich in carbon content. Such materials not only align with green chemistry principles but also open a floodgate of applications due to their tailored functionalities.

In Chapter 1, we lay the groundwork for our exploration of biomass-derived carbon materials, beginning with an overview that sets the stage for understanding their relevance, synthesis, and application. This primer introduces the fundamental concepts that

DOI: 10.1201/9781003520566-1

form the scaffold of subsequent sections, addressing the hierarchies of structure and the nuances of chemical properties that define the utility of the resulting materials.

Porous carbon materials exhibit a constellation of remarkable properties, chiefly influenced by their micro- and macrostructures, which can be intricately designed through controlled fabrication processes using different biomass precursors. Meticulously controlled porosity—one of the essential features of these materials—catapults them into the forefront for applications in gas storage, separation technologies, catalysis, and energy conversion devices. The nuances of surface area, pore volume, and functional group chemistry go hand-in-hand to tailor these carbons for specific functionalities, ranging from selective adsorption of pollutants to facilitating electrochemical reactions.

Diving into the roots of material sources, this overview emphasizes the transformative journey of biomass, rich in cellulose, hemicellulose, lignin, and sometimes proteins and fats, into carbon frameworks. Each step of the process, from the selection of the biomass type to the carbonization and activation methods employed, fundamentally influences the microarchitecture of the final carbon product. In doing so, this chapter underlines the pivotal role of precursor choice and pretreatment strategies in defining the resultant carbon's texture, porosity, and chemical character.

Moreover, the chapter introduces the reader to the interplay between physicochemical properties and the theories that describe porous carbon behavior. The discussion delves into absorption models and the integration of heteroatoms into the carbon matrix, shedding light on how these modifications can significantly augment the material's performance in particular applications. This sets a foundation for understanding how dopants such as oxygen, nitrogen, sulphur, and phosphorus can be strategically introduced to fine-tune the electronic and chemical properties of carbon materials.

Lastly, this overview sets the tone for deeply analytical content to follow, predicated upon comprehensive characterization methods. Each technique, including X-ray Diffraction (XRD), Inductively Coupled Plasma (ICP) analysis, electron microscopy (SEM/TEM), Brunauer–Emmett–Teller (BET) surface area analysis, X-ray Photoelectron Spectroscopy (XPS), and Raman Spectroscopy, provides a unique window into the material's structure and composition. Discussion of these methods prepares readers to engage with the intricacies of how these carbon materials are studied and understood at a fundamental level.

Beginning with this exploratory overview, readers are driven through a meticulous, methodical journey from rudimentary biomass to sophisticated, functional carbon constructs—an exciting traverse from organic precursor to engineered material.

1.2 PRECURSORS

1.2.1 Plant Biomass

Biomass is a renewable natural product produced by plant photosynthesis. It is mainly composed of cellulose (40–50%), hemicellulose (20–35%), lignin (10–35%), and other macromolecular substances[1,2], as shown in Figure 1.1. The world's inedible lignocellulosic materials exceed 40 million tons/year, in the form of straw, trees, fibers, etc. The mineral element distribution and microstructure networks of different plants are different, and

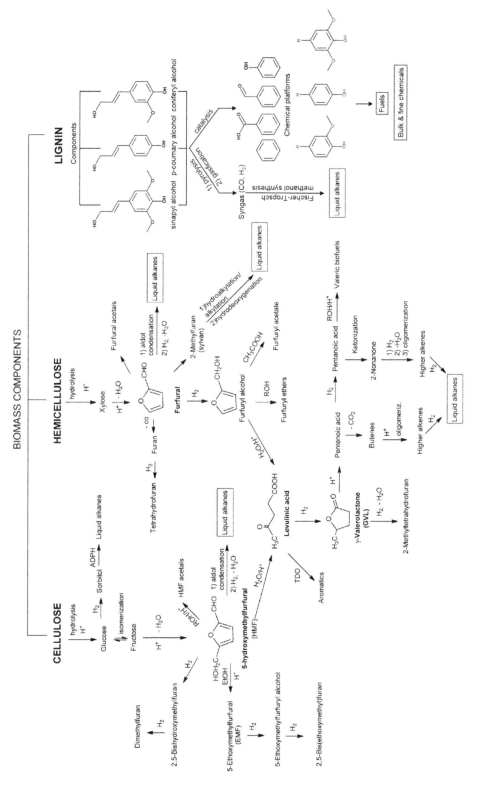

FIGURE 1.1 Classification of plant biomass.

the water and nutrient transport channels are very complex[3]. After the carbonization of these plants, the obtained carbon materials usually retain the structure and channel of the plants[4]. According to different raw materials and activation methods, carbon materials with superior morphological structure can be obtained, which has been paid attention to in industrial production.

1.2.1.1 Cellulose

(i) Introduction   Cellulose stands as the most widely distributed and abundant polysaccharide in nature, constituting over 50% of the carbon content in the plant kingdom[5]. It serves as a ubiquitous macromolecule in plants and forms the principal structural component of the plant cell wall, where it creates a cytoskeletal framework in the form of fiber bundles. Cotton boasts nearly 100% cellulose content in its natural state, making it the purest cellulose source in nature[6]. Conversely, in other wood sources like wheat straw, rice straw, and bagasse, cellulose typically accounts for approximately 40% to 50% of the composition. Notably, cellulose serves as a traditional renewable energy source and finds applications as a raw material in industries such as papermaking, pharmaceutical production, and food processing[7]. Moreover, cellulose serves as a widely utilized biomass precursor for the synthesis of carbon materials.

(ii) Structure and Properties   Cellulose is a complex linear polysaccharide composed of repeated glucose residues connected in an orderly manner through 1,4-glycosidic bonds. The molecular formula can be written as $(C_6H_{10}O_5)n$, and the molecular weight varies from 50000 to 2500000 according to different degrees of polymerization[8]. Natural cellulose is a kind of tasteless white filament that is very stable under normal temperature and pressure. Indeed, the limited flexibility of cellulose stems from the hydrogen bonds and Van der Waals forces among its molecules. Nonetheless, cellulose exhibits exceptional mechanical strength and properties[5]. Cellulose's insolubility in water and resistance to dilute acid, dilute alkali, and organic solvents like ethanol, ether, and acetone are well-known. However, it can dissolve in certain complex solutions like copper ammonia $Cu(NH_3)_4(OH)_2$ and copper ethylenediamine$[NH_2CH_2CH_2NH_2]Cu(OH)_2$. Additionally, under heating conditions, it undergoes hydrolysis when exposed to acid[9]. During the reaction, the oxygen bridge within the cellulose molecule breaks, allowing water molecules to add to the structure. This process leads to the cleavage of the long-chain cellulose molecules, resulting in the formation of shorter-chain molecules. Eventually, these shorter chains break down further to yield glucose molecules[10]. A common method for preparing cellulose-derived carbon is pyrolysis.

(iii) Sources and Applications   At present, the raw materials for cellulose production come from trees, cotton, reeds, straw, and bagasse. However, due to the lack of forest resources in China, 70% of the raw materials for cellulose are derived from non-forest resources[1].

Cellulose, the most abundant natural polymer on Earth, stands as a vital renewable resource for humanity, playing a pivotal role in both daily life and industrial production [11]. Cellulose cannot be degraded and utilized in the human body, but it can adsorb a large amount of water, promote intestinal peristalsis, and reduce adverse stimuli to the intestine [12]. It was called the seventh nutrient in 2013. In the paper industry, lignin in wood raw materials is often removed by alkali solution or sulfate solution cooking, and then the obtained alkali pulp and sulfite pulp are bleached. The resulting slurry can be used for papermaking, mainly for pulp addition and surface sizing[13]. Up to 8 million tons of cellulose are used for papermaking and textiles every year. With the separated and purified cellulose as raw material, rayon can be made. Cellulose and its derivatives are also widely used in electronics, food, architectural coatings, agriculture, oil drilling, and scientific research equipment[6].

1.2.1.2 Hemicellulose

(i) Introduction   Hemicellulose is a renewable natural polysaccharide, and its content in lignocellulose is second only to cellulose[14]. Hemicellulose serves as the primary component of the plant cell wall's primary structure, excluding pectin, and lacks a specific structure[15]. It readily dissolves in alkali solutions and undergoes facile hydrolysis in acidic environments. Typically, a plant harbors multiple hemicelluloses comprised of two or three sugar groups, each exhibiting distinct chemical structures[16]. Moreover, the content and composition of hemicellulose vary across plant components such as stems, branches, roots, and barks. Thus, hemicellulose represents a broad class of substances.

(ii) Structure and Properties   Unlike cellulose, hemicellulose is a heterogeneous polymer comprising various types of monosaccharides, predominantly pentose and hexose sugars, such as D-xylose, D-mannose, D-glucose, D-galactose, L-arabinose, galacturonic acid, and glucuronic acid[14]. These monosaccharide units are linked together via covalent bonds, hydrogen bonds, ether bonds, and ester bonds.[17]. Of these components, xylan makes up 50% of the total content. It binds to the surface of cellulose microfibers, forming connections with each other. Together with other structural proteins, wall enzymes, and pectin, xylan contributes to the formation of a cell wall with specific hardness and elasticity[18].

The hydrophilicity of hemicellulose causes the swelling of the cell wall, which gives the fiber elasticity and makes the fiber tough to cut off[10]. The molecular stiffness and multi-side chain structure of hemicellulose make it less polymerized than cellulose, and the hydrogen bonds between molecules are reduced, so it is easier to hydrolyze.

(iii) Sources and Applications   Hemicellulose is widely present in plants and varies greatly according to plant species, maturity, and morphological parts[19]. For example, the main hemicellulose of coniferous wood is polygalactose grape mannose, while that of broad-leaved wood and grass is polyxylose. The ray cells of needle and hardwood contained more xylan than tracheid and fiber cells. In the middle layer of the secondary wall of the coniferous wood cells, the content of xylans was the lowest, but it was higher in the outer and inner layers of the secondary wall, while the distribution of polygalactose grape mannose was the opposite[20]. The method of extracting hemicellulose from plant fiber is as follows: the plant fiber, alkali, and water are mixed and put into a reactor with a stirring device and a heating system, and the temperature of the reactor is raised to 35 °C ~ 85 °C. At the same time, the mixture is stirred for 10 s ~ 10 min at a speed of 300 rpm ~ 2000 rpm and filtered or centrifuged by conventional methods. The filtrate or supernatant obtained is the extract of hemicellulose[3]. 2 ~ 3 times the volume of 80% ~ 95% ethanol was added to the extraction solution of hemicellulose to precipitate soluble hemicellulose[2]. The soluble hemicellulose was filtered, collected, and dried to obtain hemicellulose.

In the pulping industry, the waste liquid obtained from pulping can be used to make yeast again, and the yeast can be further purified and processed. The obtained product can be used as a flavoring agent and an anti-cancer agent[21]. The hydrolysate obtained in the process of hydrolyzing fiber raw materials can be separated into pentose and hexose components, which can be made into sweeteners, adhesives, and surfactants after treatment. Using biotechnology, alcohol can be produced from renewable plant fiber raw materials, which is of great significance for solving the energy crisis, food shortage, and environmental pollution problems that human beings will face.

1.2.1.3 Lignin

(i) Introduction   Lignin, an amorphous polymer, consists of phenylpropane units linked by carbon-carbon and ether bonds. As one of the principal polymer constituents of lignocellulose and wood, it possesses aromatic properties and a carbon content exceeding 60%[21,3]. Recognized as an exceptionally sustainable precursor for carbonaceous materials, lignin is abundantly available from various sources. Approximately 10% to 20% of its production is sourced from agricultural residues, while 20% to 30% originates from forest biomass materials. With a global annual production ranging from 5 to 6×10^8 tons per year, lignin represents a substantial and sustainable reservoir of bio-based carbon materials.

(ii) Structure and Properties   Lignin, a complex phenolic polymer, exhibits a highly intricate and amorphous disordered structure. Comprising phenylpropanoid units linked by C-C bonds (such as biphenyl and pinoresinol) and ether bonds (such as aryl and phenyl ether), it features a heterogeneous aromatic composition[1]. Primarily derived from three aromatic alcohol precursors—coumarin, coniferyl, and sinapyl alcohol—these alcohols, or monolignols, respectively contribute to lignin units of p-hydroxyphenyl (H), guaiacol (G), and syringyl (S), as depicted in Figure 1.2. Within lignin macromolecules, these monomeric units are interconnected by various carbon–oxygen and carbon–carbon bonds, including β-O-4, 5-5, 4-O-5, and β-β linkages, with β-O-4 being the most prevalent[21], as illustrated in Figure 1.3.

There are aromatic, phenolic hydroxyl, carbonyl conjugated double bonds, and other active groups in the molecular structure of lignin, which can carry out a series of chemical reactions, such as oxidation, reduction, hydrolysis, photolysis, vulcanization, alkylation, halogenation, nitrification, polycondensation, and so on. According to the different extraction processes, plant lignin is mainly divided into three categories: softwood, hardwood, and grass. Among them, cork lignin and grass lignin contain a large number of H units, and the S/G ratios are 1: 2, 2: 1 and 1: 1, respectively. These lignins exhibit thermosplasticity, and the glass transition temperature (Tg) is about 50–150°C within the melt-processable temperature range[21].

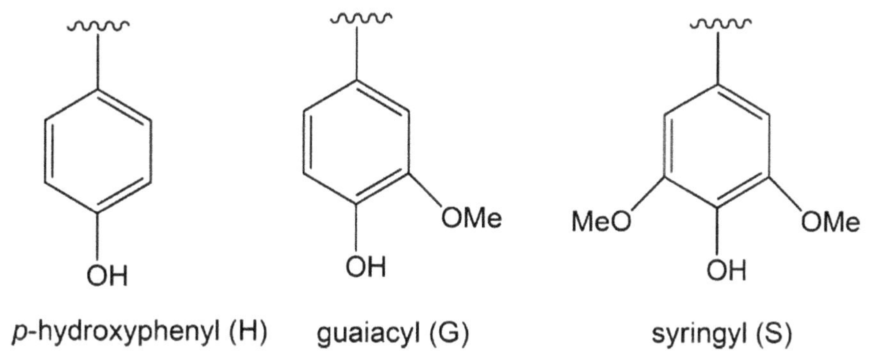

FIGURE 1.2   Structural units of lignin[21].

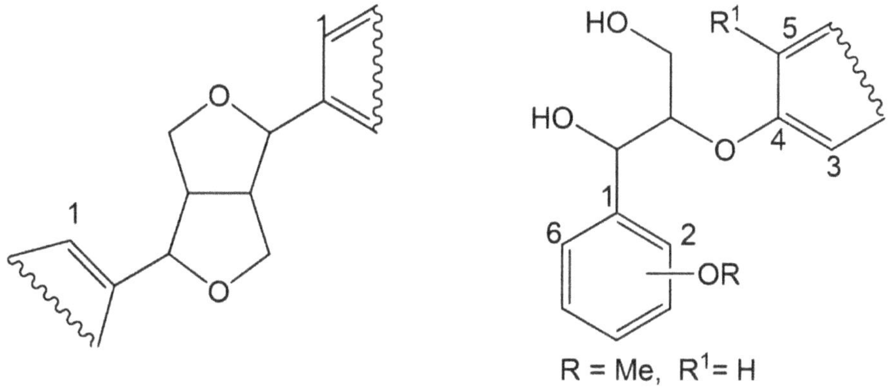

FIGURE 1.3   Common chemical bonds in lignin[21].

(iii) Sources and Applications   Lignin stands as a primary by-product of both the pulp industry and cellulosic ethanol biorefinery processes. Presently, lignin primarily originates from plant extraction through carbonization. However, owing to the intricate structure of plant cell walls and the interplay among various components, the extraction process encounters inherent challenges.

The tightly integrated structure of lignin and carbohydrates within plant materials necessitates chemical-mediated cleavage as the sole method for the separation and purification of lignin. Throughout this separation process, condensation and oxidation reactions introduce considerable complexity. Currently, two primary methods are employed for lignin separation from plants: enzymatic hydrolysis of carbohydrates, leading to the separation of insoluble lignin, and chemical-based separation of lignin from cellulose fibers.

Enzymatic hydrolysis, typically employed for carbohydrate breakdown, cannot be directly applied to lignin and often necessitates raw material pretreatment. Various pretreatment methods are utilized to disrupt the bonds between lignin and carbohydrates in biomass, including mechanical grinding, acid/alkali treatment, and hydrogen peroxide treatment. Given the higher stability of carbohydrate bonds in lignin compared to other bonds, these pretreatment methods not only cleave lignin-carbohydrate bonds but also, to some extent, disrupt certain lignin-lignin ether bonds. Regardless of the specific method used, all pre-treatment techniques reduce the biomass volume and disrupt its physical structure, which facilitates subsequent processing. Based on the chemical reagents employed, chemically derived lignin can be broadly categorized into two groups: (1) sulphur-containing lignin, such as sulfate lignin, and (2) sulphur-free lignin, such as organosolv lignin[21].

As a natural polymer, lignin is widely used in medicine, metallurgy and metal industry, printing and dyeing industry. At present, in the industrial field, lignin is usually modified and applied in the form of macromolecules. The modified lignin and its products have good dispersibility and surface activity, which can be used as a fuel stabilizer, cement grinding aid, insecticide and bactericidal dispersant, clay stabilizer, and so on. Lignosulfonate can also be used as emulsifiers, such as petroleum, asphalt, and wax. According to the rich carbon content and porous structure of lignin, the preparation of biochar materials is also deeply loved by researchers. A variety of carbon materials can be prepared by using lignin as a precursor, such as template carbon and carbon fiber. However, activated carbon is common in adsorption. A high specific surface area (500 ~ 3000 m²/g) can provide suffi-cient adsorption sites for the adsorption and separation of small molecules. In recent years, researchers have reported the preparation of many novel lignin-based carbon nanomateri-als and their high value-added applications, mainly including: (1) exploring the high-value utilization of heteroatom-doped lignin-based carbon materials in the field of energy stor-age; (2) preparing lignin carbon/metal nanocomposites for electrocatalytic energy conver-sion; and (3) exploring the application of lignin carbon/metal oxide nanocomposites in the field of photocatalysis[22].

1.2.1.4 Derived Carbon Materials

(i) Carbon Fiber   Carbon fiber is an ideal catalyst carrier with a carbon content of up to 90%, which is a microcrystalline graphite material obtained from organic fibers after carbonization and graphitization[21]. The molecular structure of carbon fiber is between graphite and diamond, with high stiffness, high flexibility, high fatigue resistance, and high temperature resistance. These properties have made carbon fiber an important part of advanced composites in various industries, including automotive, civil infrastructure, aerospace, and others. The performance of carbon fibers and reinforcements depends largely on the precursor material[23]. For example, carbon fibers are currently produced mainly from polyacrylonitrile and, to a lesser extent, from asphalt, and the high price of these petroleum-based raw materials makes carbon fiber materials expensive. Therefore, the conversion of carbon fibers using lignin, a low-cost precursor material, holds great promise[9].

Two types of lignin carbon fibers were prepared based on the precursor materials: (1) unmodified lignin carbon fibers, where no additives were used; and (2) lignin and addi-tives coextruded carbon fibers.

The process of converting purified lignin into carbon fibers involves three main stages (Scheme 3): (1) spinning carbon fiber precursors, (2) thermally stabilizing the precursors, and (3) carbonizing the stabilized fibers. Depending on the intended use, these carbon fibers may undergo graphitization or activation to achieve specific structural or functional properties. High-purity lignin is essential for producing carbon fibers of high quality. Moreover, it is advantageous for the lignin to possess a narrow molecular weight distribution, which could potentially reduce the need for diverse lignin sources, as discussed earlier. This property helps ensure uniform molecular weight enhancement throughout the fiber during thermal stabilization, resulting in a more homogeneous structure during carbonization. In certain cases, the lignin undergoes vacuum washing and heat treatment for about an hour to remove volatile compounds before the spinning process[24].

(ii) Activated Carbon   Activated carbon, a non-graphitized porous carbon material, possesses a large internal surface area ranging from 500 to 3000 m^2/g. These carbon materials exhibit high thermal and chemical stability, making them suitable for various applications, including gas separation, catalyst carriers, and water filtration systems. Lignin, due to its high carbon content and functionalized phenolic structure, is recognized as a crucial raw material for activated carbon production. The lignin content in the feedstock significantly influences the quantity and quality of micropores in activated carbon. As the primary component in carbon production, feedstocks with higher lignin content yield greater amounts of activated carbon, thereby increasing the total pore volume and BET surface area[16]. The porosity of activated carbon is determined by the type of surface functional groups in the feedstock and the activation method employed. Consequently, modifications to surface functional groups and activation procedures are often implemented to alter the porosity and properties of activated carbon. While lignin activation primarily results in the formation of micropores, cellulose activation yields a mixture of pore sizes, including macropores and micropores. Generally, the pore sizes of activated carbon fall within the range of 1.5 nm to 5 μm[25], as shown in Figure 1.4.

Activated carbon can be classified into two types: powdered activated carbon (PAC) and granular activated carbon (GAC). PAC contains fine particles with diameters of less than 0.2 mm and therefore has a large external surface area and relatively low diffusion resistance. These properties of PAC lead to a very high adsorption capacity. On the other hand, GAC has a large particle size, about 5 mm in diameter, and a small external surface area and adsorption capacity relative to PAC. However, GAC is more stable during continuous contact[15].

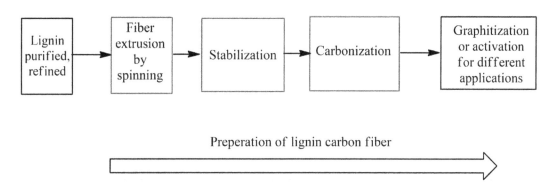

FIGURE 1.4   Preparation of carbon fiber from lignin[25].

Activated charcoal is predominantly produced via physical and chemical activation methods. In both approaches, the amount of char obtained depends on the lignin content of the initial material. Physical activation, also termed gas activation, comprises two successive stages: carbonization and activation. During carbonization, volatile components are expelled at moderate to high temperatures, enriching the carbon and forming carbon-rich char. Subsequent interaction with water vapor, carbon dioxide (CO_2), and other oxidizing gases triggers a water gas reaction, generating small-molecule gases. This reaction expands and opens up pores, creating pores of varying shapes and sizes on both the surface and within the interior of the carbonized material, ultimately yielding activated carbon. Given the limited reactivity of lignocellulosic carbon, the physical activation process requires prolonged activation times, resulting in activated carbon with relatively high microporosity and a wide distribution of micropore sizes[26].

Chemical activation involves a complex pyrolytic reaction between a carbon precursor and an activator in an inert gas, resulting in the formation of pores and ultimately activated carbon. Unlike the physical activation method, the charring and activation processes in the chemical activation method can be combined into a single stage, albeit requiring subsequent treatment to remove the activator[10]. Initially, lignin infiltrates the carbon precursor's organic matter via impregnation with an activating reagent (e.g., $ZnCl_2$), causing a portion of the carbon precursor to react away and generating high surface area and porosity in the resulting carbon. The crucial parameters in chemical activation are the ratio of lignin to impregnating reagents and the activation temperature[27].

(iii) Mesoporous Carbon   The carbon precursor is impregnated in the pores of the templating agent, and the templating agent mesoporous pores are replicated in reverse, and the mesoporous carbon can be obtained by removing the templating agent after the carbon source is carbonized at a high temperature[4]. The key to the template method is the selection of the template agent, which is often categorized into hard template method, soft template method, and double template method. Porous materials are often used as hard or soft templates to control the structure of nanostructured carbon materials. Hard template agents are mainly silica, zeolite molecular sieves, and other silicon-based template agents, as well as some metal salts and metal oxides[18]. Zeolite is an ideal template agent for synthesizing mesoporous carbon because its wall thickness is less than 1 nm. Lignin can not only penetrate into the micropores of zeolite but also be deposited or covered on the outer surface of zeolite (mesoporous wall surface), resulting in layered porous carbon after carbonization and the removal of zeolite templating agent[2]. Compared with the hard template method, the soft template method uses surfactants as template agents, utilizes intermolecular hydrogen bonding and electrostatic interactions to achieve self-assembly, and can eliminate the template agent at high temperatures, making the process easier. However, the soft template method requires higher requirements for raw materials and templates. The relative molecular weight of raw materials directly affects the final pore size, and the high complexity of lignin makes it more difficult to control than small-molecule monomers. The ability of templates to be removed at low temperatures after completing pore modification is also crucial, which determines the complexity of the process. The relative molecular weight of lignin is an important factor affecting the ordered mesoporous carbon to obtain highly ordered channels and high specific surface area. As the relative molecular weight of lignin decreases, the specific surface area and pore volume of mesoporous carbon gradually increase. The dual template method combines the advantages of two processes and proposes the simultaneous use of two template agents in the preparation of mesoporous carbon, providing a comprehensive method for preparing mesoporous carbon[21].

1.2.2 Animal Biomass

Animal biomass refers to organic matter derived from animals, including tissues, bones, fur, and other parts of the animal body. Animal biomass usually contains protein, fat, cholesterol, and other components. The complexity of animal biomass composition exceeds that of plant biomass[28]. Zhou et al.[29] explored nitrogen conversion discrepancies between animal and plant-derived biochar during the composting of papermaking sludge. Their findings revealed that animal-derived biochar exhibited reduced nitrogen loss compared to plant-derived biochar in the control group. Additionally, Chen et al.[30] assessed the impact of pig carcass and woody biochar on soil nutrients, properties, and enzyme activity in a potted cabbage experiment. Their research indicated that pig carcass-derived biochar mitigated the risk of soil chromium pollution.

Animal biomass is a rich biological resource and has wide application value. Exactly. Carbonization is the initial step where carbonaceous materials are subjected to high temperatures in the absence of oxygen, leading to the removal of volatile components and leaving behind a carbon-rich residue. Activation, on the other hand, enhances the porous structure of the carbon material by creating or enlarging pores, typically through processes like chemical activation using activating agents like KOH or physical activation via exposure to high temperatures and steam. This dual-step process is crucial for tailoring the porosity and surface properties of the resulting porous carbon material for various applications[31]. Omotayo et al.[32] employed direct pyrolysis and activation techniques to convert shrimp waste into carbon materials characterized by abundant micropores and mesopores. These carbon materials exhibit effectiveness in removing low concentrations of ciprofloxacin and copper ions from water.

In recent years, biomass carbon electrode materials have attracted extensive attention due to their wide availability, renewability, and low cost. Moreover, porous carbon derived from animal biomass possesses nitrogen-rich characteristics and exhibits superior electrochemical performance compared to pure carbon materials[33]. Yang et al.[34] synthesized nitrogen-doped carbon materials from pig hearts, demonstrating comparable oxygen reduction electrocatalytic activity to commercial Pt/C materials in alkaline solutions. N-doped carbon materials effectively catalyze oxygen reduction reactions, and are also widely used in capacitors. Xie et al.[35] subjected animal bones to calcination in a nitrogen atmosphere at 500 °C, resulting in the production of multi-level porous carbon through ball milling. They successfully incorporated the rare metal selenium into this carbon structure. This material, utilized as a positive electrode in lithium-selenium batteries, exhibits a high specific capacity, with a reversible capacity of 705 mAh/g at 0.1 °C. Qin et al.[36] acquired porous carbon materials from demineralized and calcified livestock teeth. This supercapacitor exhibits an impressive specific capacitance of 131.7 mF/cm^2, surpassing its previously reported counterparts. Moreover, the gel electrolyte retains a fundamental tubular hierarchical structure post-demineralization, facilitating ion transport.

1.3 STRUCTURE

Benefiting from the abundant precursors and various preparation methods, the carbon materials could be designed into diverse structures. According to the quantum confinement in each spatial direction, carbon materials are structurally divided into zero-dimension (0D), one-dimension (1D), two-dimension (2D), and three-dimension (3D), which separately possess three, two, one, and zero dimensions within the nanoscale. Decreasing the extended length of carbon materials in each dimension within the nano range brings about distinct physicochemical properties as compared to bulky materials. For example, the stability, surface and interface properties, mechanical properties, electrical properties, and optical properties of low-dimension carbon materials could be significantly enhanced. Therefore, the performance of carbon materials for diverse applications is largely dependent on their dimensions. This section summarized the representative carbon materials with different dimensional structures, highlighting their properties, applications, and synthesis, especially from biomass precursors.

1.3.1 Zero-Dimensional Structure (0D)

0D carbon materials are small in size, large in specific surface area (SSA), and high in aspect ratio. Combined with the featured quantum size effects, they show superiorities in terms of mobility and conductivity, and thus the utilization potentiality in the fields of electronics and energy. Among various 0D materials, carbon nanodots (CNDs), carbon nanocages (CNCs), and carbon onions (CNOs) have been frequently investigated.

Carbon nanodots (CNDs) are spherical carbon materials with an average size of <10 nm[37]. Differing in structure, CNDs can be classified into graphitic CNDs and amorphous CNDs. The first mentioned of two are crystalline carbon nanoparticles, which are composed of a graphene lattice with no more than ten layers thick (Figure 1.5(a, b))[38]. Benefiting from quantum confinement and edge effects, graphitic CNDs are considered to possess a bandgap[39]. The amorphous CNDs are amorphous carbon nanoparticles made up of both graphite and turbostratic carbon[40]. Notably, graphitic and amorphous CNDs are mainly sp^2- and sp^3-hybridized carbon, respectively. Besides, graphitic CNDs contain more carbon and less oxygen than amorphous CNDs[41]. The advantages of CNDs include flexible modifiability, biocompatibility, low toxicity, decent emission tenability, broad absorption spectra, ultra-photo stability, lasting fluorescent life, high luminescence, large quantum yield, and chemical inertness[42,43], leading to wide applications in drug delivery, bioimaging, chemical sensors, and energy devices[44]. Xu et al.[45] first discovered CNDs and reported their preparation from single-walled carbon nanotubes via gel electrophoresis in 2004. Diverse top-down approaches of chemical ablation or oxidation cracking, electrochemical oxidation and exfoliation, arc-discharge deposition, encompassing laser ablation, plasma treatment, and so on, have been reported for preparing CNDs. For the bottom-up

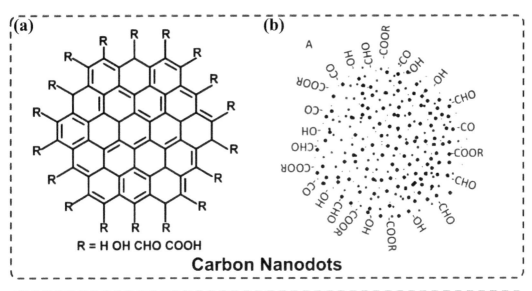

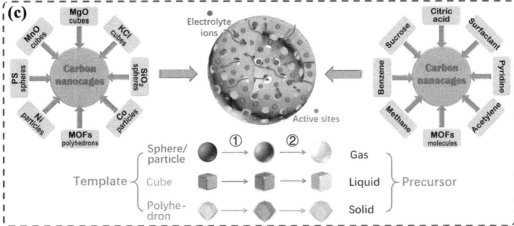

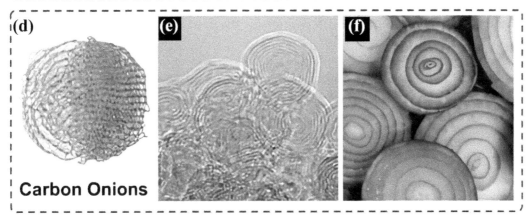

FIGURE 1.5   Examples of structures of (a) graphitic CNDs and (b) amorphous CNDs. (c) Schematic diagram for formation of CNCs[15]. (d) Examples of structures[79]. (e) TEM image and (f) SEM image of CNOs[58].

strategy, biomass-based materials are renewable carbon precursors, which could be transferred into CNDs via the hydrothermal carbonization process and electric or microwave-assisted pyrolysis method[46].

Carbon nanocages (CNCs) consist of a hollow core surrounded by sp^2-hybridized carbon shells in the shapes of quasi-spheres or cubes[47,48]. The carbon shells are enriched with interconnected nanopores (1–10 nm). The unique structure of CNCs brings integrated features of high SAA, large porosity, and flexible modifiability[49]. Additionally, the 0D CNCs can be further assembled into 3D architecture with abundant micro-, meso-, and macropores, providing channels for mass or charge transfer[49,50]. Benefiting from these structural characteristics, CNCs have been applied as promising electrode materials and electrocatalysts in the electronic and energy fields[51–53]. Moreover, CNCs could also serve as nanoscale reactors or containers to encapsulate functional guest materials. The confinement effects of the porous carbon shell open new avenues in size-dependent procedures[54–56]. To realize the efficient synthesis of CNCs with tailored architecture, a variety of template-free and template-based methods have been developed (Figure 1.5(c)). Co, Ni, MnO, MgO, SiO_2, PS, MOFs, and so on are typical templates. The precursor for carbon shells varied from solid (surfactant, biomass, MOFs), and liquid (pyridine) to gas phase (benzene vapor, CH_4, CO), as determined by the synthetic techniques[49].

Carbon onions (CNOs) contain multilayer fullerene-like carbon shells, in which smaller ones are nested with larger ones, forming a specific onion-like structure. The carbon atoms in CNOs interact with each other via one double bond and two single bonds. The distance between two carbon layers (3.335 Å) is comparable with that between two graphene (3.334 Å)[57], while the size of CNOs is within the range of 2 to 100 nm[46]. Typical features of carbon onions are high SAA, low density, well-organized carbon arrangement, high conductivity, and relatively inert surface with limited functionalities (Figure 1.5(d–f)), which allows for application in the fields of physical, biological, medical, and chemical[58]. CNOs were first reported by Lijima in 1980[59]. Up to now, diverse synthetic methods have been developed, among which, thermal annealing, pyrolysis, arc-discharge, electron-beam irradiation, ion implantation, and chemical vapor deposition are the most widely applied[57].

1.3.2 One-Dimensional Structure (1D)

1D carbon materials show superiority in terms of aspect ratio, SSA, conductivity, and flexibility. Hence, they find extensive applications in electronics and energy sectors. Besides, 1D carbon materials are also typical supports for nano-sized metal species. The shape of 1D carbon materials mainly includes nanotubes and nanofibers.

Carbon nanotubes (CNTs) are tubular carbon materials with nanoscale diameter and millimeter-sized length[60–62]. According to the number of cylindroid graphene layers, CNTs could be classified into single-wall (SWNTs) and multiwall nanotubes (MWNTs). SWNTs contain one layer of graphene, whose diameter is about 1.4 nm. MWNTs consist of multilayer graphene cylindroids (4–24). Their diameter is 10–20 nm, and the distance between them is 0.34 nm (Figure 1.6(a–c)). Benefit from the unique tubular structure, CNTs show outstanding SAA, thermal conductivity, electrical properties, and so on and thus have been widely used in the fields of electronics[63], biosensors[64], bio-medicals[65], air

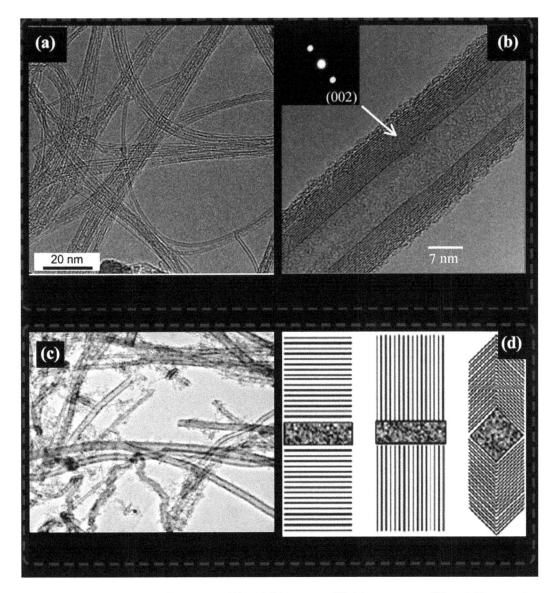

FIGURE 1.6   TEM image of (a) SWNT[60] and (b) MWNTs[61]. (c) TEM image[62] and (d) examples of structures of CNFs[73].

and water purification[66,67], and environmental remediation[68]. The most popular methods for obtaining CNTs are chemical vapor deposition[69], arc discharge[70], and pyrolysis [71]. Popular biomass carbon precursors include grass, cotton fiber, and potato peels[72].

Carbon nanofibers (CNFs) are 1D carbon materials with cylindrical structures, whose diameters and lengths range from a few to hundred nanometers and micrometers to millimeters, respectively. CNFs are made up of graphene, which is stacked and curved into the 1D structure. Based on the different inclination angles of graphene layers relative to the fiber axis, the morphology of CNFs could be classified into plate, ribbon, and herringbone (Figure 1.6(d))[73]. CNFs feature nanosized diameter, a high ratio between surface area and volume, and strong mechanical capacity, leading to great utilization potential in the fields of materials synthesis, chemical industry, and energy storage[74]. CNFs were first reported over 120 years ago and were obtained in a metallic crucible under a carbon-containing gas atmosphere[75]. Nowadays, CNFs can be prepared via chemical vapor dispersion[76] and carbonization[77]. The electrospinning technique and the templating routes are generally used to obtain the filamentous carbon precursors for carbonization[78]. Many biomasses with fibrous shapes (silk fibers, cotton fibers, and wool fibers) are thus suitable templates[28]. Notably, since their sizes are on the micrometer scale, biomass as raw materials is always treated by electrospinning technology to produce nanoscale CNFs[79].

1.3.3 Two-Dimensional Structure (2D)

2D carbon materials are known for their large SSA, abundant surface edge, high conductivity, and rich defect sites, and thus are explored for promising utilizations in the field of energy storage. 2D carbon materials generally possess layer, lamellar, and ribbon-like morphology.

Carbon nanosheets as the typical 2D carbon materials, are made up of monolayer carbon or stacked multilayer carbon (< 10). Based on the structure of carbon, carbon nanosheets can be divided into graphene and graphene-like carbon. Graphene is a planar sheet with a single-atom thickness. The carbon atoms in graphene are sp^2-hybridized and are arrayed into a honeycomb crystal lattice via σ bonds (Figure 1.7). Graphene shows large SAA, strong mechanical stability as well as high Young's modulus and thermal conductivity[80]. The π orbitals perpendicular to the plane of graphene allow for the transfer of electrons between graphene or between graphene and substrate, leading to high carrier mobility[74]. Benefiting from these remarkable properties, graphene is widely applied in the fields of physics, catalysis, supercapacitors, and so on[81]. Mechanical exfoliation is the most famous method for obtaining graphene from graphite, as proposed by Andre Geim and Kostya Novoselov in 2004. Other typical methods include chemical exfoliation, epitaxial growth, and so on. Carbonization of biomass is also an effective approach for preparing graphene, in which sp^3-hybridized carbon is transformed to aromatic sp^2-hybridized carbon at a high temperature to achieve the layered microstructures[72]. However, it is difficult to avoid the formation of amorphous carbon, namely graphene-like materials. The disordered arrangement of carbon leads to the existence of vacancies, sp^3-hybridized carbon, oxygenated functions, etc.

1.3.4 Three-Dimensional Structure (3D)

3D carbon materials are mainly constructed from graphene, CNTs, or CNFs, forming versatile architectures of microsphere[82,] hollow fiber[83], and sponge (Figure 1.8)[84]. In addition to the high conductivity inherited from graphene and CNTs, 3D carbon materials also possess large SSA, low density, good interconnectivity, controlled porous structures, and mechanical and chemical stability[85,86]. These advantages could improve their performance in the environment[86], electronics[87], and energy-related applications[88]. Normally, the synthesis of 3D carbon materials could be realized through a solution-based self-assembly approach[89,90], hydrothermal method[91,92], chemical vapor deposition[93,94], pyrolysis[95,96],

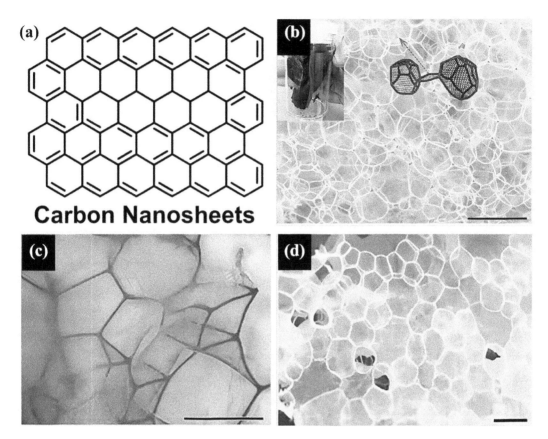

FIGURE 1.7 (a) Examples of structures. (b, c) optical photo and (d) SEM images of carbon nanosheets[81].

etc. Pyrolysis features procedure simplicity and cost-effectiveness, which call for carbon-rich raw materials of polymers, MOFs, and biomass[88]. Notably, the pyrolysis of biomass-based precursors usually requires extra processes, like freeze-drying[97].

1.4 PHYSICOCHEMICAL PROPERTIES AND RELATED THEORIES

1.4.1 Porosity

Porosity stands as one of the pivotal characteristics of porous carbon. A detailed examination and analysis of the pores within porous carbon aid in the more effective utilization and modulation of its properties. Categorized by pore size, pores measuring less than 2 nm are commonly referred to as micropores; those ranging from 2 nm to 50 nm are termed mesopores; and pores exceeding 50 nm in size are classified as macropores.

To investigate the pore properties of activated carbon, we can use adsorption as a measurement method. Physical adsorption offers a method to assess the material's surface area, average pore size, and distribution of pore sizes.

Adsorption refers to the process where molecules of a substance stick to the surface of a solid or liquid material, called the adsorbent. This occurs as the adsorbate (the substance being adsorbed) forms weak bonds with the surface of the adsorbent. It's a surface-based

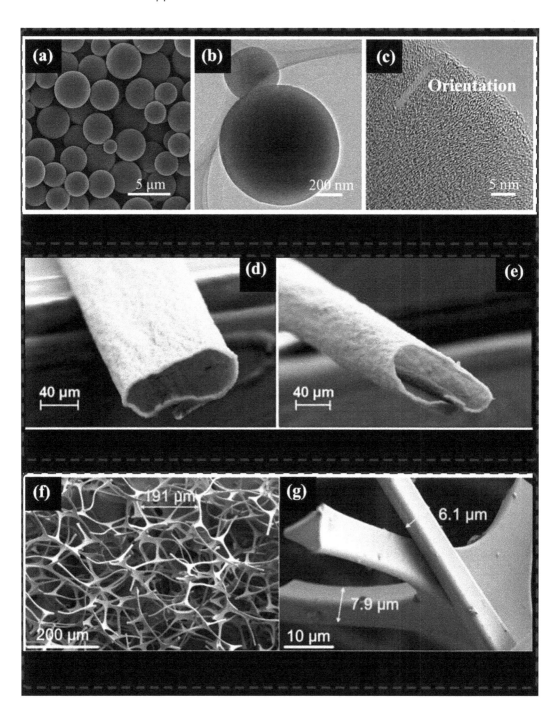

FIGURE 1.8   (a) SEM image and (b, c) TEM images of carbon microspheres[82]. (d, e) SEM images of carbon hollow fibers[83]. (f, g) SEM images of carbon sponges[84].

phenomenon that can take place on various materials like activated carbon, silica gel, and metal oxides. Adsorption can be categorized into physical and chemical types, depending on the forces between the adsorbate and the solid surface. Physical adsorption involves Van der Waals forces, while chemical adsorption entails the formation of chemical bonds.

Van der Waals forces are typically weaker than chemical bonds, with interaction energies ranging from 10^{-1} to 10 kJ/mol, which is 2–3 orders of magnitude lower than that of ionic or covalent bonds (10^2–10^3 kJ/mol)[98].

Porous carbon usually contains numerous micropores, characterized by their small size, which is only slightly larger than that of the adsorbed molecules. This proximity of size allows the adsorbed molecules to be closely surrounded by the solid material, intensifying the adsorption potential within the micropores. Consequently, micropores exhibit a robust trapping capability for adsorbed molecules, particularly at low relative pressures. This adsorption mechanism is known as micropore filling (Figure 1.9)[99].

The adsorption capacity is the most important indicator for measuring the adsorption, and it is often expressed as the quantity of the adsorbate adsorbed per unit mass of adsorbent (mass, volume, amount of substance, etc.). Many factors affect adsorption, including the properties of the adsorbate and adsorbent themselves, temperature, equilibrium concentration, and more. When the adsorbate and adsorbent remain unchanged and the system is at a constant temperature, we can study the relationship between adsorption capacity and equilibrium pressure. This relationship curve is called an adsorption isotherm. For gas adsorption, the isotherm is often expressed as the relationship between the adsorption capacity and the relative pressure of the gas, $V = f(p/p_0)$, where V is the adsorption capacity, p is the equilibrium pressure of the gas adsorption, and p_0 is the saturated vapor pressure of the gas at the adsorption temperature (which is maintained below the critical temperature of the gas). The actual adsorption isotherm is very complex, and the International Union of Pure and Applied Chemistry (IUPAC) has identified six types

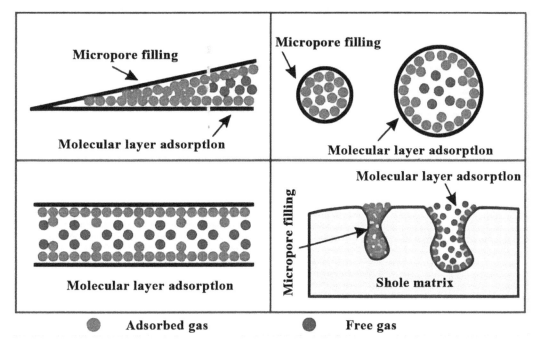

FIGURE 1.9   Schematic of micropore filling[100].

based on extensive experimental results from researchers[101]. Among them, Types I and IV are further divided into two subtypes, as shown in Figure 1.10.

Researchers have developed many mathematical models to describe adsorption isotherms. The widely used models include Langmuir and BET equations, BJH method, and D-R equation.

1.4.1.1 Langmuir Adsorption Theory

The Langmuir adsorption model is suitable for both chemisorption and physisorption and is thus most commonly applied in heterogeneous catalytic studies. It is based on the theory

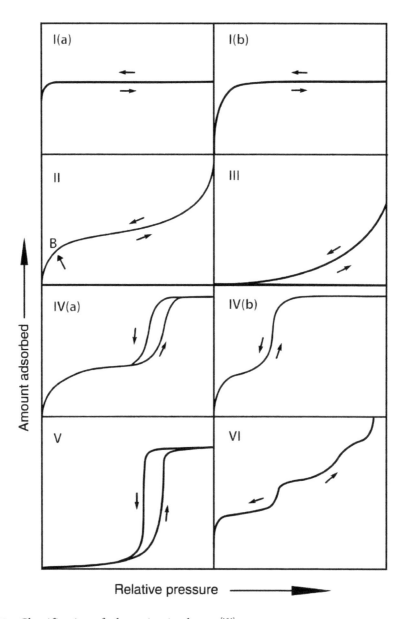

FIGURE 1.10   Classification of adsorption isotherms[101].

of monolayer adsorption, which means that only a single layer of gas molecules can be adsorbed on the surface of the adsorbent[102–104].

The assumption of Langmuir adsorption model:

i. Adsorption sites are present on the surface of the adsorbent, and gas molecules can only be adsorbed in a monolayer on these sites.

ii. The adsorption sites exhibit homogeneity both thermodynamically and kinetically, indicating that the adsorbent surface is uniform and the heat of adsorption remains constant regardless of the site occupancy.

iii. There are no interactions between adsorbed molecules and no lateral interactions.

iv. The adsorption-desorption process achieves kinetic equilibrium, where the desorption rate equals the adsorption rate once adsorption equilibrium is reached.

v. The rate of adsorption of gas molecules onto an adsorbent is directly proportional to the partial pressure of the gas phase for that component.

The rate of gas desorption is directly proportional to the adsorbent surface coverage (θ), whereas the rate of adsorption of gas molecules onto the adsorbent is proportional to the remaining adsorption area ($1 - \theta$) and the partial pressure of the gas phase (p). The rate of adsorption is equal to the rate of desorption when the adsorption develops to equilibrium, which can be expressed by the equation:

$$k_a \left(1-\theta\right) p = k_d \theta \tag{1}$$

where k_a is the adsorption rate constant and k_d is the desorption rate constant.

Setting $\alpha = k_a/k_d$ infers the Langmuir adsorption model:

$$\theta = \frac{V}{V_m} = \frac{\alpha p}{1+\alpha p} \tag{2}$$

where θ is the adsorbent surface coverage, V is the adsorbed amount, V_m denotes the monolayer adsorption capacity, p is the pressure at which the adsorbent vapor adsorption equilibrium, α is the adsorption equilibrium constant, which is related to the characteristics of the adsorbent and adsorbate, as well as temperature.

The Langmuir adsorption isotherm is reformulated into a linear form:

$$\frac{p}{V} = \frac{p}{V_m} + \frac{1}{\alpha V_m} \tag{3}$$

A graph with p/V and p as coordinates leads to a straight line with slope of $1/V_m$ and intercept of $1/\alpha V_m$, which can be used to find the monolayer adsorption capacity V_m and the adsorption equilibrium constant α.

It should be noted that the Langmuir adsorption isotherm is a type I isotherm. At very low pressure or weak adsorption, the adsorption amount is proportional to the equilibrium pressure. At higher pressure or strong adsorption, θ approximately equal to 1, and the adsorption amount is the capacity of monolayer adsorption, independent of the pressure. When the partial pressure of the adsorbate in the gas phase is too high, multilayer adsorption occurs on the surface of the adsorbent, resulting in a significant increase in the adsorption amount, which is not consistent with the assumptions of the Langmuir adsorption model. Therefore, Langmuir adsorption model is only suitable for describing adsorption isotherms in the low and medium pressure ranges.

1.4.1.2 Brunauer–Emmett–Teller (BET) Theory

Different from the Langmuir adsorption theory, S. Brunauer, P. H. Emmett, and E. Teller proposed a multimolecular layer adsorption theory in 1938 and deduced the BET equation[105]. BET adsorption theory suggests that the adsorption process depends on Van der Waals forces. The existence of interaction forces between adsorbed molecules resulted in the formation of multilayer adsorption of adsorbate molecules on the surface of the adsorbent. Therefore, the BET adsorption isotherm applies to physical adsorption.

The BET model retained the assumptions of the Langmuir model that the heat of adsorption was independent of surface coverage and that there were no interactions between the adsorbed molecules, and added:

i. The adsorption may be multi-layered, and the adsorbed molecules may not necessarily fill a single layer before being paved for other layers.

ii. In the first layer of adsorption, the gas molecules interacted directly with the solid surface and the adsorption heat (E1) was different from that of the subsequent layers of adsorption.

According to this model, an infinite multilayer adsorption BET equation was obtained:

$$\theta = \frac{V}{Vm} = \frac{C \times p}{(p_0 - p)\left[1 + (C-1)\dfrac{p}{p_0}\right]} \tag{4}$$

In the equation, V denotes the adsorbed amount, p signifies the gas adsorption equilibrium pressure, and p^0 represents the saturated vapor pressure of the gas at the adsorption temperature. Vm indicates the amount of adsorption when the solid surface is entirely covered with a monomolecular layer, while C is a constant associated with the heat of adsorption.

Transform Eq. 4 into a linear form:

$$\frac{p}{V(p_0 - p)} = \frac{C-1}{V_m C} \times \frac{p}{p_0} + \frac{1}{V_m C} \tag{5}$$

Plotting with $\dfrac{p}{V(p_0 - p)}$ and $\dfrac{p}{p_0}$ as coordinates, a straight line should be obtained with the slope of $\dfrac{C-1}{V_m C}$ and the intercept of $\dfrac{1}{V_m C}$, thus figuring out V_m and C.

The number of molecules covering the surface of the adsorbent in a single molecular layer could be calculated from V_m. If the cross-sectional area of each molecule was known, the total surface area and specific surface area of the adsorbent could be figured out.

$$S = \frac{V_m}{22400} N_A \sigma_m \tag{6}$$

where S is the total surface area of the adsorbent, σ_m is the cross-sectional area of the adsorbate molecules, and N_A is the Avogadro constant.

The BET equation has a wide applicability and can describe many types of adsorption isotherms. Limited by the multilayer physical adsorption modelling, it incurs large errors at very low or very high adsorbent partial pressures. When the relative pressure is too small, multilayer physical adsorption cannot be formed. When the relative pressure is too large, the capillary coalescence phenomenon destroys the multilayer physical adsorption.

1.4.1.3 Capillary Coalescence Phenomenon and BJH Method

In fact, adsorbents, such as biocarbons, not only have irregular surface shapes, but many also have multiple pore structures. The size, shape, and number of holes greatly affect the results of the specific surface area measurement. For example, the adsorption isotherm of an adsorbent with a mesoporous structure will show a hysteresis loop. This means that the adsorption curve does not coincide with the desorption curve within a certain relative pressure range, and the separation forms a loop. Langmuir theory and BET theory based on ideal assumptions fail to explain it.

In 1911, R. A. Zsigmondy applied the Kelvin formula (Eq. 7) to porous solids and proposed the theory of capillary coalescence, which explained the hysteresis loop phenomenon of adsorption isotherms in porous solids[106,107]. The theory of capillary coalescence suggests that if a concave liquid surface is formed in a capillary pore by adsorption, the vapor pressure on the concave surface must be less than the saturated vapor pressure on the flat surface. When the pore diameter decreases, the curvature radius of the concave liquid surface also decreases, causing a reduction in the equilibrium vapor pressure. As the relative pressure (p/p_0) increases, the pore radius capable of condensation also increases, denoted by a p/p_0 ratio corresponding to a critical pore radius (R_k). Pores with a radius smaller than R_k undergo capillary condensation, while those larger do not. During desorption, the liquid surface curvature post-capillary coalescence is always smaller, resulting in desorption pressure consistently being lower than adsorption pressure at the same adsorption level, creating a hysteresis loop.

$$\ln \frac{p}{p_0} = -\frac{2\gamma M}{RT\rho} \times \frac{1}{r} \tag{7}$$

where γ represents the surface tension of the adsorbent liquid, M represents the molar mass of the adsorbent, ρ represents the density of the adsorbent liquid, and r is the radius of the capillary pore corresponding to the p/p_0.

The pore volume and pore size distribution significantly affect the adsorption properties, catalytic performance, and stability of materials. Therefore, investigating the pore structure, size, and distribution is crucial. In 1951, E. P. Barett, L. G. Joyner, and P. P. Halenda introduced the BJH method. This technique calculates the pore size distribution of mesoporous materials using the Kelvin formula and a theoretical model of equivalent capillary coalescence for cylindrical pores closed at one end[108].

The BJH method assumed that the thickness (t) of the adsorbed layer in the capillary is only dependent on the relative pressure (p/p_0) and not on the properties of the adsorbate or the pore radius. In the process of adsorption, the adsorption amount (V) corresponding to the relative pressure not only includes the capillary condensation amount of all pores with radius (r_i) smaller than the capillary condensation but also includes the adsorption amount of liquid film formation on the pore wall with a radius larger than r_i. As the relative pressure (p/p_0) increases, the liquid film on the pore wall continues to thicken, while capillary coalescence moves toward the microporous. The equation is:

$$V_{ads}\left(x_k\right) = \sum_{i=1}^{k}\Delta V_i\left[r_i \leq r_k\left(x_k\right)\right] + \sum_{j=k+1}^{n}\Delta S_j t_j\left[r_j \geq r_k\left(x_k\right)\right] \tag{8}$$

where $V_{ads}(x_k)$ is the total adsorbed amount at $x_k = p_k/p_0$, r_k is the Kelvin radius where capillary coalescence occurs, ΔV_i is the adsorption amount in the capillary with radius r_i ($r_i \leq r_k$), ΔS_j is the surface area of the wall of the capillary with radius r_j ($r_j \geq r_k$), t_j is the thickness of the liquid film on the wall of the capillary, and $\Delta S_j t_j$ is the adsorbed amount corresponding to the liquid film on the wall of the capillary with radius r_j.

1.4.1.4 The Theory of Bulk Filling of Micropores

Contrary to the mesoporous, the micropores are too small in size to allow the adsorbed gases to form a liquid phase therein or to have a gas-liquid interface. Therefore, the theory of capillary coalescence is not applicable.

In 1947, M. M. Dubinin and L. V. Radushkevic proposed the Theory of Bulk Filling of Micropores (TBFM) and the D-R equation (Eq. 9) to describe the adsorption process on microporous adsorbents[109,110]. The theory suggested that, due to the increased adsorption energy of the micropores, they could be filled with adsorbate molecules at very low relative pressures. When the micropores are filled, the inner surface loses the ability to continue to adsorb molecules, and the adsorption capacity decreases dramatically, resulting in a plateau of isothermal adsorption lines, which are consistent with type I isotherms. The saturated adsorption value is equal to the volume of the micropores.

$$\ln V = \ln V_0 - k\left(\ln \frac{p_0}{p}\right)^2 \tag{9}$$

where V is the adsorbed amount, V_0 is the microporous saturated adsorption amount, and k is a constant that varies with adsorbent, adsorbate, and temperature.

According to the D-R equation, the microporous volume can be found with a known adsorption volume (V) and relative pressure (p/p_0). This equation is only applicable to the measurement of the microporous volume of a homogeneous pure microporous adsorbent under low relative pressure conditions.

Most adsorbents bear the combinations of microporous, mesoporous, and microporous surfaces and a significant proportion of external surfaces. To describe the inhomogeneous microporous system, M. M. Dubinin and V. A. Astakhov replaced the second power in the exponent in the D-R equation with the mth power to obtain the D-A equation[111]:

$$\ln V = \ln V_0 - k\left(\ln \frac{p_0}{p}\right)^m \tag{10}$$

In addition to specific theoretical models, several empirical methods for measuring the surface area of adsorbents have been developed, for example, the t-plot Method and the α_s-plot Method[112,113]. These two empirical graphing methods enable us to differentiate between microporous adsorption, mesoporous adsorption and capillary coalescence phenomena, and further calculate the surface area, microporous and mesoporous volumes.

1.4.2 Dopants

Surface properties play a crucial role in determining the performance of carbon materials. Heteroatom doping is a prevalent method used to modify these properties, involving the introduction of elements such as nitrogen, phosphorus, and sulphur onto the carbon surface or within its structure. Typically, there are two doping approaches: artificial and self-doping. Artificial doping involves treating biomass with chemical agents like urea, melamine, thiourea, or phosphoric acid during carbonization or activation processes. On the other hand, self-doping entails the direct carbonization or activation of biomass already containing heteroatoms[114].

Nitrogen doping is widely utilized in the fabrication of biomass-derived porous carbon materials, primarily due to the ease with which N atoms can be integrated into carbon matrices. This is owing to their higher electronegativity and close atomic radius to carbon[115]. The incorporation of nitrogen serves to adjust the pore structure of biochar, enrich nitrogen-containing functional groups within carbon materials[116], modify the charge and spin density of carbon atoms, introduce carbon defects, and foster the creation of additional catalytically active sites[117]. Common nitrogen-containing groups encompass pyridine-N, pyrrole-N, quaternary-N, and pyridine oxide-N[117], whose structures are shown in Figure 1.11.

NH_3 is often used as a nitrogen-containing gas to replace artificially doped nitrogen in the preparation of biochar. When biochar is treated with NH_3, NH_3 is decomposed into ·NH_2, ·NH, hydrogen atoms, N atoms, etc., at 500 °C[119]. The generated free radicals react with the surface active groups of carbon materials to form pyridine-N and pyrrole-N. Then, some of the generated pyridine-N is converted into quaternary-N by polymerization[120].

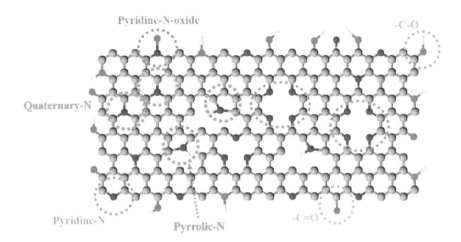

FIGURE 1.11   Types of N-doping functional groups on biochar[118].

Yang et al.[121] synthesized nitrogen-doped carbon by ammonia oxidation method using coconut shells as raw material. The prepared biochar showed high micropores and nitrogen content, which enhanced the adsorption performance of CO_2 (4.26 mmol/g). The temperature and residence time for heat treatment of biomass have a great influence on the nitrogen content and specific surface area of the prepared nitrogen-doped biochar. The specific surface area increases with the increase in ammonia modification temperature[122]. Luo et al.[123] also reported similar results and explored the preparation temperature and time of the optimal nitrogen content (10.3%).

Nitrogen-containing chemicals such as urea and melamine were also used to impregnate biochar and were then heat-treated to prepare nitrogen-doped biochar. The NH_3 produced by the thermal decomposition of urea and melamine can react with the edge sites in carbonaceous materials[124]. Chang et al.[125] prepared a nitrogen-doped porous carbon material with abundant micropores and a high specific surface area (2351 m^2/g) by co-pyrolysis of urea and Firmiana simplex fluff. Seredych et al.[126] treated wood with melamine and urea and carbonized it at 950 °C to prepare biochar with good electrochemical properties. Liu et al.[127] used rice husk as a biomass precursor and polyethyleneimine as a nitrogen source to prepare nitrogen-doped porous carbon with high adsorption capacity for CO_2 (1.90 mmol/g). However, both the ammonia-heated biochar and the doping method of nitrogen-containing chemicals impregnated original biochar will damage the pore structure of biochar to a certain extent.

The self-doping technique involves utilizing the pyrolysis or hydrothermal carbonization of nitrogen-enriched waste biomass to achieve the in-situ production of nitrogen-doped carbon. Certain waste biomasses rich in nitrogen content can serve as suitable precursors for the fabrication of nitrogen-doped carbon. For example, microalgae with rich nitrogen content (13.94%) can be pyrolyzed to obtain biochar containing pyrrolic-N, pyridinic-N, and quaternary-N[128]. Kante et al.[129] successfully prepared biochar containing nitrogen (3.5%) from spent ground coffee. The nitrogen in the prepared biochar promoted the catalytic activity for H_2S oxidation. The temperature at which nitrogen-containing biomass undergoes pyrolysis significantly affects the nitrogen content and species found in the resultant biochar. Elevated pyrolysis temperatures tend to decrease the nitrogen content due to the decomposition of nitrogen-containing organic components and functional groups[114].

Phosphorus is also often used to dope biomass carbon materials. It belongs to the same group of elements as nitrogen, and has a similar structure of electron arrangement. According to the different precursors prepared, the main ways of phosphorus doping can be divided into three types. The first is the preparation of biomass carbon precursor doping, mixing with phosphorus compounds, or chemical bonding. The second is the modification of prefabricated carbon materials with phosphorus-containing compounds. The third is the co-deposition of carbon and phosphorus in the gas phase[130–132].

The primary doping method frequently employed involves carbonizing carbon-containing precursors in conjunction with phosphoric acid, known as phosphoric acid activation. Phosphoric acid serves as an acid catalyst, facilitating bond cleavage and crosslinking through cyclization and condensation reactions. Additionally, it can create bridges, including phosphoric acid and polyphosphoric acid, by combining with organic substances, thus linking and crosslinking biopolymer fragments[133]. The incorporation of phosphate groups initiates the expansion process, expanding the matrix. Upon acid removal, the matrix adopts an expansive state with an accessible pore structure, essentially constituting the activation process. At temperatures exceeding 450 °C, the phosphate bond's thermal instability triggers secondary shrinkage. This reduction in crosslinking density facilitates the growth and arrangement of polyaromatic clusters, leading to a denser structure with reduced porosity. [134]. Additional phosphorus-containing substances, such as pyrophosphate, polyphosphoric acid, disodium hydrogen phosphate, and ammonium dihydrogen phosphate, can also serve for the activation or modification of biomass carbon[135,136]. Nahil et al.[137]. The biomass porous carbon materials were prepared by activating cotton stalks with phosphoric acid at various temperatures ranging from 500 to 800 °C. Experimental findings indicate that both the impregnation ratio and activation temperature influence the yield and porous structure of the carbon materials. Specifically, higher activation temperatures coupled with increased impregnation ratios tend to facilitate the development of mesopores.

The second doping method involves modifying pre-formed carbon with phosphorus compounds, typically at elevated temperatures. Various compounds can be used for this modification, including phosphoric acid, phosphate, phosphite, and methylphosphonic acid, PCl_3, $POCl_3$, phytic acid, ammonium hexafluorophosphate, ionic liquid, red phosphorus, or white phosphorus[138,139]. Park et al.[139]. used ammonium hexafluorophosphate to modify the surface of graphite, which showed excellent electrochemical performance in lithium-ion battery anode. This improvement can be attributed to the introduction of phosphorus on the surface of graphite, which made the surface more stable.

The third method is to prepare phosphorus-containing carbon materials by vapor deposition. The commonly used mixed gases are CH_4 and PCl_3, toluene and triphenylphosphine, pyridine and triphenylphosphine, or trichlorophosphazene, which are doped as phosphorus sources[140,141]. Liu et al.[142]. pyrolyzed toluene and triphenylphosphine at 1000 °C to prepare phosphorus-doped graphite carbon materials. This phosphorus-doped carbon material exhibited high electrocatalytic activity and stability in the oxygen reduction reaction. This metal-free catalyst can not only be used as a simple and effective alternative to platinum and platinum-based catalysts for ORR but also determine that the ORR electrocatalytic activity of heteroatom-doped carbon catalysts is caused by their unique electronic properties.

1.5 CHARACTERIZATION METHODS

In recent years, there has been significant interest in the low-cost and readily available recycling of biomass. The resulting biomass-derived porous carbon (BDPC) possesses unique properties, including a large specific surface area, well-developed pore structure, resistance to acid and alkali, good conductivity, and adjustable pore size. As a result, BDPC has found wide-ranging applications as a novel material in various fields[143–146]. The following section offers an extensive overview of various characterization techniques, including XRD, XPS, SEM, TEM, Raman spectroscopy, BET measurements, and ICP, specifically tailored for biomaterials. It also briefly examines the analysis of microstructure, physical properties, composition, and chemical structure of the resultant carbon materials through these diverse characterization methods. The aim is to advance the application of biomass-derived porous carbon materials across various domains and provide valuable insights for their further development and utilization.

1.5.1 XRD

XRD phase analysis is based on the diffraction effect of polycrystalline samples on X-rays, analyzing the existing forms of each component in the sample. By measuring the crystallinity, crystal phase, crystal structure, and bonding state, the structure and content of various crystalline components can be determined. The sensitivity is relatively low, and it can generally only determine phases with a content of more than 1% in the sample. At the same time, the accuracy of quantitative determination is also not high, usually in the order of 1%. XRD phase analysis requires a large sample size (0.1 g) to obtain more accurate results, which cannot be analyzed for amorphous samples. The main uses of X-ray diffraction analysis include XRD phase qualitative analysis, phase quantitative analysis, determination of grain size, determination of mesoporous structure (small angle X-ray diffraction), multi-layer film analysis (small angle XRD method), and identification of material state (distinguishing crystalline and amorphous states).

For example, Jalalah et al.[145] reported that a kind of non-metallic nitrogen-doped carbon material was prepared by using *Bambusa vulgaris* leaves as the precursor. For this investigation, the crystallinity of the synthetic metal-free nitrogen-doped porous carbon framework derived from *B. vulgaris* leaves at a specific carbonization temperature was assessed using P-XRD. Analysis of the porous carbon matrix devoid of metal nitrogen doping did not exhibit any distinct or prominent peaks, consistent with the amorphous characteristics of the synthesized *B. vulgaris* material. As can be seen from Figure 1.12(a), The BV-700, BV-800, and BV-900 display two peaks at approximately $2\theta = 24°$ and 44°, aligning with the reflections of graphitized carbon (002) and (100) crystal faces, respectively[147–151]. The P-XRD pattern of the BV-800 undergoes slight changes due to its lower pyrolysis temperature, enabling biomass activation in the presence of a KOH activator. With increasing activation temperature, there is a slight enhancement in the graphitization degree of carbon framework, and the diffraction patterns (900) of the resulting BV-002 and BV-800 are larger than those of BV-700. These observed peaks suggest that the frameworks synthesized at various pyrolysis temperatures contain a higher proportion of disordered carbon atoms[152]. The skeleton BV-800 exhibits a broader XRD peak, suggesting a higher level of disorder in its carbon matrix compared to other synthesized carbon materials. Apart from these diffraction peaks, the XRD pattern does not indicate any signs of disruption. Pre-activated *B. vulgaris* leaves at 800°C display the most pronounced disturbance in the carbon matrix, facilitating further processing[153].

1.5.2 Raman

When a beam of photons of excited light interacts with the molecule as the scattering center, most of the photons only change direction, scattering, and the frequency of the light is still the same as the excitation source, which is called Rayleigh scattering. However, there are also very small amounts of photons that not only change the direction of light propagation but also change the frequency of light waves; this scattering is called Raman scattering, and the intensity of the scattered light accounts for about 10^{-6} to 10^{-10} of the total scattered light intensity. Raman scattering is caused by the energy exchange between the photon and the molecule, which changes the energy of the photon. There are many mechanisms of Raman activation in solid materials, and the range of reflection is also very wide, such as molecular vibration, various elementary excitations (electrons, phonons, plasma, etc.), impurities, defects. Raman spectroscopy can be used to analyze the molecular structure, physical and chemical properties, and qualitative identification of materials, which can reveal information on vacancy, gap atom, dislocation, grain boundary, and phase boundary in materials.

Figure 1.12(b) displays the Raman spectra for the activated carbon skeleton derived from green leaves at three distinct pyrolysis temperatures. Notably, 1352 and 1581 cm^{-1}

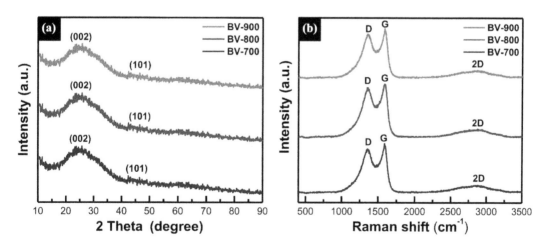

FIGURE 1.12   (a) Powder XRD pattern and (b) Raman spectroscopy of the synthesized metal-free nitrogen-doped porous carbon materials from green and raw leaves of *Bambusa vulgaris* at various activation temperatures, BV-700, BV-800, and BV-900[145].

correspond to the defect (D) band and the graphitic (G) band, respectively[154]. The level of disorder in the carbon matrices was assessed by calculating the ratio of the intensities of the two peaks (D and G)[155,156]. Among these, BV-800 stands out for exhibiting the highest degree of graphitization, indicating a more disordered carbon structure[157,158]. Reducing electron communication resistance during electrocatalytic reactions enhances electro-chemical performance through graphitization. Intercalation of potassium during the KOH activation step elongates the lattices, leading to carbon etching and pore generation[159,160]. Consequently, activation leads to a higher concentration of defects and structural disor-der in the carbon material. Conversely, high-temperature pyrolysis fosters both structural alignment and disorder in the carbon matrices. The 2D band (2800 cm^{-1}) suggests the occurrence of complete graphitization during the activation process[146,148].

1.5.3 SEM

SEM analysis offers morphological images spanning from a few nanometers to millime-ters, featuring a broad observation field and a resolution typically around 6 nm. Field emis-sion scanning electron microscopy (FESEM) achieves spatial resolutions on the order of 0.5 nm. It primarily provides insights into the geometric morphology of materials, the dis-persion state of powders, the size and distribution of nanoparticles, and the elemental com-position and phase structure of specific morphology regions. SEM has relatively lenient sample requirements, allowing for direct observation of both powder and bulk samples for morphology analysis. Liu et al.[161] reported that the water lettuce, known as the most pro-lific invasive aquatic plant globally, served as the source for synthesizing nitrogen-doped porous carbon nanosheets (N-doped PCNs) through an economical and scalable method devoid of any activators, as illustrated in Figure 1.13(a). SEM imaging of the N-doped PCNs reveals a porous nanosheet structure characterized by thin and stacked layered nanosheets (Figure 1.13(b)), while TEM images depict multi-level interconnected porous ultra-thin nanosheets (Figure 1.13(c)). These nanosheets boast a thickness close to 1.5 nm, compris-ing approximately 5 layers of graphite. Notably, the sample exhibits a higher pore density at 8000 due to NH and C activation, resulting in the formation of more interconnected pores. Furthermore, carbon (C), nitrogen (N), and oxygen (O) are uniformly distributed in the N-doped sample, with no presence of other metals.

1.5.4 TEM

Transmission electron microscopy (TEM) boasts high spatial resolution, making it particu-larly suited for analyzing nanopowder materials. Its key feature lies in its ability to analyze samples with minimal usage, allowing for the determination of morphology, particle size, dis-tribution, elemental composition, and phase structure of specific regions. TEM is most suit-able for morphology analysis of nanopowder samples, although the particle size should not exceed 300 nm to ensure proper penetration of the electron beam. For bulk sample analysis, thinning treatment of samples is typically required. TEM enables observation of particle size, morphology, particle distribution, and range of particle size distribution, with particle size cal-culated using statistical averaging methods. Typically, electron microscopy focuses on product particle size rather than grain size. High-resolution electron microscopy (HRTEM) provides

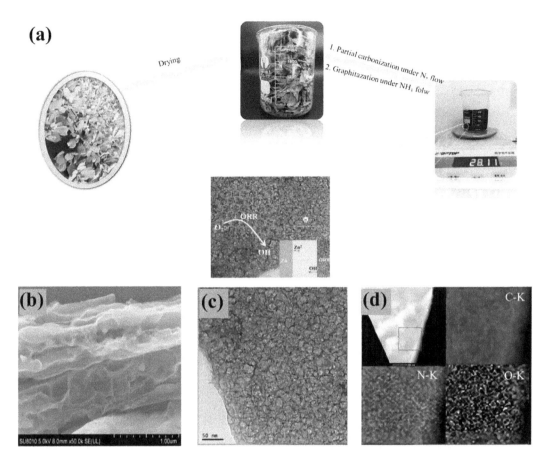

FIGURE 1.13 (a) Schematic details of the preparation of N-doped PCNs from water lettuce leaves. (b) SEM. (c) TEM image and (d) C, N, and O mapping of N-doped PCN sample synthesized at 800 °C[161].

direct observation of microcrystalline structure, offering an effective means for interface atomic structure analysis. It facilitates observation of the solid appearance of small particles and enables the study of crystal growth direction based on crystal morphology, corresponding diffraction patterns, and high-resolution images. Quan et al.[161] *Osmanthus fragrans* and *Sterculia lychnophora* were chosen to produce OPC and SPC porous carbons via a two-step method involving hydrothermal carbonization and KOH activation. In Figure 1.14, SEM and TEM images reveal distinctive structures for OPC and SPC samples. OPC (Figure 1.14 (a₁)) displayed a tremella-like structure, while SPC exhibited a honeycomb-like morphology (Figure 1.14 (b₁)). Further examination of the morphological and structural details of OPC and SPC using TEM revealed numerous pores (Figure 1.14 (a₂)) and Figure 1.14 (b₂)). These pores, ranging from tens of nanometers to sub-micrometers, intricately intertwined to form a cross-linked carbon framework that was challenging to separate. Additionally, the disordered structure observed in the high-resolution TEM images (Figure 1.14 (a₃) and (b₃)) indicated the amorphous nature of the OPC and SPC, aligning with previous reports on porous carbons derived from other biomass sources[162–166].

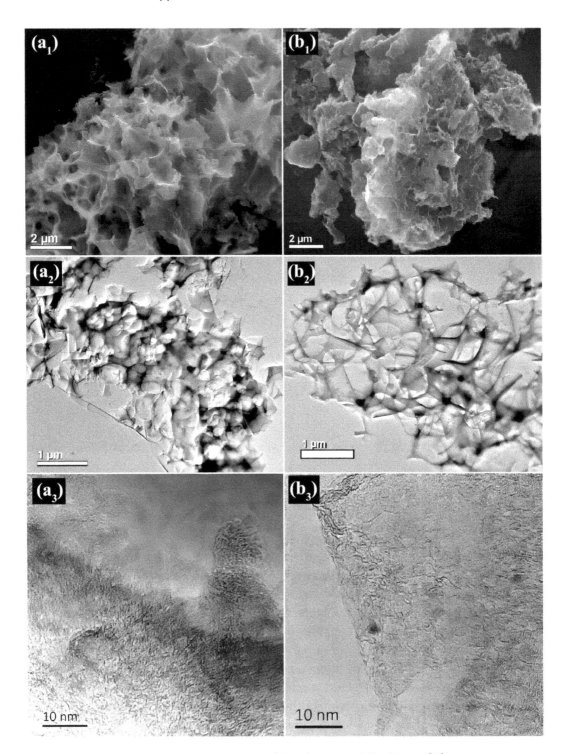

FIGURE 1.14   Typical SEM and TEM images of (a$_1$–a$_3$) OPC and (b$_1$–b$_3$) SPC[161].

1.5.5 ICP

ICP is a method that uses inductively coupled plasma as the excitation source to analyze the measured element based on the characteristic spectral lines of the atoms returning to the ground state when the element is excited. It can perform multi-element analysis and is suitable for the analysis of nearly 70 elements. The detection limit is extremely low and can reach 10^{-1}~10^{-5} $\mu g/cm^3$. It has good stability, high precision, a relative deviation of less than 1%, and a good quantitative analysis effect. The linear range can reach 4 to 6 orders of magnitude, but the detection sensitivity for non-metallic elements is relatively low.

1.5.6 XPS

X-ray photoelectron spectroscopy (XPS) utilizes X-rays to irradiate a sample's surface, exciting the electrons of its atoms or molecules, and then measures the energy distribution of these photoelectrons to gather essential information. Over time, XPS has seen advancements, including the development of small-area XPS, significantly enhancing spatial resolution. By scanning the sample comprehensively, XPS can detect most elements in a single measurement, making it a potent surface analysis instrument. It offers functions such as surface element analysis, chemical state and band structure analysis, and microregion chemical state imaging analysis. Based on Einstein's formula for photoelectron divergence, XPS serves as a crucial method for studying the electronic and atomic structures of material surfaces and interfaces. It can theoretically measure all elements on the periodic table except for hydrogen and helium. Its primary functions and applications encompass qualitative, quantitative, and chemical state information of surface layers, depth distribution analysis of heterogeneous cover layers, and imaging of elements and their chemical states, providing insight into surface element distribution. Zhu et al.[167] introduced a novel approach that combines nitrogen functionalization and pore structure control to produce nitrogen-doped hierarchical porous carbons (NHPCs) from a biodegraded product. Humic acid (HA), a natural polymer derived from decaying organic matter, served as the precursor for NHPCs. HA is recognized as a blend of various organic compounds and polymers, although its exact molecular formula remains a topic of debate[150]. The additional XPS analysis depicted in Figure 1.15 reveals that oxygen (O) primarily appears in the forms of C=O, C-O, and O-C=O. These diverse

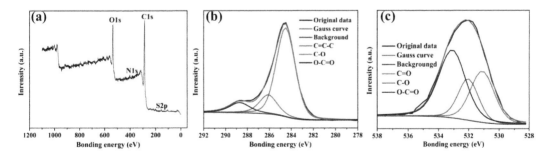

FIGURE 1.15 (a) XPS survey spectra of HA and high-resolution XPS scan of the C1s (b) and O1s (c) for HA[167].

oxygen-containing groups play a crucial role in facilitating the homogeneous blending of humic acid (HA), K_2CO_3, and urea. This blending not only enhances activation efficiency but also facilitates nitrogen (N) doping.

1.5.7 BET

BET testing is a widely employed technique for assessing the surface properties of materials, primarily aimed at evaluating their specific surface area and pore structure. These parameters are pivotal indicators of material performance and suitability for various applications within the realm of materials science. The methodology of BET testing relies on gas adsorption onto the material's surface to characterize its specific surface area and pore structure. Fundamentally, BET testing operates on the principle that it correlates specific surface area with the adsorption isotherm, typically an extension of the Langmuir isotherm. In BET testing, the material is usually in the form of a solid powder or thin film, while the gas is usually nitrogen. In the test, the material is first placed in a vacuum to remove moisture and other impurities from the surface and pores of the material. Then, slowly introduce nitrogen into the material until it reaches equilibrium with the surface and pores of the material. In this equilibrium state, nitrogen has completely adsorbed on the surface and pores of the material, forming a certain amount of adsorption. Then, the specific surface area and pore structure of the material can be calculated based on the amount of nitrogen added and the shape of the adsorption isotherm. Luo et al.[168] investigated the preparation of a series of N, S co-doped porous carbons through a two-step carbonization process, utilizing waste distiller's grains as the raw material and thiourea as the N and S source. To elucidate the textural properties of these N, and S co-doped porous carbons synthesized under varying conditions, nitrogen adsorption-desorption at −196°C was conducted. The resulting nitrogen adsorption-desorption isotherms and pore size distributions are depicted in Figure 1.16. Analysis revealed that the isotherms exhibited a combination of reversible types I and IV, characterized by substantial nitrogen adsorption at low relative pressures (p/p0 < 0.1), indicating well-developed microporous structures alongside a limited presence of mesopores (Figure 1.16(a–c)). Correspondingly, the average pore size distributions showcased systematic alterations in micropore and mesopore volumes across the porous carbons prepared under different conditions (Figure 1.16(d–f)).

1.6 CONCLUSION

In conclusion, when summarizing the basic principles of biomass-derived porous carbon, the following can be included:

(i) Raw material selection—The selection of biomass as a raw material is the first step in the preparation of porous carbon. Biomass usually includes renewable resources such as wood, plant fibers, and fruit husks.

(ii) Carbonization process—Biomass is carbonized at a high temperature, that is, heated in an anoxic or inert atmosphere, to convert organic matter into carbon.

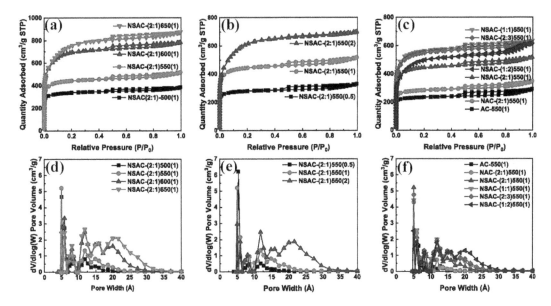

FIGURE 1.16 (a–c) N_2 adsorption–desorption isotherms and (d–f) average pores size distributions of the N,S co-doped porous carbons prepared under different conditions[168].

(iii) Activation treatment—The carbonized product is activated and the pore structure is introduced by chemical or physical methods to improve the specific surface area and pore volume of porous carbon. Common activation methods include physical activation (such as thermal activation) and chemical activation (such as alkali metal activation).

(iv) Pore structure—The properties of porous carbon are closely related to its pore structure, including micropores, mesoporous pores, and macropores. These pore structures have important implications for applications such as adsorption, energy storage, and mass transfer.

(v) Surface functionalization—In order to further improve the specific application performance of porous carbon, its surface is often functionalized and specific functional groups or modifiers are introduced.

(vi) Applications—Porous carbon materials have a wide range of applications in energy storage, adsorption, catalysis, sensing and other fields, such as supercapacitors, lithium-sulphur batteries, adsorbents, and so on. In general, the preparation principle of biomass-derived porous carbon involves key steps such as raw material selection, carbonization process, activation treatment, pore structure control, and surface functionalization. Through reasonable design and optimization of these steps, porous carbon materials with specific structure and properties can be obtained to meet the needs of different application fields.

NOTE

* Corresponding author

REFERENCES

[1] Jian Gan, Lizhen Chen, Zhijun Chen, Jilei Zhang, Wenji Yu, Caoxing Huang, Yan Wu, Kai Zhang, Lignocellulosic biomass-based carbon dots: Synthesis processes, properties, and applications, Small, **2023**, 19, 2304066. DOI:10.1002/smll.202304066.

[2] Siji Chen, Yuhan Xia, Bolun Zhang, Huan Chen, Guang Chen, Shanshan Tang, Disassembly of lignocellulose into cellulose, hemicellulose, and lignin for preparation of porous carbon materials with enhanced performances, Journal of Hazardous Materials, **2021**, 408, 124956. DOI:10.1016/j.jhazmat.2020.124956.

[3] Zhongqing Ma, Youyou Yang, Youlong Wu, Jiajia Xu, Hehuan Peng, Xiaohuan Liu, Wenbiao Zhang, Shurong Wang, In-depth comparison of the physicochemical characteristics of biochar derived from biomass pseudo components: Hemicellulose, cellulose, and lignin, Journal of Analytical and Applied Pyrolysis, **2019**, 140, 195. DOI:10.1016/j.jaap.2019.03.015.

[4] Jiang Deng, Tianyi Xiong, Haiyan Wang, Anmin Zheng, Yong Wang, Effects of cellulose, hemicellulose, and lignin on the structure and morphology of porous carbons, ACS Sustainable Chemistry & Engineering, **2016**, 4, 3750. DOI:10.1021/acssuschemeng.6b00388.

[5] Jonatan Henschen, Dongfang Li, Monica Ek, Preparation of cellulose nanomaterials via cellulose oxalates, Carbohydrate Polymers, **2019**, 213, 208. DOI:10.1016/j.carbpol.2019.02.056.

[6] Monika Österberg, Juan José Valle-Delgado, Surface forces in lignocellulosic systems, Current Opinion in Colloid & Interface Science, **2017**, 27, 33. DOI:10.1016/j.cocis.2016.09.005.

[7] Jessica Lucenius, Juan José Valle-Delgado, Kirsti Parikka, Monika Österberg, Understanding hemicellulose-cellulose interactions in cellulose nanofibril-based composites, Journal of Colloid and Interface Science, **2019**, 555, 104. DOI:10.1016/j.jcis.2019.07.053.

[8] Snehasish Basu, Okako Omadjela, David Gaddes, Srinivas Tadigadapa, Jochen Zimmer, Jeffrey M. Catchmark, Cellulose microfibril formation by surface-tethered cellulose synthase enzymes, ACS Nano, **2016**, 10, 1896. DOI:10.1021/acsnano.5b05648.

[9] D. Siva Priya, L. John Kennedy, G. Theophil Anand, Emerging trends in biomass-derived porous carbon materials for energy storage application: A critical review, Materials Today Sustainability, **2023**, 21, 100320. DOI:10.1016/j.mtsust.2023.100320.

[10] Luis Reyes, Lokmane Abdelouahed, Chetna Mohabeer, Jean-Christophe Buvat, Bechara Taouk, Energetic and exergetic study of the pyrolysis of lignocellulosic biomasses, cellulose, hemicellulose and lignin, Energy Conversion and Management, **2021**, 244, 114459. DOI:10.1016/j.enconman.2021.114459.

[11] Ngo Anh Dao Ho, C. P. Leo, A review on the emerging applications of cellulose, cellulose derivatives and nanocellulose in carbon capture, Environmental Research, **2021**, 197, 111100. DOI:10.1016/j.envres.2021.111100.

[12] Jiliang Liu, Herbert Sixta, Yu Ogawa, Michael Hummel, Michael Sztucki, Yoshiharu Nishiyama, Manfred Burghammer, Multiscale structure of cellulose microfibrils in regenerated cellulose fibers, Carbohydrate Polymers, **2024**, 324, 121512. DOI:10.1016/j.carbpol.2023.121512.

[13] Yuxin Liu, Bing Sun, Xuefan Zheng, Lingfang Yu, Jianguo Li, Integrated microwave and alkaline treatment for the separation between hemicelluloses and cellulose from cellulosic fibers, Bioresource Technology, **2018**, 247, 859. DOI:10.1016/j.biortech.2017.08.059.

[14] Jun Rao, Ziwen Lv, Gegu Chen, Feng Peng, Hemicellulose: Structure, chemical modification, and application, Progress in Polymer Science, **2023**, 140, 101675. DOI:10.1016/j.progpolymsci.2023.101675.

[15] Zesheng Li, Bolin Li, Changlin Yu, Hongqiang Wang, Qingyu Li, Recent progress of hollow carbon nanocages: General design fundamentals and diversified electrochemical applications, Advanced Science, **2023**, 10, 2206605. DOI:10.1002/advs.202206605.

[16] Thamarys Scapini, Maicon S. N. dos Santos, Charline Bonatto, João H. C. Wancura, Jéssica Mulinari, Aline F. Camargo, Natalia Klanovicz, Giovani L. Zabot, Marcus V. Tres, Gislaine Fongaro, Helen Treichel, Hydrothermal pretreatment of lignocellulosic biomass for hemicellulose recovery, Bioresource Technology, **2021**, 342, 126033. DOI:10.1016/j.biortech.2021.126033.

[17] Oliver P. Sarosi, Robert H. Bischof, Antje Potthast, Tailoring pulp cellulose with electron beam irradiation: Effects of lignin and hemicellulose, ACS Sustainable Chemistry & Engineering, **2020**, 8, 7235. DOI:10.1021/acssuschemeng.0c02165.

[18] Haimiao Yu, Zilu Wu, Geng Chen, Catalytic gasification characteristics of cellulose, hemicellulose and lignin, Renewable Energy, **2018**, 121, 559. DOI:10.1016/j.renene.2018.01.047.

[19] Lindokuhle Precious Magagula, Clinton Michael Masemola, Muhammed As'ad Ballim, Zikhona Nobuntu Tetana, Nosipho Moloto, Ella Cebisa Linganiso, Lignocellulosic biomass waste-derived cellulose nanocrystals and carbon nanomaterials: A review, International Journal of Molecular Sciences, **2022**, 23, 4310. DOI:10.3390/ijms23084310.

[20] Younghyun Lee, Eilhann E. Kwon, Jechan Lee, Polymers derived from hemicellulosic parts of lignocellulosic biomass, Reviews in Environmental Science and Bio/Technology, **2019**, 18, 317. DOI:10.1007/s11157-019-09495-z.

[21] Sabornie Chatterjee, Tomonori Saito, Lignin-derived advanced carbon materials, ChemSusChem, **2015**, 8, 3941. DOI:10.1002/cssc.201500692.

[22] Rodrigo Navia, Lignin valorization into biocarbon materials, Waste Management & Research, **2020**, 38, 109. DOI:10.1177/0734242x20902892.

[23] Yuting Zhu, Yuhe Liao, Luying Lu, Wei Lv, Jing Liu, Xiangbo Song, Jingcheng Wu, Lei Li, Chenguang Wang, Longlong Ma, Bert F. Sels, Oxidative catalytic fractionation of lignocellulose to high-yield aromatic aldehyde monomers and pure cellulose, ACS Catalysis, **2023**, 13, 7929. DOI:10.1021/acscatal.3c01309.

[24] Tian Li, DanDan Zhi, ZiHao Guo, JinZhe Li, Yao Chen, FanBin Meng, 3D porous biomass-derived carbon materials: Biomass sources, controllable transformation and microwave absorption application, Green Chemistry, **2022**, 64, 647. DOI:10.1039/d1gc02566j.

[25] Tan Yi, Hanyu Zhao, Qi Mo, Donglei Pan, Yang Liu, Lijie Huang, Hao Xu, Bao Hu, Hainong Song, From cellulose to cellulose nanofibrils—A comprehensive review of the preparation and modification of cellulose nanofibrils, Materials, **2020**, 13, 5062. DOI:10.3390/ma13225062.

[26] Arjeta Kryeziu, Václav Slovák, Alžběta Parchaňská, Liquefaction of cellulose for production of advanced porous carbon materials, Polymers, **2022**, 14, 1621. DOI:10.3390/polym14081621.

[27] Eldho Abraham, David E. Weber, Sigal Sharon, Shaul Lapidot, Oded Shoseyov, Multifunctional cellulosic scaffolds from modified cellulose nanocrystals, ACS Applied Materials & Interfaces, **2017**, 9, 2010. DOI:10.1021/acsami.6b13528.

[28] Hongzhe He, Ruoqun Zhang, Pengcheng Zhang, Ping Wang, Ning Chen, Binbin Qian, Lian Zhang, Jianglong Yu, Baiqian Dai, Functional carbon from nature: Biomass-derived carbon materials and the recent progress of their applications, Advanced Science, **2023**, 10, 2205557. DOI:10.1002/advs.202205557.

[29] Xiaoming Liu, Yueqiang Wang, Shaoqi Zhou, Peng Cui, Weiwu Wang, Wenfeng Huang, Zhen Yu, Shungui Zhou, Differentiated strategies of animal-derived and plant-derived biochar to reduce nitrogen loss during paper mill sludge composting, Bioresource Technology, **2022**, 360, 127583. DOI:10.1016/j.biortech.2022.127583.

[30] Hanbo Chen, Xing Yang, Hailong Wang, Binoy Sarkar, Sabry M. Shaheen, Gerty Gielen, Nanthi Bolan, Jia Guo, Lei Che, Huili Sun, Jörg Rinklebe, Animal carcass- and wood-derived biochars improved nutrient bioavailability, enzyme activity, and plant growth in metal-phthalic acid ester co-contaminated soils: A trial for reclamation and improvement of degraded soils, Journal of Environmental Management, **2020**, 261, 110246. DOI:10.1016/j.jenvman.2020.110246.

[31] Changlei Xia, Sheldon Q. Shi, Self-activation for activated carbon from biomass: Theory and parameters, Green Chemistry, **2016**, 18, 2063. DOI:10.1039/C5GC02152A.

[32] Omobayo A. Salawu, Ziwei Han, Adeyemi S. Adeleye, Shrimp waste-derived porous carbon adsorbent: Performance, mechanism, and application of machine learning, Journal of Hazardous Materials, **2022,** 437, 129266. DOI:10.1016/j.jhazmat.2022.129266.

[33] Da-Wei Wang, Feng Li, Li-Chang Yin, Xu Lu, Zhi-Gang Chen, Ian R. Gentle, Gao Qing Lu, Hui-Ming Cheng, Nitrogen-doped carbon monolith for alkaline supercapacitors and understanding nitrogen-induced redox transitions, Chemistry–A European Journal, **2012,** 18, 5345. DOI:10.1002/chem.201102806.

[34] Bingye Yang, Jianping Gao, Minhui Xie, Saisai Zuo, Huiying Kang, Yu Sun, Xiaoyang Xu, Wei Wang, Chunjuan Gao, Yu Liu, Jing Yan, N-self-doped porous carbon derived from animal-heart as an electrocatalyst for efficient reduction of oxygen, Journal of Colloid and Interface Science, **2020,** 579, 832. DOI:10.1016/j.jcis.2020.06.113.

[35] Li-Sheng Xie, Sheng-Xue Yu, Hui-Jun Yang, Jun Yang, Jian-Lan Ni, Jiu-Lin Wang, Hierarchical porous carbon derived from animal bone as matric to encapsulated selenium for high performance Li–Se battery, Rare Metals, **2017,** 36, 434. DOI:10.1007/s12598-017-0910-0.

[36] Chaoran Qin, Xiaoyi Wu, Cui Huang, Bo Duan, Jinping Zhou, Hongye Yang, Ang Lu, Tooth-derived flexible supercapacitor, Journal of Energy Storage, **2022,** 52, 104728. DOI:10.1016/j.est.2022.104728.

[37] Rashmita Das, Rajib Bandyopadhyay, Panchanan Pramanik, Carbon quantum dots from natural resource: A review, Materials Today Chemistry, **2018,** 8, 96. DOI:10.1016/j.mtchem.2018.03.003.

[38] Dan Wang, Jian-Feng Chen, Liming Dai, Recent advances in graphene quantum dots for fluorescence bioimaging from cells through tissues to animals, Particle & Particle Systems Characterization, **2015,** 32, 515. DOI:10.1002/ppsc.201400219.

[39] Aumber Abbas, Lim Tuti Mariana, Anh N. Phan, Biomass-waste derived graphene quantum dots and their applications, Carbon, **2018,** 140, 77. DOI:10.1016/j.carbon.2018.08.016.

[40] Abu Bakar Siddique, Ashit Kumar Pramanick, Subrata Chatterjee, Mallar Ray, Amorphous carbon dots and their remarkable ability to detect 2,4,6-Trinitrophenol, Scientific Reports, **2018,** 8, 9770. DOI:10.1038/s41598-018-28021-9.

[41] Chee Shan Lim, Katerina Hola, Adriano Ambrosi, Radek Zboril, Martin Pumera, Graphene and carbon quantum dots electrochemistry, Electrochemistry Communications, **2015,** 52, 75. DOI:10.1016/j.elecom.2015.01.023.

[42] Haitao Li, Xiaodie He, Zhenhui Kang, Hui Huang, Yang Liu, Jinglin Liu, Suoyuan Lian, Chi Him A. Tsang, Xiaobao Yang, Shuit-Tong Lee, Water-soluble fluorescent carbon quantum dots and photocatalyst design, Angewandte Chemie International Edition, **2010,** 49, 4430. DOI:10.1002/anie.200906154.

[43] Jin Zhou, Pei Lin, Juanjuan Ma, Xiaoyue Shan, Hui Feng, Congcong Chen, Jianrong Chen, Zhaosheng Qian, Facile synthesis of halogenated carbon quantum dots as an important intermediate for surface modification, RSC Advances, **2013,** 3, 9625. DOI:10.1039/C3RA41243A.

[44] Baskar Thangaraj, Pravin Raj Solomon, Surawut Chuangchote, Nutthapon Wongyao, Werasak Surareungchai, Biomass-derived carbon quantum dots—A Review. Part 1: Preparation and characterization, ChemBioEng Reviews, **2021,** 8, 265. DOI:10.1002/cben.202000029.

[45] Xiaoyou Xu, Robert Ray, Yunlong Gu, Harry J. Ploehn, Latha Gearheart, Kyle Raker, Walter A. Scrivens, Electrophoretic analysis and purification of fluorescent single-walled carbon nanotube fragments, Journal of the American Chemical Society, **2004,** 126, 12736. DOI:10.1021/ja040082h.

[46] Zheng Yang, Tiantian Xu, Hui Li, Mengyao She, Jiao Chen, Zhaohui Wang, Shengyong Zhang, Jianli Li, Zero-dimensional carbon nanomaterials for fluorescent sensing and imaging, Chemical Reviews, **2023,** 123, 11047. DOI:10.1021/acs.chemrev.3c00186.

[47] Takeo Oku, Takanori Hirano, Katsuaki Suganuma, Satoru Nakajima, Formation and structure of carbon nanocage structures produced by polymer pyrolysis and electron-beam irradiation, Journal of Materials Research, **1999,** 14, 4266. DOI:10.1557/JMR.1999.0578.

[48] Yahachi Saito, Takehisa Matsumoto, Carbon nanocages created as cubes, Nature, **1998,** 392, 237. DOI:10.1038/32555.

[49] Qiang Wu, Lijun Yang, Xizhang Wang, Zheng Hu, Carbon-based nanocages: A new platform for advanced energy storage and conversion, Advanced Materials, **2020,** 32, 1904177. DOI:10.1002/adma.201904177.

[50] Qiang Wu, Lijun Yang, Xizhang Wang, Zheng Hu, From carbon-based nanotubes to nanocages for advanced energy conversion and storage, Accounts of Chemical Research, **2017,** 50, 435. DOI:10.1021/acs.accounts.6b00541.

[51] Yufei Jiang, Lijun Yang, Tao Sun, Jin Zhao, Zhiyang Lyu, Ou Zhuo, Xizhang Wang, Qiang Wu, Jing Ma, Zheng Hu, Significant contribution of intrinsic carbon defects to oxygen reduction activity, ACS Catalysis, **2015,** 5, 6707. DOI:10.1021/acscatal.5b01835.

[52] Sheng Chen, Jiyu Bi, Yu Zhao, Lijun Yang, Chen Zhang, Yanwen Ma, Qiang Wu, Xizhang Wang, Zheng Hu, Nitrogen-doped carbon nanocages as efficient metal-free electrocatalysts for oxygen reduction reaction, Advanced Materials, **2012,** 24, 5593. DOI:10.1002/adma.201202424.

[53] Zhiyang Lyu, Lijun Yang, Dan Xu, Jin Zhao, Hongwei Lai, Yufei Jiang, Qiang Wu, Yi Li, Xizhang Wang, Zheng Hu, Hierarchical carbon nanocages as high-rate anodes for Li- and Na-ion batteries, Nano Research, **2015,** 8, 3535. DOI:10.1007/s12274-015-0853-4.

[54] Lingyu Du, Xueyi Cheng, Fujie Gao, Youbin Li, Yongfeng Bu, Zhiqi Zhang, Qiang Wu, Lijun Yang, Xizhang Wang, Zheng Hu, Electrocatalysis of S-doped carbon with weak polysulfide adsorption enhances lithium–sulfur battery performance, Chemical Communications, **2019,** 55, 6365. DOI:10.1039/C9CC02134E.

[55] Zhiyang Lyu, Dan Xu, Lijun Yang, Renchao Che, Rui Feng, Jin Zhao, Yi Li, Qiang Wu, Xizhang Wang, Zheng Hu, Hierarchical carbon nanocages confining high-loading sulfur for high-rate lithium–sulfur batteries, Nano Energy, **2015,** 12, 657. DOI:10.1016/j.nanoen.2015.01.033.

[56] Liming Shen, Tao Sun, Ou Zhuo, Renchao Che, Danqin Li, Yucheng Ji, Yongfeng Bu, Qiang Wu, Lijun Yang, Qiang Chen, Xizhang Wang, Zheng Hu, Alcohol-tolerant platinum electrocatalyst for oxygen reduction by encapsulating platinum nanoparticles inside nitrogen-doped carbon nanocages, ACS Applied Materials & Interfaces, **2016,** 8, 16664. DOI:10.1021/acsami.6b03482.

[57] Olena Mykhailiv, Halyna Zubyk, Marta E. Plonska-Brzezinska, Carbon nano-onions: Unique carbon nanostructures with fascinating properties and their potential applications, Inorganica Chimica Acta, **2017,** 468, 49. DOI:10.1016/j.ica.2017.07.021.

[58] Marta E. Plonska-Brzezinska, Carbon nano-onions: A Review of recent progress in synthesis and applications, ChemNanoMat, **2019,** 5, 568. DOI:10.1002/cnma.201800583.

[59] Sumio Iijima, Direct observation of the tetrahedral bonding in graphitized carbon black by high resolution electron microscopy, Journal of Crystal Growth, **1980,** 50, 675. DOI:10.1016/0022-0248(80)90013-5.

[60] W. L. Wang, X. D. Bai, K. H. Liu, Z. Xu, D. Golberg, Y. Bando, E. G. Wang, Direct synthesis of B-C-N single-walled nanotubes by bias-assisted hot filament chemical vapor deposition, Journal of the American Chemical Society, **2006,** 128, 6530. DOI:10.1021/ja0606733.

[61] Andrews Rodney, Jacques David, Qian Dali, Rantell Terry, Multiwall carbon nanotubes: Synthesis and application, Accounts of Chemical Research, **2002,** 35. DOI:1008. 10.1021/ar010151m.

[62] Wei Xia, Dangsheng Su, Alexander Birkner, Lars Ruppel, Yuemin Wang, Christof Wöll, Jun Qian, Changhai Liang, Gabriela Marginean, Waltraut Brandl, Chemical vapor deposition and synthesis on carbon nanofibers: Sintering of ferrocene-derived supported iron nanoparticles and the catalytic growth of secondary carbon nanofibers, Chemistry of Materials, **2005,** 17, 5737. DOI:10.1021/cm051623k.

[63] M. Nagatsu, M. Miyake, J. Maeda, Plasma CVD reactor with two-microwave oscillators for diamond film synthesis, Thin Solid Films, **2006,** 506–507, 617. DOI:10.1016/j.tsf.2005.08.041.

[64] Albert G. Nasibulin, Anna Moisala, David P. Brown, Esko I. Kauppinen, Carbon nanotubes and onions from carbon monoxide using $Ni(acac)_2$ and $Cu(acac)_2$ as catalyst precursors, Carbon, **2003,** 41, 2711. DOI:10.1016/S0008-6223(03)00333-6.

[65] L. G. Bulusheva, A. V. Okotrub, V. L. Kuznetsov, D. V. Vyalikh, Soft X-ray spectroscopy and quantum chemistry characterization of defects in onion-like carbon produced by nanodiamond annealing, Diamond and Related Materials, **2007,** 16, 1222. DOI:10.1016/j.diamond.2006.11.064.

[66] Sorin Vizireanu, Leona Nistor, Michael Haupt, Verena Katzenmaier, Christian Oehr, Gheorghe Dinescu, Carbon nanowalls growth by radiofrequency plasma-beam-enhanced chemical vapor deposition, Plasma Processes and Polymers, **2008,** 5, 263. DOI:10.1002/ppap.200700120.

[67] Antonio de Lucas, Prado B. García, Agustín Garrido, Amaya Romero, J. L. Valverde, Catalytic synthesis of carbon nanofibers with different graphene plane alignments using Ni deposited on iron pillared clays, Applied Catalysis A: General, **2006,** 301, 123. DOI:10.1016/j.apcata.2005.11.026.

[68] Marc Monthioux, Vladimir L. Kuznetsov, Who should be given the credit for the discovery of carbon nanotubes?, Carbon, **2006,** 44, 1621. DOI:10.1016/j.carbon.2006.03.019.

[69] J. F. Colomer, C. Stephan, S. Lefrant, G. Van Tendeloo, I. Willems, Z. Kónya, A. Fonseca, Ch Laurent, J. B. Nagy, Large-scale synthesis of single-wall carbon nanotubes by Catalytic Chemical Vapor Deposition (CCVD) method, Chemical Physics Letters, **2000,** 317, 83. DOI:10.1016/S0009-2614(99)01338-X.

[70] Jieshan Qiu, Yongfeng Li, Yunpeng Wang, Wen Li, Production of carbon nanotubes from coal, Fuel Processing Technology, **2004,** 85, 1663. DOI:10.1016/j.fuproc.2003.12.010.

[71] Minglong Yang, Ye Yuan, Ying Li, Xianxian Sun, Shasha Wang, Lei Liang, Yuanhao Ning, Jianjun Li, Weilong Yin, Renchao Che, Yibin Li, Dramatically enhanced electromagnetic wave absorption of hierarchical CNT/Co/C fiber derived from cotton and metal-organic-framework, Carbon, **2020,** 161, 517. DOI:10.1016/j.carbon.2020.01.073.

[72] Yucheng Zhou, Jiajun He, Ruoxi Chen, Xiaodong Li, Recent advances in biomass-derived graphene and carbon nanotubes, Materials Today Sustainability, **2022,** 18, 100138.

[73] Jianshe Huang, Yang Liu, Tianyan You, Carbon nanofiber based electrochemical biosensors: A review, Analytical Methods, **2010,** 2, 202. DOI:10.1039/B9AY00312F.

[74] Hongyan Song, Wenzhong Shen, Carbon nanofibers: Synthesis and applications, Journal of Nanoscience and Nanotechnology, **2014,** 14, 1799. DOI:10.1166/jnn.2014.9005.

[75] Gary G. Tibbetts, Vapor-grown carbon fibers: Status and prospects, Carbon, **1989,** 27, 745. DOI:10.1016/0008-6223(89)90208-x.

[76] Lichao Feng, Ning Xie, Jing Zhong, Carbon nanofibers and their composites: A Review of synthesizing, properties and applications, Materials, **2014,** 7, 3919. DOI:10.3390/ma7053919.

[77] Elfina Azwar, Wan Adibah Wan Mahari, Joon Huang Chuah, Dai-Viet N. Vo, Nyuk Ling Ma, Wei Haur Lam, Su Shiung Lam, Transformation of biomass into carbon nanofiber for supercapacitor application—A review, International Journal of Hydrogen Energy, **2018,** 43, 20811. DOI:10.1016/j.ijhydene.2018.09.111.

[78] Andreas Greiner, Joachim H. Wendorff, Electrospinning: A fascinating method for the preparation of ultrathin fibers, Angewandte Chemie International Edition, **2007,** 46, 5670. DOI:10.1002/anie.200604646.

[79] Ji-Won Jung, Cho-Long Lee, Sunmoon Yu, Il-Doo Kim, Electrospun nanofibers as a platform for advanced secondary batteries: A comprehensive review, Journal of Materials Chemistry A, **2016,** 4, 703. DOI:10.1039/C5TA06844D.

[80] A. H. Castro Neto, F. Guinea, N. M. R. Peres, K. S. Novoselov, A. K. Geim, The electronic properties of graphene, Reviews of Modern Physics, **2009,** 81, 109. DOI:10.1103/RevModPhys.81.109.

[81] Xiao Kong, Yifeng Zhu, Hanwu Lei, Chenxi Wang, Yunfeng Zhao, Erguang Huo, Xiaona Lin, Qingfa Zhang, Moriko Qian, Wendy Mateo, Rongge Zou, Zhen Fang, Roger Ruan, Synthesis

of graphene-like carbon from biomass pyrolysis and its applications, Chemical Engineering Journal, **2020,** 399, 125808. DOI:10.1016/j.cej.2020.125808.

[82] Yong Qian, Yang Li, Zhen Pan, Jie Tian, Ning Lin, Yitai Qian, Hydrothermal "disproportionation" of biomass into oriented carbon microsphere anode and 3d porous carbon cathode for potassium ion hybrid capacitor, Advanced Functional Materials, **2021,** 31, 2103115. DOI:10.1002/adfm.202103115.

[83] Kozlov M. E., Capps R.C., Sampson W.M., Ebron V.H., Ferraris J.P., Baughman R.H., Spinning solid and hollow polymer-free carbon nanotube fibers, Advanced Materials, **2005,** 17, 614. DOI:10.1002/adma.200401130.

[84] Fu Min, Lv Ruitao, Lei Yu, Terrones Mauricio, Ultralight flexible electrodes of nitrogen-doped carbon macrotube sponges for high-performance supercapacitors, Small, **2021,** 17, 2004827. DOI:10.1002/smll.202004827.

[85] Yi Wang, Mingmei Wu, Kun Wang, Junwei Chen, Tongwen Yu, Shuqin Song, $Fe_3O_4@$ N-Doped interconnected hierarchical porous carbon and its 3D integrated electrode for oxygen reduction in acidic media, Advanced Science, **2020,** 7, 2000407. DOI:10.1002/advs.202000407.

[86] Shaofeng Zhou, Lihua Zhou, Yaping Zhang, Jian Sun, Junlin Wen, Yong Yuan, Upgrading earth-abundant biomass into three-dimensional carbon materials for energy and environmental applications, Journal of Materials Chemistry A, **2019,** 7, 4217. DOI:10.1039/C8TA12159A.

[87] Hongtao Sun, Jian Zhu, Daniel Baumann, Lele Peng, Yuxi Xu, Imran Shakir, Yu Huang, Xiangfeng Duan, Hierarchical 3D electrodes for electrochemical energy storage, Nature Reviews Materials, **2019,** 4, 45. DOI:10.1038/s41578-018-0069-9.

[88] Bo Chen, Qinglang Ma, Chaoliang Tan, Teik-Thye Lim, Ling Huang, Hua Zhang, Carbon-based sorbents with three-dimensional architectures for water remediation, Small, **2015,** 11, 3319. DOI:10.1002/smll.201570161.

[89] Zhihong Tang, Shuling Shen, Jing Zhuang, Xun Wang, Noble-metal-promoted three-dimensional macroassembly of single-layered graphene oxide, Angewandte Chemie International Edition, **2010,** 122, 4707. DOI:10.1002/anie.201000270.

[90] Duc Dung Nguyen, Nyan-Hwa Tai, San-Boh Lee, Wen-Shyong Kuo, Superhydrophobic and superoleophilic properties of graphene-based sponges fabricated using a facile dip coating method, Energy & Environmental Science, **2012,** 5, 7908. DOI:10.1039/C2EE21848H.

[91] Robin J. White, Nicolas Brun, Vitaly L. Budarin, James H. Clark, Maria-Magdalena Titirici, Always look on the "light" side of life: Sustainable carbon aerogels, ChemSusChem, **2014,** 7, 670. DOI:10.1002/cssc.201300961.

[92] Maria-Magdalena Titirici, Robin J. White, Nicolas Brun, Vitaliy L. Budarin, Dang Sheng Su, Francisco Del Monte, James H. Clark, Mark J. MacLachlan, Sustainable carbon materials, Chemical Society Reviews, **2015,** 44, 250. DOI:10.1039/C4CS00232F.

[93] Xuchun Gui, Jinquan Wei, Kunlin Wang, Anyuan Cao, Hongwei Zhu, Yi Jia, Qinke Shu, Dehai Wu, Carbon nanotube sponges, Advanced Materials, **2010,** 22, 617. DOI:10.1002/adma.200902986.

[94] L. Camilli, C. Pisani, E. Gautron, M. Scarselli, P. Castrucci, F. D'Orazio, M. Passacantando, D. Moscone, M. De Crescenzi, A three-dimensional carbon nanotube network for water treatment, Nanotechnology, **2014,** 25, 065701. DOI:10.1088/0957-4484/25/6/065701.

[95] Ning Chen, Qinmin Pan, Versatile fabrication of ultralight magnetic foams and application for oil-water separation, ACS Nano, **2013,** 7, 6875–6883. DOI:10.1021/nn4020533.

[96] Erik Frank, Lisa M. Steudle, Denis Ingildeev, Johanna M. Spörl, Michael R. Buchmeiser, Carbon fibers: Precursor systems, processing, structure, and properties, Angewandte Chemie International Edition, **2014,** 53, 5262. DOI:10.1002/anie.201306129.

[97] Daniel Kobina Sam, Ebenezer Kobina Sam, Arulappan Durairaj, Xiaomeng Lv, Zijing Zhou, Jun Liu, Synthesis of biomass-based carbon aerogels in energy and sustainability, Carbohydrate Research, **2020,** 491, 107986. DOI:10.1016/j.carres.2020.107986.

[98] Peiqi Wang, Chuancheng Jia, Yu Huang, Xiangfeng Duan, Van der Waals heterostructures by design: From 1D and 2D to 3D, Matter, **2021**, 4, 552. DOI:10.1016/j.matt.2020.12.015.

[99] ShangWen Zhou, HongYan Wang, HuaQing Xue, Wei Guo, XiaoBo Li, Supercritical methane adsorption on shale gas: Mechanism and model, Chinese Science Bulletin, **2017**, 62, 4189. DOI:10.1360/n972017-00151.

[100] Weidong Xie, Meng Wang, Veerle Vandeginste, Si Chen, Zhenghong Yu, Jiyao Wang, Hua Wang, Huajun Gan, Adsorption behavior and mechanism of CO_2 in the longmaxi shale gas reservoir, RSC Advance, **2022**, 12, 25947. DOI:10.1039/d2ra03632k.

[101] Matthias Thommes, Katsumi Kaneko, Alexander V. Neimark, James P. Olivier, Francisco Rodriguez-Reinoso, Jean Rouquerol, Kenneth S. W. Sing, Physisorption of gases, with special reference to the evaluation of surface area and pore size distribution (IUPAC technical report), Pure and Applied Chemistry, **2015**, 87, 1051. DOI:10.1515/pac-2014-1117.

[102] Irving Langmuir, The constitution and fundamental properties of solids and liquids, Journal of the Franklin Institute, **1917**, 183, 102. DOI:10.1016/s0016-0032(17)90938-x.

[103] Irving Langmuir, The adsorption of gases on plane surfaces of glass, mica and platinum, Journal of the American Chemical Society, **2002**, 40, 1361. DOI:10.1021/ja02242a004.

[104] H. Swenson, N. P. Stadie, Langmuir's theory of adsorption: A centennial review, Langmuir, **2019**, 35, 5409. DOI:10.1021/acs.langmuir.9b00154.

[105] Stephen Brunauer, P. H. Emmett, Edward Teller, Adsorption of gases in multimolecular layers, Journal of the American Chemical Society, **2002**, 60, 309. DOI:10.1021/ja01269a023.

[106] Kenneth S. W. Sing, Ruth T. Williams, Historical aspects of capillarity and capillary condensation, Microporous and Mesoporous Materials, **2012**, 154, 16. DOI:10.1016/j.micromeso.2011.09.022.

[107] Richard Zsigmondy, Über die Struktur des Gels der Kieselsäure. Theorie der Entwässerung, Zeitschrift für anorganische Chemie, **2004**, 71, 356. DOI:10.1002/zaac.19110710133.

[108] Elliott P. Barrett, Leslie G. Joyner, Paul P. Halenda, The determination of pore volume and area distributions in porous substances. I. computations from nitrogen isotherms, Journal of the American Chemical Society, **2002**, 73, 373. DOI:10.1021/ja01145a126.

[109] M. M. Dubinin, Contemporary status of the theory of the volume filling of the micropores of carbonaceous adsorbents, Bulletin of the Academy of Sciences of the USSR Division of Chemical Science, **1991**, 40, 1. DOI:10.1007/bf00959621.

[110] M. M. Dubinin, The equation of the characteristic curve of activated charcoal, Proceedings of the USSR Academy of Sciences, **1947**, 55, 327.

[111] A. Dąbrowski, Adsorption-its development and application for practical purposes. In Adsorption and its applications in industry and environmental protection-Vol. I: Applications in industry, edited by A. Dąbrowski, **1999**, Elsevier.

[112] B. Lippens, Studies on pore systems in catalysts: V. The t method, Journal of Catalysis, **1965**, 4, 319. DOI:10.1016/0021-9517(65)90307-6.

[113] K. S. W. Sing, Reporting physisorption data for gas/solid systems with special reference to the determination of surface area and porosity (provisional), Pure and Applied Chemistry, **1982**, 54, 2201. DOI:10.1351/pac198254112201.

[114] Dan Li, Wenhua Chen, Jianping Wu, Charles Qiang Jia, Xia Jiang, The preparation of waste biomass-derived N-doped carbons and their application in acid gas removal: Focus on N functional groups, Journal of Materials Chemistry A, **2020**, 8, 24977. DOI:10.1039/D0TA07977D.

[115] Qing Lv, Wenyan Si, Jianjiang He, Lei Sun, Chunfang Zhang, Ning Wang, Ze Yang, Xiaodong Li, Xin Wang, Weiqiao Deng, Yunze Long, Changshui Huang, Yuliang Li, Selectively nitrogen-doped carbon materials as superior metal-free catalysts for oxygen reduction, Nature Communications, **2018**, 9, 3376. DOI:10.1038/s41467-018-05878-y.

[116] Yuanting Qiao, Chunfei Wu, Nitrogen enriched biochar used as CO_2 adsorbents: A brief review, Carbon Capture Science & Technology, **2022**, 2, 100018. DOI:10.1016/j.ccst.2021.100018.

[117] Xiaomin Yang, Huihui He, Ting Lv, Jieshan Qiu, Fabrication of biomass-based functional carbon materials for energy conversion and storage, Materials Science and Engineering: R: Reports, **2023,** 154, 100736. DOI:10.1016/j.mser.2023.100736.

[118] Arthi Gopalakrishnan, Sushmee Badhulika, Effect of self-doped heteroatoms on the performance of biomass-derived carbon for supercapacitor applications, Journal of Power Sources, **2020,** 480, 228830. DOI:10.1016/j.jpowsour.2020.228830.

[119] Ke Zhu, Wenlei Qin, Yaping Gan, Yizhe Huang, Zhiwei Jiang, Yuwen Chen, Xin Li, Kai Yan, Acceleration of Fe^{3+}/Fe^{2+} cycle in garland-like MIL-101(Fe)/MoS_2 nanosheets to promote peroxymonosulfate activation for sulfamethoxazole degradation, Chemical Engineering Journal, **2023,** 470, 144190. DOI:10.1016/j.cej.2023.144190.

[120] Luc Moens, Robert J. Evans, Michael J. Looker, Mark R. Nimlos, A comparison of the Maillard reactivity of proline to other amino acids using pyrolysis-molecular beam mass spectrometry, Fuel, **2004,** 83, 1433. DOI:10.1016/j.fuel.2004.01.020.

[121] Mingli Yang, Liping Guo, Gengshen Hu, Xin Hu, Leqiong Xu, Jie Chen, Wei Dai, Maohong Fan, Highly cost-effective nitrogen-doped porous coconut shell-based CO_2 sorbent synthesized by combining ammoxidation with KOH activation, Environmental Science & Technology, **2015,** 49, 7063. DOI:10.1021/acs.est.5b01311.

[122] Xiong Zhang, Jing Wu, Haiping Yang, Jingai Shao, Xianhua Wang, Yingquan Chen, Shihong Zhang, Hanping Chen, Preparation of nitrogen-doped microporous modified biochar by high temperature CO_2–NH_3 treatment for CO_2 adsorption: Effects of temperature, RSC Advances, **2016,** 6, 98157. DOI:10.1039/C6RA23748G.

[123] Wei Luo, Bao Wang, Christopher G. Heron, Marshall J. Allen, Jeff Morre, Claudia S. Maier, William F. Stickle, Xiulei Ji, Pyrolysis of cellulose under ammonia leads to nitrogen-doped nanoporous carbon generated through methane formation, Nano Letters, **2014,** 14, 2225. DOI:10.1021/nl500859p.

[124] Naoto Tsubouchi, Megumi Nishio, Yuuki Mochizuki, Role of nitrogen in pore development in activated carbon prepared by potassium carbonate activation of lignin, Applied Surface Science, **2016,** 371, 301. DOI:10.1016/j.apsusc.2016.02.200.

[125] Chengshuai Chang, Miao Li, He Wang, Shulan Wang, Xuan Liu, Huakun Liu, Li Li, A novel fabrication strategy for doped hierarchical porous biomass-derived carbon with high microporosity for ultrahigh-capacitance supercapacitors, Journal of Materials Chemistry A, **2019,** 7, 19939. DOI:10.1039/C9TA06210F.

[126] Mykola Seredych, Denisa Hulicova-Jurcakova, Gao Qing Lu, Teresa J. Bandosz, Surface functional groups of carbons and the effects of their chemical character, density and accessibility to ions on electrochemical performance, Carbon, **2008,** 46, 1475. DOI:10.1016/j.carbon.2008.06.027.

[127] Xin Liu, Chenggong Sun, Hao Liu, Wei Herng Tan, Wenlong Wang, Colin Snape, Developing hierarchically ultra-micro/mesoporous biocarbons for highly selective carbon dioxide adsorption, Chemical Engineering Journal, **2019,** 361, 199. DOI:10.1016/j.cej.2018.11.062.

[128] Kristina Maliutina, Arash Tahmasebi, Jianglong Yu, Pressurized entrained-flow pyrolysis of microalgae: Enhanced production of hydrogen and nitrogen-containing compounds, Bioresource Technology, **2018,** 256, 160. DOI:10.1016/j.biortech.2018.02.016.

[129] Karifala Kante, Cesar Nieto-Delgado, J. Rene Rangel-Mendez, Teresa J. Bandosz, Spent coffee-based activated carbon: Specific surface features and their importance for H_2S separation process, Journal of Hazardous Materials, **2012,** 201–202, 141. DOI:10.1016/j.jhazmat.2011.11.053.

[130] Dimple P. Dutta, Brindaban Modak, Balaji Rao Ravuri, Phosphorous/fluorine Co-doped biomass-derived carbon for enhanced sodium-ion and lithium-ion storage, ChemNanoMat, **2023,** 9, e202300077. DOI:10.1002/cnma.202300077.

[131] Chang Hyuck Choi, Min Wook Chung, Sung Hyeon Park, Seong Ihl Woo, Additional doping of phosphorus and/or sulfur into nitrogen-doped carbon for efficient oxygen reduction

reaction in acidic media, Physical Chemistry Chemical Physics, **2013**, 15, 1802. DOI:10.1039/C2CP44147K.

[132] Fatemeh Razmjooei, Kiran Pal Singh, Eun Jin Bae, Jongsung Yu, A new class of electroactive Fe- and P-functionalized graphene for oxygen reduction, Journal of Materials Chemistry A, **2015**, 3, 11031. DOI:10.1039/C5TA00970G.

[133] Yade Zhu, Ying Huang, Chen Chen, Mingyue Wang, Panbo Liu, Phosphorus-doped porous biomass carbon with ultra-stable performance in sodium storage and lithium storage, Electrochimica Acta, **2019**, 321, 134698. DOI:10.1016/j.electacta.2019.134698.

[134] Fatemeh Razmjooei, Kiran Pal Singh, Min Young Song, Jong-Sung Yu, Enhanced electrocatalytic activity due to additional phosphorous doping in nitrogen and sulfur-doped graphene: A comprehensive study, Carbon, **2014**, 78, 257. DOI:10.1016/j.carbon.2014.07.002.

[135] Arjunan Ariharan, Balasubramanian Viswanathan, Vaiyapuri Nandhakumar, Heteroatom doped multi-layered graphene material for hydrogen storage application, Graphene, **2016**, 05, 39. DOI:10.4236/graphene.2016.52005.

[136] Sumio Iijima, Toshinari Ichihashi, Single-shell carbon nanotubes of 1-nm diameter, Nature, **1993**, 363, 603. DOI:10.1038/363603a0.

[137] Mohamad Anas Nahil, Paul T. Williams, Pore characteristics of activated carbons from the phosphoric acid chemical activation of cotton stalks, Biomass and Bioenergy, **2012**, 37, 142. DOI:10.1016/j.biombioe.2011.12.019.

[138] Rong Li, Zidong Wei, Xinglong Gou, Wei Xu, Phosphorus-doped graphene nanosheets as efficient metal-free oxygen reduction electrocatalysts, RSC Advances, **2013**, 3, 9978. DOI:10.1039/C3RA41079J.

[139] Min-Sik Park, Jae-Hun Kim, Yong-Nam Jo, Seung-Hyun Oh, Hansu Kim, Young-Jun Kim, Incorporation of phosphorus into the surface of natural graphite anode for lithium ion batteries, Journal of Materials Chemistry, **2011**, 21, 17960. DOI:10.1039/C1JM13158C.

[140] Chenzhen Zhang, Nasir Mahmood, Han Yin, Fei Liu, Yanglong Hou, Synthesis of phosphorus-doped graphene and its multifunctional applications for oxygen reduction reaction and lithium ion batteries, Advanced Materials, **2013**, 25, 4932. DOI:10.1002/adma.201301870.

[141] Xueni Zhao, Hejun Li, Mengdi Chen, Kezhi Li, Bin Wang, Zhanwei Xu, Sheng Cao, Leilei Zhang, Hailiang Deng, Jinhua Lu, Strong-bonding calcium phosphate coatings on carbon/carbon composites by ultrasound-assisted anodic oxidation treatment and electrochemical deposition, Applied Surface Science, **2012**, 258, 5117. DOI:10.1016/j.apsusc.2012.01.144.

[142] Ziwu Liu, Feng Peng, Hongjuan Wang, Hao Yu, Wenxu Zheng, Jian Yang, Phosphorus-doped graphite layers with high electrocatalytic activity for the O_2 reduction in an alkaline medium, Angewandte Chemie International Edition, **2011**, 50, 3257. DOI:10.1002/anie.201006768.

[143] Dong Xie, Xinhui Xia, Wanjia Tang, Yu Zhong, Yadong Wang, Donghuang Wang, Xiuli Wang, Jiangping Tu, Novel carbon channels from loofah sponge for construction of metal sulfide/carbon composites with robust electrochemical energy storage, Journal of Materials Chemistry A, **2017**, 5, 7578. DOI:10.1039/C7TA01154G.

[144] Yangke Long, Sifan Bu, Yixuan Huang, Yueqi Shao, Ling Xiao, Xiaowen Shi, N-doped hierarchically porous carbon for highly efficient metal-free catalytic activation of peroxymonosulfate in water: A non-radical mechanism, Chemosphere, **2019**, 216, 545. DOI:10.1016/j.chemosphere.2018.10.175.

[145] Mohammed Jalalah, HyukSu Han, Arpan Kumar Nayak, Farid A. Harraz, Biomass-derived metal-free porous carbon electrocatalyst for efficient oxygen reduction reactions, Journal of the Taiwan Institute of Chemical Engineers, **2023**, 147, 104905. DOI:10.1016/j.jtice.2023.104905.

[146] Wei Ren, Gang Nie, Peng Zhou, Hui Zhang, Xiaoguang Duan, Shaobin Wang, The intrinsic nature of persulfate activation and N-doping in carbocatalysis, Environmental Science & Technology, **2020**, 54, 6438. DOI:10.1021/acs.est.0c01161.

[147] Anqi Wang, Zhikeng Zheng, Ruiqi Li, Di Hu, Yiran Lu, Huixia Luo, Kai Yan, Biomass-derived porous carbon highly efficient for removal of Pb(II) and Cd(II), Green Energy and Environment **2019,** 4, 414–423. DOI:10.1016/j.gee.2019.05.002.

[148] Wei Ren, Gang Nie, Peng Zhou, Hui Zhang, Xiaoguang Duan, Shaobin Wang, The intrinsic nature of persulfate activation and N-doping in carbocatalysis, Environmental Science & Technology, **2020,** 54, 6438. DOI:10.1021/acs.est.0c01161.

[149] Jian Su, Changqing Fang, Mannan Yang, Youliang Cheng, Zhen Wang, Zhigang Huang, Caiyin You, A controllable soft-templating approach to synthesize mesoporous carbon micro-spheres derived from d-xylose via hydrothermal method, Journal of Materials Science & Technology, **2020,** 38, 183. DOI:10.1016/j.jmst.2019.03.050.

[150] Ali Mostofizadeh, Yanwei Li, Bo Song, Yudong Huang, Synthesis, properties, and applications of low-dimensional carbon-related nanomaterials, Journal of Nanomaterials, **2011, 2011,** 685081. DOI:10.1155/2011/685081.

[151] Ruiqi Li, Di Hu, Kang Hu, Hao Deng, Man Zhang, Anqi Wang, Rongliang Qiu, Kai Yan, Coupling adsorption-photocatalytic reduction of Cr(VI) by metal-free N-doped carbon, Science of The Total Environment, **2020,** 704, 135284. DOI:10.1016/j.scitotenv.2019.135284.

[152] Qinghua Liang, Ling Ye, Zheng-Hong Huang, Qiang Xu, Yu Bai, Feiyu Kang, Quan-Hong Yang, A honeycomb-like porous carbon derived from pomelo peel for use in high-performance supercapacitors, Nanoscale, **2014,** 6, 13831. DOI:10.1039/C4NR04541F.

[153] Changyu Leng, Kang Sun, Jihui Li, Jianchun Jiang, From dead pine needles to O, N Codoped activated carbons by a one-step carbonization for high rate performance supercapacitors, ACS Sustainable Chemistry & Engineering, **2017,** 5, 10474. DOI:10.1021/acssuschemeng.7b02481.

[154] Jiang Deng, Mingming Li, Yong Wang, Biomass-derived carbon: Synthesis and applications in energy storage and conversion, Green Chemistry, **2016,** 18, 4824. DOI:10.1039/C6GC01172A.

[155] Yuanfeng Qi, Jing Li, Yanqing Zhang, Qi Cao, Yanmei Si, Zhiren Wu, Muhammad Akram, Xing Xu, Novel lignin-based single atom catalysts as peroxymonosulfate activator for pollutants degradation: Role of single cobalt and electron transfer pathway, Applied Catalysis B: Environmental, **2021,** 286, 119910. DOI:10.1016/j.apcatb.2021.119910.

[156] Yuanfeng Qi, Jing Li, Yanqing Zhang, Qi Cao, Yanmei Si, Zhiren Wu, Muhammad Akram, Xing Xu, Novel lignin-based single atom catalysts as peroxymonosulfate activator for pollutants degradation: Role of single cobalt and electron transfer pathway, Applied Catalysis B: Environmental, **2021,** 286, 119910. DOI:10.1016/j.apcatb.2021.119910.

[157] Akshay Jain, Sundaramurthy Jayaraman, Rajasekhar Balasubramanian, M. P. Srinivasan, Hydrothermal pre-treatment for mesoporous carbon synthesis: Enhancement of chemical activation, Journal of Materials Chemistry A, **2014,** 2, 520. DOI:10.1039/C3TA12648J.

[158] Akshay Jain, Rajasekhar Balasubramanian, M. P. Srinivasan, Hydrothermal pre-treatment for mesoporous carbon synthesis: Enhancement of chemical activation, Journal of Materials Chemistry A, **2014,** 2, 520. DOI:10.1039/c3ta12648j.

[159] Yongqing Zhao, Min Lu, Pengyu Tao, Yunjie Zhang, Xiaoting Gong, Zhi Yang, Guoqing Zhang, Hulin Li, Hierarchically porous and heteroatom doped carbon derived from tobacco rods for supercapacitors, Journal of Power Sources, **2016,** 307, 391. DOI:10.1016/j.jpowsour.2016.01.020.

[160] Miao Liang, Jing Wang, Min Zhao, Jiaxiao Cai, Mingjian Zhang, Nan Deng, Bing Wang, Bin Li, Ke Zhang, Borax-assisted hydrothermal carbonization to fabricate monodisperse carbon spheres with high thermostability, Materials Research Express, **2019,** 6, 065615. DOI:10.1088/2053-1591/ab109f.

[161] L. Liu, G. Zeng, J. Chen, L. Bi, L. Dai, Z. Wen, N-doped porous carbon nanosheets as pH-universal ORR electrocatalyst in various fuel cell devices, Nano Energy. **2018,** 49, 393. DOI:10.1016/j.nanoen.2018.04.061.

[162] Cuixian Liu, Gaoyi Han, Yunzhen Chang, Yaoming Xiao, Miaoyu Li, Wen Zhou, Dongying Fu, Wenjing Hou, Properties of porous carbon derived from cornstalk core in high-performance electrochemical capacitors, ChemElectroChem, **2016,** 3, 323. DOI:10.1002/celc.201500376.

[163] Akshay Jain, Rajasekhar Balasubramanian, M. P. Srinivasan, Hydrothermal conversion of biomass waste to activated carbon with high porosity: A review, Chemical Engineering Journal, **2016,** 283, 789. DOI:10.1016/j.cej.2015.08.014.

[164] Padmaker Pandey, Anamika Pandey, Shruti Singh, Nikhil Kant Shukla, Self assembled mono-layers and carbon nanotubes: A significant tool's for modification of electrode surface, Sensor Letters, **2020,** 18, 669–685. DOI:10.1166/sl.2020.4280.

[165] Ning Fu, Ying Liu, Rui Liu, Xiaodong Wang, Zhenglong Yang, Metal cation-assisted synthesis of amorphous B, N Co-doped carbon nanotubes for superior sodium storage, Small **2020,** 16, 2001607. DOI:10.1002/smll.202001607.

[166] Ke Zhu, Yaqian Shen, Junming Hou, Jie Gao, Dongdong He, Jin Huang, Hongmei He, Lele Lei, Wenjin Chen, One-step synthesis of nitrogen and sulfur co-doped mesoporous graphite-like carbon nanosheets as a bifunctional material for tetracycline removal via adsorption and catalytic degradation processes: Performance and mechanism, Chemical Engineering Journal, **2021,** 412, 128521. DOI:10.1016/j.cej.2021.128521.

[167] Youyu Zhu, Mingming Chen, Yang zhang, Wenxuan Zhao, Chengyang Wang, A biomass-derived nitrogen-doped porous carbon for high-energy supercapacitor, Carbon, **2018,** 140, 404. DOI:10.1016/j.carbon.2018.09.009.

[168] Lan Luo, Chunliang Yang, Fei Liu, Tianxiang Zhao, Heteroatom-N,S co-doped porous carbons derived from waste biomass as bifunctional materials for enhanced CO_2 adsorption and conversion, Separation and Purification Technology, **2023,** 320, 124090. DOI:10.1016/j.seppur.2023.124090.

CHAPTER 2

Fabrication of Biomass-Derived Porous Carbon

Zhenhao Xu, Yutong Zhang, Ke Li, Ke Zhu,
Wanrong Bu, and Yuchen Wang*

ABSTRACT

Since the physicochemical properties of biomass-derived porous carbon are crucial to the specific application, the mechanism and parameters of fabrication processes should be well studied and understood. Up to date, common methods for preparing biomass-derived porous carbon include physical activation methods (e.g. direct pyrolysis, CO_2-activated pyrolysis, and steam-activated pyrolysis) and chemical activation methods (e.g. KOH, $ZnCl_2$, and H_3PO_4). In this chapter, the pore formation mechanism of these traditional methods is explicitly described. Meanwhile, the corresponding examples of biomass-derived porous carbon using these methods are listed. Furthermore, the advances of other new fabrication methods including hydrothermal method and microwave-assisted pyrolysis are discussed.

2.1 INTRODUCTION

Biomass, especially plant biomass, are promising alternative resources to fossil fuels for synthesizing carbon materials and high-value chemicals[1,2]. For the fabrication of porous carbon materials, lignocellulosic biomass with the composition of cellulose, hemicellulose, and lignin are main precursors[3]. The fabrication processes involve the thermal decomposition and structural reorganization of cellulose, hemicellulose, and lignin at certain temperature and anaerobic environment[4].

2.2 PHYSICAL ACTIVATION METHODS

2.2.1 Direct Pyrolysis

Direct pyrolysis is a traditional technology for the conversion of biomass to porous carbon that is conducted under anaerobic or anoxic conditions at high temperatures[5]. Most

DOI: 10.1201/9781003520566-2

of the porous carbon in our daily life is obtained from the direct pyrolysis of plants such as coconut husk, bamboo, and rice husk (Figure 2.1)[6]. The pyrolysis temperature of the components of biomass (cellulose, hemicellulose, and lignin) have significant differences. Specifically, after water evaporation at the temperature below 100 °C, hemicellulose and cellulose are decomposed at the temperature range of 220–315 °C and 315–400 °C, respectively[4]. Differently, lignin is decomposed at a wide temperature range of 250–550 °C[7,8]. During the pyrolysis of biomass, the main reaction involves cross-linking polymerization, which is accompanied by the release of small molecule volatiles, e.g., H_2O, CO_2, CH_4, CO, and H_2. Direct pyrolysis method is simple and the obtained pore structure is composed of micropores and macropores with relatively small specific surface area.

The physicochemical properties of porous carbon obtained from direct pyrolysis are greatly affected by the type of precursors. Gan et al.[9] performed direct pyrolysis on three kinds of biomass (walnut shell, cypress sawdust, and rice straw) to synthesize porous carbon materials. Walnut shell-derived carbon afforded the largest specific surface area of 555.0 m^2/g and mesopore volume of 0.116 cm^3/g. Moreover, the pyrolysis parameters, including heating rate, holding temperature, and holding time, greatly affect the properties of porous carbon. Supriya et al.[10] prepared a series of porous carbon materials from *Caesalpinia sappan* waste pods via direct pyrolysis at a varied temperature of 400–1000 °C and demonstrated that the specific surface area and pore volume were positively dependent on pyrolysis temperature. Fan et al.[11] reported that heating rate (5, 10, 15, and 20 °C/min) of cellulose pyrolysis could influence

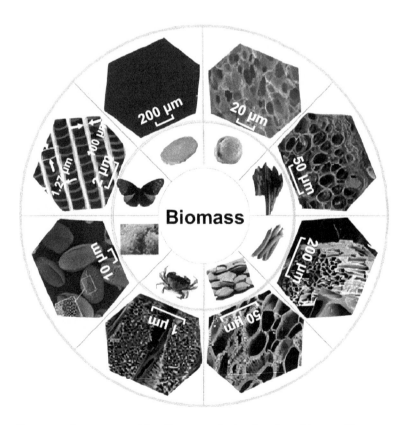

FIGURE 2.1   Porous carbon prepared by direct pyrolysis of various biomass[6].

the surface functional groups (-OH, C-H, C-O, C=C, and C=O), and thus the hydrophobic/hydrophilic nature of porous carbon. Al-Rumaihi et al.[12] claimed that long residence time is beneficial for the increased yield of biomass-derived porous carbon due to high possibility of repolymerization.

2.2.2 CO$_2$-Activated Pyrolysis

CO$_2$ is a cheap and abundant natural activation agent to react with carbon in the skeleton of biomass precursors for the creation of pores at high temperatures in CO$_2$ atmosphere[13,14]. The activation of CO$_2$ proceeds through the Boudouard reaction (C + CO$_2$ → 2CO), which thermodynamically occurs at temperatures higher than 710 °C[15]. Fang et al.[16] investigated the effects of activation time and temperature on the structure of CO$_2$-activated porous carbon from hickory and peanut hull. Undoubtedly, the specific surface area and pore volume of the as-prepared porous carbon increased with increasing time and temperature. Nabais et al.[17] prepared porous carbon material from coffee endocarp via steam- and CO$_2$-activated pyrolysis. The specific surface area and pore volume of CO$_2$-activated porous carbon could reach 424 m^2/g and 0.22 cm^3/g, respectively, which were higher than those of steam-activated porous carbon. Jung et al.[18] mixed CO$_2$ with N$_2$ to improve the porous structure of oak-derived porous carbon. Concretely, the specific surface area of porous carbon via CO$_2$/N$_2$ activation (1126 m^2/g) was greater than that of porous carbon via direct CO$_2$ activation (800 m^2/g). Furthermore, Xiong et al.[19] developed a novel cyclic pressure switching (CPS) technique to enrich the porous structure of mushroom-derived CO$_2$-activated carbon. As seen in Figure 2.2, the introduction of intermittent vacuumization accelerates the gas transport of CO$_2$ and CO to break the diffusion balance of the reaction. Thus the specific surface area and pore volume of the porous carbon reached 1175 m^2/g and 0.52 cm^3/g, respectively.

2.2.3 Steam-Activated Pyrolysis

The steam-activated pyrolysis employs water vapor to activate biomass precursors with the advantages of facile operation and environmental friendliness[20]. The overall reaction

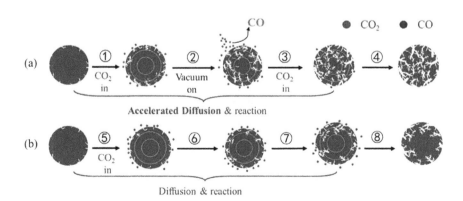

FIGURE 2.2 The schematic illustration of (a) cyclic pressure switching and (b) atmospheric pressure processes[19].

process is described as follows[21,22]: (1) the precursor is carbonized at suitable temperature; (2) the adsorbed water vapor on the carbon surface reacts with carbon to release CO/CO_2 at elevated temperature; and (3) CO_2 further reacts with carbon to generate pores. Qin et al.[23] investigated the influences of steam flow rate, activation temperature, and time on the porosities of pine nut shell-derived porous carbon. The optimal specific surface area was derived as high as 956 m^2/g with a mesoporous volume ratio of 37.1%. Ima et al.[24] applied a two-stage steam-activated pyrolysis to convert needle coke into porous carbon materials, which possessed a specific surface area of 1134 m^2/g with a mesoporous volume ratio of 78%.

2.3 CHEMICAL ACTIVATION METHODS

2.3.1 KOH

KOH is an efficient alkali activating agent for promoting the formation of plenty of micropores due to the enhanced intercalation ability of potassium element[25,26]. According to detailed analysis of previous works, the activation mechanism of KOH is summarized as three stages (Figure 2.3)[27,28]: (1) KOH and generated K_2CO_3, K_2O etch the carbon matrix; (2) the formation of water vapor and CO_2 during the redox reactions further etch the carbon; (3) the lattice of carbon expands through the intercalation of potassium vapor at high temperature (> 750 °C). Thus, the porosity of KOH-activated porous carbon can be modulated by adjusting activation parameters, such as activation time, activation temperature, and the amount of KOH. Apart from KOH, other potassium salts have also been used as activating agent to prepare biomass-based porous carbon through similar activation processes[29,30].

Concretely, Feng et al.[31] found that the specific surface area increased by more than 4 times to 2839 m^2/g as the ratio of KOH to carbon precursors increased from 1 to 3, indicating a significant enhancement in porosity. Nevertheless, further addition of KOH

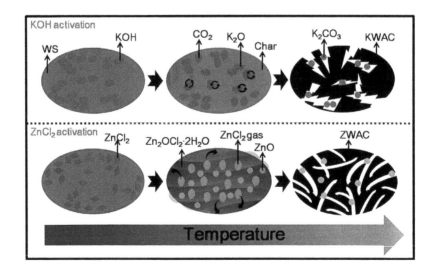

FIGURE 2.3   The chemical activation mechanisms of KOH and $ZnCl_2$[27].

decreased the specific surface area due to excessive activation. Also, when KOH amount was very high, small micropores disappeared and large micropores with pore size of 1–2 nm were dominated[32]. Moreover, Wang et al.[33] demonstrated the effect of activation temperature on the porosity of KOH-activated willow catkins-derived porous carbon. As the temperature gradually elevated from 600 to 800 °C, the specific surface area and total pore volume of activated carbon increased from 645 to 1586 m^2/g, and from 0.31 to 0.78 m^3/g, respectively, which is consistent with KOH activation mechanism.

2.3.2 $ZnCl_2$

$ZnCl_2$ is another common chemical activator for the production of biomass-derived porous carbon[34]. Similar to KOH, the activation mechanism of $ZnCl_2$ is dependent on temperature (Figure 2.3). Due to the Lewis acid property, $ZnCl_2$ could promote the dehydrogenation of plant precursors at low temperature (200–350 °C) and limit tar formation at medium temperature (350–450 °C)[35,36]. At high temperature (450–600 °C), the decomposition of pre-formed $Zn_2OCl_2 \cdot 2H_2O$ in the carbon framework releases vaporized $ZnCl_2$ molecules[27], resulting in a large specific surface area. Ma et al.[37] studied the activation temperature on $ZnCl_2$-activated potato waste residue-derived porous carbon. After comparing the specific surface area of as-prepared porous carbon at different temperature (600–900 °C), 700 °C was determined to be the optimal activation temperature at 1052 m^2/g. In order to further improve the specific surface area, $ZnCl_2$ is usually used with other chemical activators. Zhao et al.[38] employed $ZnCl_2$ and $Mg(NO_3)_2$ dual-activators to fabricate peanut meal-derived porous carbon. Except for the activation effect of $ZnCl_2$, the introduction of $Mg(NO_3)_2$ greatly increased the specific surface area from 1098 to 2090 m^2/g.

2.3.3 H_3PO_4

Owing to the potential environmental hazards of vaporized $ZnCl_2$, a medium-strength acid H_3PO_4 with relatively low activation temperature (~450 °C) is considered a promising alternative[39]. The activation mechanism of H_3PO_4 generally includes dehydration, aromatization, and cross-linking processes (Figure 2.4)[28,40]. First, the dehydration of the polysaccharides with abundant hydroxyl groups is accelerated by H_3PO_4. Second, most carbon is transformed from sp^3 to sp^2 hybridization in the carbonization process. Thirdly, the cross-linking process proceeds by interacting the hydroxyl groups of H_3PO_4 with polysaccharides to generate composites. After removing the extra H_3PO_4 and its composites, a porous structure is created.

Yagmur et al.[41] confirmed the essential role of H_3PO_4 activation. As compared with direct pyrolysis, the specific surface area of waste tea-derived porous carbon with H_3PO_4 activation increased from 3.2 to 1135 m^2/g. Meanwhile, the yield of porous carbon was also elevated from 36.3% to 51.8%. Zakaria et al.[42] used mangrove leftover-derived porous carbon to investigate the influence of the H_3PO_4 impregnation ratio and activation temperature. After control experiments, porosity results showed that a combined effect of suitable impregnation ratio and moderate activation temperature not only generates a high yield of porous carbon but also enables the formation of abundant micropores and mesopores. The specific surface area was optimized as 1011.8 m^2/g with the impregnation ratio of 4 and the

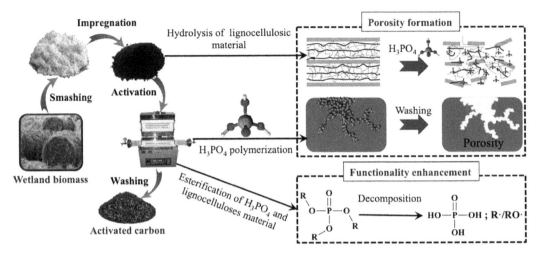

FIGURE 2.4   Principle scheme of biomass-derived porous carbon production with H_3PO_4 activation[40].

activation temperature of 300 °C. More H_3PO_4 and higher activation temperature (500 °C) would cause a shrinkage in the porous carbon structure and thus lower the specific surface area.

2.4 OTHER METHODS

2.4.1 Hydrothermal Method

Hydrothermal method is to carbonize biomass precursors in a sealed autoclave at relatively low temperature (150–350 °C)[43]. There are mainly five steps during hydrothermal process: hydrolysis, dehydration, decarboxylation, polymerization, and aromatization, which convert biomass into porous carbon materials with abundant oxygen-containing functional groups and aromatic rings[44]. Wang et al.[45] reported hydrothermal and subsequent annealing treatments to obtain three-dimensional porous carbon from kiwifruit. The optimal specific surface area was derived as 379 m²/g. Direct hydrothermal carbonization usually leads to low degree of aromatization and unsatisfied specific surface area. Therefore, chemical activators are required to further expand the pore structure of hydrothermal carbon. Zhao et al.[46] combined hydrothermal and KOH activation methods to realize tobacco rods-derived porous carbon with the high specific surface area of 1761–2115 m²/g.

2.4.2 Microwave-Assisted Pyrolysis

The conventional pyrolysis method is to heat the biomass precursor from the surface to the core through thermal convection or conduction. The thermal gradient along the whole biomass precursors always requires the additional holding step, giving rise to high energy consumption[47]. In contrast, as illustrated in Figure 2.5, microwave-assisted heating is proceeded through the homogenous penetration of electromagnetic energy into biomass precursors, resulting in an efficient and controllable heating process[48,49]. Therefore, in recent years, microwave-assisted pyrolysis has received tremendous attention in preparation of biomass-derived porous carbon materials.

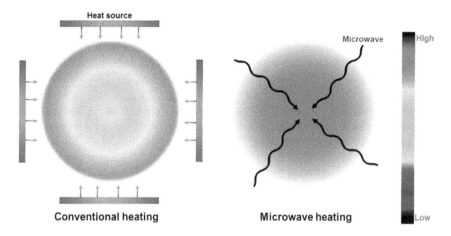

FIGURE 2.5   The difference of heating principle between conventional heating and microwave-assisted heating[49].

Huang et al.[50] compared the two heating methods for preparation of rice straw-derived porous carbon as CO_2 capture adsorbent. The maximum temperature to obtain porous carbon using microwave-assisted pyrolysis was 300 °C with the microwave power of 200 W, whereas the value of the traditional pyrolysis was 550 °C. The microwave-assisted pyrolysis is also combined with physical activation methods to upgrade the quality of porous carbon materials. Yek et al.[51] reported an innovative microwave-steam activation process to prepare microporous waste palm shell-derived porous carbon. The resultant porous carbon showed uniform porosity and a specific surface area of 570.8 m^2/g. In addition, chemical activators such as KOH[52], $ZnCl_2$[53], and H_3PO_4[54], have been extensively used to facilitate the formation of porous structure during the microwave-assisted pyrolysis of biomass.

2.5 CONCLUSION

In comparison with physical activation methods, chemically activated biomass-derived porous carbon generally possesses larger specific surface area, which is suitable for application in subsequent chapters. Nevertheless, the utilization of chemically activated agents could cause potential environmental risks. Thus, the balance between performance and economic costs should be seriously considered. Moreover, more efforts should be devoted to the mechanism of new fabrication methods.

NOTE

* Corresponding author

REFERENCES

[1] Babasaheb M. Matsagar, Ren-Xuan Yang, Saikat Dutta, Yong Sik Ok, Kevin C. W. Wu, Recent progress in the development of biomass-derived nitrogen-doped porous carbon, Journal of Materials Chemistry A, **2021,** 9, 3703. DOI:10.1039/D0TA09706C.

[2] Valentina G. Matveeva, Lyudmila M. Bronstein, From renewable biomass to nanomaterials: Does biomass origin matter?, Progress in Materials Science, **2022,** 130, 100999. DOI:10.1016/j.pmatsci.2022.100999.

[3] Yu-Te Liao, Babasaheb M. Matsagar, Kevin C. W. Wu, Metal–organic framework (MOF)-derived effective solid catalysts for valorization of lignocellulosic biomass, ACS Sustainable Chemistry & Engineering, **2018,** 6, 13628. DOI:10.1021/acssuschemeng.8b03683.

[4] Nannan Guo, Su Zhang, Luxiang Wang, Dianzeng Jia, Application of plant-based porous carbon for supercapacitors, Acta Physico-Chimica Sinica, **2020,** 36, 1903055. DOI:10.3866/pku.Whxb201903055.

[5] Xuqin Pan, Zhepei Gu, Weiming Chen, Qibin Li, Preparation of biochar and biochar composites and their application in a Fenton-like process for wastewater decontamination: A review, Science of the Total Environment, **2021,** 754, 142104. DOI:10.1016/j.scitotenv.2020.142104.

[6] Huanqin Zhao, Yan Cheng, Wei Liu, Lieji Yang, Baoshan Zhang, Luyuan Paul Wang, Guangbin Ji, Zhichuan J. Xu, Biomass-derived porous carbon-based nanostructures for microwave absorption, Nano-Micro Letters, **2019,** 11, 24. DOI:10.1007/s40820-019-0255-3.

[7] Anqi Wang, Zhikeng Zheng, Ruiqi Li, Di Hu, Yiran Lu, Huixia Luo, Kai Yan, Biomass-derived porous carbon highly efficient for removal of Pb(II) and Cd(II), Green Energy & Environment, **2019,** 4, 414. DOI:10.1016/j.gee.2019.05.002.

[8] Yufeng Yin, Qianjun Liu, Jing Wang, Yiting Zhao, Recent insights in synthesis and energy storage applications of porous carbon derived from biomass waste: A review, International Journal of Hydrogen Energy, **2022,** 47, 39338. DOI:10.1016/j.ijhydene.2022.09.121.

[9] Fengli Gan, Bowen Cheng, Ziheng Jin, Zhongde Dai, Bangda Wang, Lin Yang, Xia Jiang, Hierarchical porous biochar from plant-based biomass through selectively removing lignin carbon from biochar for enhanced removal of toluene, Chemosphere, **2021,** 279, 130514. DOI:10.1016/j.chemosphere.2021.130514.

[10] S. Supriya, Vinay S. Bhat, Titilope John Jayeoye, Thitima Rujiralai, Kwok Feng Chong, Gurumurthy Hegde, An investigation on temperature-dependent surface properties of porous carbon nanoparticles derived from biomass, Journal of Nanostructure in Chemistry, **2022,** 12, 495. DOI:10.1007/s40097-021-00427-4.

[11] Mengjiao Fan, Chao Li, Yifan Sun, Lijun Zhang, Shu Zhang, Xun Hu, In situ characterization of functional groups of biochar in pyrolysis of cellulose, Science of the Total Environment, **2021,** 799. DOI:10.1016/j.scitotenv.2021.149354.

[12] Aisha Al-Rumaihi, Muhammad Shahbaz, Gordon McKay, Hamish Mackey, Tareq Al-Ansari, A review of pyrolysis technologies and feedstock: A blending approach for plastic and biomass towards optimum biochar yield, Renewable & Sustainable Energy Reviews, **2022,** 167, 112715. DOI:10.1016/j.rser.2022.112715.

[13] Adekunle Moshood Abioye, Farid Nasir Ani, Recent development in the production of activated carbon electrodes from agricultural waste biomass for supercapacitors: A review, Renewable and Sustainable Energy Reviews, **2015,** 52, 1282. DOI:10.1016/j.rser.2015.07.129.

[14] Hanfang Zhang, Yihe Zhang, Liqi Bai, Yingge Zhang, Li Sun, Effect of physiochemical properties in biomass-derived materials caused by different synthesis methods and their electrochemical properties in supercapacitors, Journal of Materials Chemistry A, **2021,** 9, 12521. DOI:10.1039/D1TA00790D.

[15] Xiao-fei Tan, Shao-bo Liu, Yun-guo Liu, Yan-ling Gu, Guang-ming Zeng, Xin-jiang Hu, Xin Wang, Shao-heng Liu, Lu-hua Jiang, Biochar as potential sustainable precursors for activated carbon production: Multiple applications in environmental protection and energy storage, Bioresource Technology, **2017,** 227, 359. DOI:10.1016/j.biortech.2016.12.083.

[16] June Fang, Bin Gao, Andrew R. Zimmerman, Kyoung S. Ro, Jianjun Chen, Physically (CO_2) activated hydrochars from hickory and peanut hull: Preparation, characterization, and sorption of methylene blue, lead, copper, and cadmium, RSC Advances, **2016,** 6, 24906. DOI:10.1039/C6RA01644H.

[17] João Valente Nabais, Peter Carrott, M. M. L. Ribeiro Carrott, Vânia Luz, Angel L. Ortiz, Influence of preparation conditions in the textural and chemical properties of activated

carbons from a novel biomass precursor: The coffee endocarp, Bioresource Technology, **2008,** 99, 7224. DOI:10.1016/j.biortech.2007.12.068.

[18] Su-Hwa Jung, Joo-Sik Kim, Production of biochars by intermediate pyrolysis and activated carbons from oak by three activation methods using CO_2, Journal of Analytical and Applied Pyrolysis, **2014,** 107, 116. DOI:10.1016/j.jaap.2014.02.011.

[19] Jingjing Xiong, Guancong Jiang, Yu Qian, Liwen Mu, Xin Feng, Xiaohua Lu, Jiahua Zhu, Cycling pressure-switching process enriches micropores in activated carbon by accelerating reactive gas internal diffusion in porous channels, Sustainable Materials and Technologies, **2021,** 28, e00248. DOI:10.1016/j.susmat.2021.e00248.

[20] Zijiong Li, Dongfang Guo, Yanyue Liu, Haiyan Wang, Lingli Wang, Recent advances and challenges in biomass-derived porous carbon nanomaterials for supercapacitors, Chemical Engineering Journal, **2020,** 397, 125418. DOI:10.1016/j.cej.2020.125418.

[21] Rishika Chakraborty, Vilya K, Mukul Pradhan, Arpan Kumar Nayak, Recent advancement of biomass-derived porous carbon based materials for energy and environmental remediation applications, Journal of Materials Chemistry A, **2022,** 10, 6965. DOI:10.1039/D1TA10269A.

[22] Lu Luo, Yuling Lan, Qianqian Zhang, Jianping Deng, Lingcong Luo, Qinzhi Zeng, Haili Gao, Weigang Zhao, A review on biomass-derived activated carbon as electrode materials for energy storage supercapacitors, Journal of Energy Storage, **2022,** 55, 105839. DOI:10.1016/j.est.2022.105839.

[23] Liyuan Qin, Zhiwei Hou, Shuang Lu, Shuang Liu, Zhongyuan Liu, Enchen Jiang, Porous carbon derived from pine nut shell prepared by steam activation for supercapacitor electrode material, International Journal of Electrochemical Science, **2019,** 14, 8907. DOI:10.20964/2019.09.20.

[24] Ui-Su Im, Jiyoung Kim, Seon Ho Lee, Song mi Lee, Byung-Rok Lee, Dong-Hyun Peck, Doo-Hwan Jung, Preparation of activated carbon from needle coke via two-stage steam activation process, Materials Letters, **2019,** 237, 22. DOI:10.1016/j.matlet.2018.09.171.

[25] Damilola Momodu, Moshawe Madito, Farshad Barzegar, Abdulhakeem Bello, Abubakar Khaleed, Okikiola Olaniyan, Julien Dangbegnon, Ncholu Manyala, Activated carbon derived from tree bark biomass with promising material properties for supercapacitors, Journal of Solid State Electrochemistry, **2017,** 21, 859. DOI:10.1007/s10008-016-3432-z.

[26] Zuo Chen, Man Zhang, Yuchen Wang, Zhiyu Yang, Di Hu, Yetao Tang, Kai Yan, Controllable synthesis of nitrogen-doped porous carbon from metal-polluted miscanthus waste boosting for supercapacitors, Green Energy & Environment, **2021,** 6, 929. DOI:10.1016/j.gee.2020.07.015.

[27] Yuhui Ma, Comparison of activated carbons prepared from wheat straw via $ZnCl_2$ and KOH activation, Waste and Biomass Valorization, **2017,** 8, 549. DOI:10.1007/s12649-016-9640-z.

[28] Guosai Jiang, Raja Arumugam Senthil, Yanzhi Sun, Thangvelu Rajesh Kumar, Junqing Pan, Recent progress on porous carbon and its derivatives from plants as advanced electrode materials for supercapacitors, Journal of Power Sources, **2022,** 520, 230886. DOI:10.1016/j.jpowsour.2021.230886.

[29] Yuchen Wang, Yaoyu Liu, Zuo Chen, Man Zhang, Biying Liu, Zhenhao Xu, Kai Yan, In situ growth of hydrophilic nickel-cobalt layered double hydroxides nanosheets on biomass waste-derived porous carbon for high-performance hybrid supercapacitors, Green Chemical Engineering, **2022,** 3, 55. DOI:10.1016/j.gce.2021.09.001.

[30] Xin Li, Wanrong Bu, Ke Zhu, Yuwen Chen, Xiaoying Liang, Bin Wang, Yuchen Wang, Kai Yan, Fe-nanocluster embedded biomass-derived carbon for efficient photo-Fenton-like activity in water purification, Separation and Purification Technology, **2024,** 337, 126382. DOI:10.1016/j.seppur.2024.126382.

[31] Haobin Feng, Mingtao Zheng, Hanwu Dong, Yong Xiao, Hang Hu, Zhongxin Sun, Chao Long, Yijin Cai, Xiao Zhao, Haoran Zhang, Bingfu Lei, Yingliang Liu, Three-dimensional honeycomb-like hierarchically structured carbon for high-performance supercapacitors derived from high-ash-content sewage sludge, Journal of Materials Chemistry A, **2015,** 3, 15225. DOI:10.1039/C5TA03217B.

[32] M. Genovese, K. Lian, Polyoxometalate modified pine cone biochar carbon for supercapacitor electrodes, Journal of Materials Chemistry A, **2017**, 5, 3939. DOI:10.1039/C6TA10382K.

[33] Kai Wang, Ning Zhao, Shiwen Lei, Rui Yan, Xiaodong Tian, Junzhong Wang, Yan Song, Defang Xu, Quangui Guo, Lang Liu, Promising biomass-based activated carbons derived from willow catkins for high performance supercapacitors, Electrochimica Acta, **2015**, 166, 1. DOI:10.1016/j.electacta.2015.03.048.

[34] Lichao Ge, Can Zhao, Mingjin Zuo, Jie Tang, Wen Ye, Xuguang Wang, Yuli Zhang, Chang Xu, Review on the preparation of high value-added carbon materials from biomass, Journal of Analytical and Applied Pyrolysis, **2022**, 168, 105747. DOI:10.1016/j.jaap.2022.105747.

[35] Jun'ichi Hayashi, Atsuo Kazehaya, Katsuhiko Muroyama, A. Paul Watkinson, Preparation of activated carbon from lignin by chemical activation, Carbon, **2000**, 38, 1873. DOI:10.1016/S0008-6223(00)00027-0.

[36] Zhi-Long Yu, Guan-Cheng Li, Nina Fechler, Ning Yang, Zhi-Yuan Ma, Xin Wang, Markus Antonietti, Shu-Hong Yu, Polymerization under hypersaline conditions: A robust route to phenolic polymer-derived carbon aerogels, Angewandte Chemie International Edition, **2016**, 55, 14623. DOI:10.1002/anie.201605510.

[37] Guofu Ma, Qian Yang, Kanjun Sun, Hui Peng, Feitian Ran, Xiaolong Zhao, Ziqiang Lei, Nitrogen-doped porous carbon derived from biomass waste for high-performance supercapacitor, Bioresource Technology, **2015**, 197, 137. DOI:10.1016/j.biortech.2015.07.100.

[38] Guangzhen Zhao, Yanjiang Li, Guang Zhu, Junyou Shi, Ting Lu, Likun Pan, Biomass-based N, P, and S self-doped porous carbon for high-performance supercapacitors, ACS Sustainable Chemistry & Engineering, **2019**, 7, 12052. DOI:10.1021/acssuschemeng.9b00725.

[39] Yuanyuan Sun, Hong Li, Guangci Li, Baoyu Gao, Qinyan Yue, Xuebing Li, Characterization and ciprofloxacin adsorption properties of activated carbons prepared from biomass wastes by H_3PO_4 activation, Bioresource Technology, **2016**, 217, 239. DOI:10.1016/j.biortech.2016.03.047.

[40] Hai Liu, Cheng, Haiming Wu, Sustainable utilization of wetland biomass for activated carbon production: A review on recent advances in modification and activation methods, Science of the Total Environment, **2021**, 790, 148214. DOI:10.1016/j.scitotenv.2021.148214.

[41] Emine Yagmur, I. Isil Gurten Inal, Yavuz Gokce, T. Gamze Ulusoy Ghobadi, Tugce Aktar, Zeki Aktas, Examination of gas and solid products during the preparation of activated carbon using phosphoric acid, Journal of Environmental Management, **2018**, 228, 328. DOI:10.1016/j.jenvman.2018.09.046.

[42] Ridzuan Zakaria, Nur Azimah Jamalluddin, Mohamad Zailani Abu Bakar, Effect of impregnation ratio and activation temperature on the yield and adsorption performance of mangrove based activated carbon for methylene blue removal, Results in Materials, **2021**, 10, 100183. DOI:10.1016/j.rinma.2021.100183.

[43] Tanveer Ahmed Khan, Anisah Sajidah Saud, Saidatul S. Jamari, Mohd Hasbi Ab Rahim, Ji-Won Park, Hyun-Joong Kim, Hydrothermal carbonization of lignocellulosic biomass for carbon rich material preparation: A review, Biomass and Bioenergy, **2019**, 130, 105384. DOI:10.1016/j.biombioe.2019.105384.

[44] Axel Funke, Felix Ziegler, Hydrothermal carbonization of biomass: A summary and discussion of chemical mechanisms for process engineering, Biofuels, Bioproducts and Biorefining, **2010**, 4, 160. DOI:10.1002/bbb.198.

[45] Chao Wang, Ye Xiong, Hanwei Wang, Chunde Jin, Qingfeng Sun, Naturally three-dimensional laminated porous carbon network structured short nano-chains bridging nanospheres for energy storage, Journal of Materials Chemistry A, **2017**, 5, 15759. DOI:10.1039/C7TA04178K.

[46] Yong-Qing Zhao, Min Lu, Peng-Yu Tao, Yun-Jie Zhang, Xiao-Ting Gong, Zhi Yang, Guo-Qing Zhang, Hu-Lin Li, Hierarchically porous and heteroatom doped carbon derived from tobacco rods for supercapacitors, Journal of Power Sources, **2016**, 307, 391. DOI:10.1016/j.jpowsour.2016.01.020.

[47] Wenya Ao, Jie Fu, Xiao Mao, Qinhao Kang, Chunmei Ran, Yang Liu, Hedong Zhang, Zuopeng Gao, Jing Li, Guangqing Liu, Jianjun Dai, Microwave assisted preparation of activated carbon from biomass: A review, Renewable and Sustainable Energy Reviews, **2018,** 92, 958. DOI:10.1016/j.rser.2018.04.051.

[48] Madhuchhanda Bhattacharya, Tanmay Basak, A review on the susceptor assisted microwave processing of materials, Energy, **2016,** 97, 306. DOI:10.1016/j.energy.2015.11.034.

[49] Sabzoi Nizamuddin, Humair Ahmed Baloch, M. T. H. Siddiqui, N. M. Mubarak, M. M. Tunio, A. W. Bhutto, Abdul Sattar Jatoi, G. J. Griffin, M. P. Srinivasan, An overview of microwave hydrothermal carbonization and microwave pyrolysis of biomass, Reviews in Environmental Science and Bio/Technology, **2018,** 17, 813. DOI:10.1007/s11157-018-9476-z.

[50] Yu-Fong Huang, Pei-Te Chiueh, Chun-Hao Shih, Shang-Lien Lo, Liping Sun, Yuan Zhong, Chunsheng Qiu, Microwave pyrolysis of rice straw to produce biochar as an adsorbent for CO_2 capture, Energy, **2015,** 84, 75. DOI:10.1016/j.energy.2015.02.026.

[51] Peter Nai Yuh Yek, Rock Keey Liew, Mohammad Shahril Osman, Chern Leing Lee, Joon Huang Chuah, Young-Kwon Park, Su Shiung Lam, Microwave steam activation, an innovative pyrolysis approach to convert waste palm shell into highly microporous activated carbon, Journal of Environmental Management, **2019,** 236, 245. DOI:10.1016/j.jenvman.2019.01.010.

[52] Shiela Marie Villota, Hanwu Lei, Elmar Villota, Moriko Qian, Jeffrey Lavarias, Victorino Taylan, Ireneo Agulto, Wendy Mateo, Marvin Valentin, Melba Denson, Microwave-assisted activation of waste cocoa pod husk by H_3PO_4 and KOH—Comparative insight into textural properties and pore development, ACS Omega, **2019,** 4, 7088. DOI:10.1021/acsomega.8b03514.

[53] Osvaldo Pezoti Junior, André L. Cazetta, Ralph C. Gomes, Érica O. Barizão, Isis P. A. F. Souza, Alessandro C. Martins, Tewodros Asefa, Vitor C. Almeida, Synthesis of $ZnCl_2$-activated carbon from macadamia nut endocarp (Macadamia integrifolia) by microwave-assisted pyrolysis: Optimization using RSM and methylene blue adsorption, Journal of Analytical and Applied Pyrolysis, **2014,** 105, 166. DOI:10.1016/j.jaap.2013.10.015.

[54] Qing-Song Liu, Tong Zheng, Peng Wang, Liang Guo, Preparation and characterization of activated carbon from bamboo by microwave-induced phosphoric acid activation, Industrial Crops and Products, **2010,** 31, 233. DOI:10.1016/j.indcrop.2009.10.011.

CHAPTER 3

Biomass-Derived Porous Carbon for Supercapacitors

Sunaina Saini, Aman Joshi, and Prakash Chand*

ABSTRACT

This chapter examines the various forms and modifications of biomass-derived porous carbon for supercapacitors in order to improve its electrochemical properties. The importance of porous carbon with a large surface area, heteroatom doping as a performance enhancer, and the development of hybrid composites for synergistic effects are all covered in this chapter. Activation techniques are often used after the carbonization of biomass feedstock to customize the pore size distribution and surface area. Various factors collectively contribute to making supercapacitors with such materials attractive for various energy storage needs in both portable electronics and emerging technologies like electric vehicles and renewable energy systems.

3.1 INTRODUCTION

Supercapacitors, often referred to as electrochemical capacitors or ultracapacitors, have become a potential energy storage technology because of their outstanding cycling stability, high power density, and quick charge/discharge characteristics[1]. Supercapacitors, as opposed to conventional electrochemical energy storage systems like batteries, store energy by the physical separation of charges at the electrode-electrolyte interface[2,3]. They have a number of benefits as a result of this fundamental difference, including almost infinite charge-discharge cycles, rapid energy release, and the capacity to deliver high bursts of power. Supercapacitors with increased energy storage capacity are still being developed, which is a major challenge[4].

The electrode material is one of the essential elements of supercapacitors and is crucial in determining the overall performance of the device[5]. In supercapacitor research, electrode materials with a large surface area and excellent electrical conductivity are highly desired. Due to their exceptional qualities, porous carbon materials have become

DOI: 10.1201/9781003520566-3

the preferred choice for the fabrication of electrodes in this situation[6]. These substances have an extensive specific surface area and a three-dimensional network of pores that speed up ion transport and adsorption at the electrode-electrolyte interface, enhancing the capacity of the material for energy storage[7]. Due to its well-known porous nature, activated carbon has often been the preferred material for supercapacitor electrodes. However, in recent years, scientists have concentrated on biomass-derived porous carbon as a viable and affordable substitute. Because biomass-derived carbon compounds are made from renewable resources, including agricultural waste, wood, and plant-derived precursors, as shown in Figure 3.1, they are both environmentally and commercially viable[8]. The synthesis, characterization, and use of porous carbon materials produced from biomass for supercapacitors are the subjects of this developing field of research. After the carbonization of biomass feedstock, activation techniques are frequently employed to tailor the pore size distribution and surface area[9]. The resulting porous carbon materials have a special combination of characteristics, such as highly specialized surface areas, customizable pore architectures, and superior electrical conductivity, all of which are necessary for supercapacitors to achieve high energy densities and power densities[10].

In this chapter, we will go into the realm of porous carbon made from biomass for supercapacitors, looking at its different forms and alterations to enhance its electrochemical

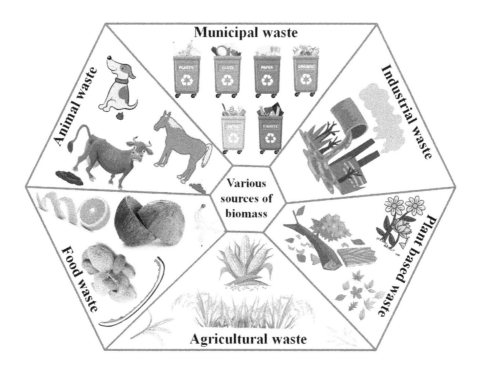

FIGURE 3.1   Various sources of biomass to produce porous activated carbon.

capabilities. We will go through the significance of porous carbon with a high surface area, heteroatom doping as a performance booster, and the creation of hybrid composites for synergistic effects.

3.2 HIGH SURFACE AREA POROUS CARBON

High surface area porous carbon materials have attracted a lot of interest in a number of scientific and industrial applications because of their remarkable properties, such as high surface area, tunable porosity, superior electrical conductivity, and chemical durability. These materials have uses in a variety of fields, including gas separation, environmental cleanup, catalysis, energy storage (such as supercapacitors and batteries), and adsorption[11,12]. This section will discuss the fundamentals of high surface area porous carbon compounds and give an outline of the processes used to create them. Porous carbon materials are distinguished by a large surface area per unit volume and a three-dimensional network of linked pores. The abundance of micro- and mesopores inside the carbon structure is principally responsible for the high surface area[13]. They gain a number of beneficial qualities because of this distinctive quality, including:

  i. **High absorbing power:** They are ideal for gas and liquid adsorption applications because the pores are plentiful and provide a lot of surface area for molecules to adsorb onto.

 ii. **High electrical conductivity:** Due to their superior electrical conductivity, porous carbons are frequently utilized as electrode materials in energy storage systems like supercapacitors and batteries.

iii. **Chemical stability:** Carbon is chemically stable, which makes porous carbon materials resistant to chemical reactions and degradation under various environmental conditions.

 iv. **Tailorable properties:** The porosity, pore size distribution, and surface functionalization of porous carbons can be tuned to meet specific application requirements.

3.2.1 Synthesis of High Surface Area Porous Carbon

There are several methods for synthesizing high surface area porous carbon materials[8]. The choice of method depends on the desired properties and intended applications, as shown in Figure 3.2. Here are some common synthesis techniques:

  i. **Activation:** Activation is a widely used method that involves the chemical or physical treatment of carbon precursors (such as coconut shells, wood, or polymers) to create pores. Physical activation typically uses high temperatures and an inert atmosphere, while chemical activation involves impregnating the precursor with activating agents (e.g., KOH, $ZnCl_2$) and heating. The precursor is then carbonized to form porous carbon.

ii. *Template synthesis:* In this method, a sacrificial template (such as silica spheres or block copolymers) is used to create pores within the carbon structure. After carbonization, the template is removed, leaving behind a porous carbon material.

iii. *Carbonization of organic precursors:* Organic precursors, such as biomass, polymers, or resins, can be directly carbonized at elevated temperatures in an inert atmosphere. The resulting carbon material can have inherent porosity depending on the precursor used.

iv. *Chemical vapor deposition (CVD):* Carbon can be deposited onto a substrate from gaseous carbon-containing precursors using CVD techniques. By controlling the deposition conditions, porous carbon films or structures can be synthesized.

v. *Hydrothermal carbonization (HTC):* HTC is a green synthesis method that involves the hydrothermal treatment of biomass or organic precursors at elevated temperatures and pressures. It results in the formation of porous carbon materials.

vi. *Electrochemical synthesis:* Porous carbons can also be synthesized by electrochemical methods, such as electro spraying or electrospinning of carbon-containing precursors followed by carbonization.

The choice of synthesis method depends on factors like desired pore size distribution, surface area requirements, scalability, and cost-effectiveness. The resulting porous carbon

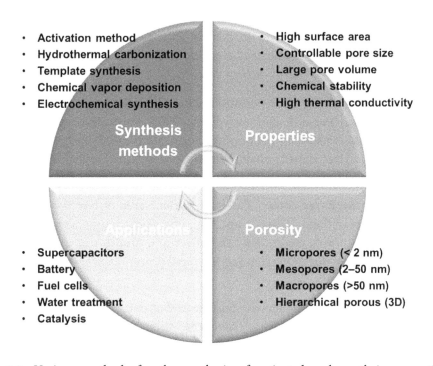

FIGURE 3.2 Various methods for the synthesis of activated carbon, their properties, and applications.

materials can be further altered using post-processing techniques like chemical functionalization or doping to improve their properties for particular purposes. With continued efforts to enhance their performance in many applications, high surface area porous carbons remain a fascinating area of research and development.

3.2.2 Importance of Surface Area

The surface area of porous carbon is a critical factor affecting supercapacitor performance. More electrochemical processes can take place at places with a large surface area, which improves capacitance. Porous carbon materials made from biomass frequently have surface areas per gram that range from hundreds to thousands of square meters, making them very desirable for supercapacitor applications. The high surface area in porous carbon made from biomass is important for supercapacitor applications because it can increase capacitance, energy density, charge/discharge rates, cycle life, sustainability, and affordability[7,14]. These factors collectively contribute to making supercapacitors with such materials attractive for various energy storage needs in both portable electronics and emerging technologies like electric vehicles and renewable energy systems.

i. ***Enhanced electrode-electrolyte interaction:*** Greater contact between the porous carbon electrode and the electrolyte is made possible by a high surface area. Supercapacitors' high capacitance and quick charge/discharge rates depend on efficient ion adsorption and desorption at the electrode-electrolyte interface, which is made possible by the larger contact area.

ii. ***Increased capacitance:*** The capacitance of a supercapacitor is directly correlated with the surface area of the electrode material. Higher surface area enables the electrode-electrolyte interface to hold more charge, increasing the capacitance values. This, in turn, leads to improved energy storage capacity.

iii. ***Improved energy density:*** Supercapacitors with high surface area electrodes can store more energy per unit mass or volume. This is crucial for applications where space and weight constraints are important, such as portable electronics and electric vehicles. High energy density enables supercapacitors to complement or replace traditional energy storage devices like batteries.

iv. ***Faster charge/discharge rates:*** The large surface area of biomass-derived porous carbon enables rapid ion diffusion and high charge/discharge rates. Supercapacitors with high surface area electrodes can deliver and absorb energy quickly, making them suitable for applications that require rapid power delivery, such as regenerative braking in electric vehicles.

v. ***Long cycle life:*** Biomass-derived porous carbon with a high surface area often exhibits excellent cyclic stability. This means that the supercapacitor can undergo many charge/discharge cycles without significant degradation in performance. This durability is a desirable characteristic for energy storage systems in various applications.

vi. *Sustainable material source:* Biomass-derived porous carbon is typically produced from renewable and sustainable sources, such as agricultural residues, wood, or algae. This eco-friendly aspect aligns with the growing demand for sustainable energy storage solutions.

vii. *Cost-effectiveness:* Biomass-derived porous carbon is often cost-effective to produce, especially when compared to some other high-surface-area carbon materials like activated carbon. This cost-effectiveness makes supercapacitors more affordable and accessible for a wide range of applications.

3.2.3 Tunable Porosity

Tunable porosity in biomass-derived porous carbon materials is a key feature that can be controlled and optimized for various applications, such as adsorption, energy storage, catalysis, and more[15]. The porosity of these materials can be adjusted at different scales, including micropores (less than 2 nm), mesopores (2–50 nm), and macropores (greater than 50 nm), depending on the specific requirements of the application[16]. Here are some ways in which porosity can be tuned in biomass-derived porous carbon:

i. *Choice of biomass source:* The starting biomass material plays a significant role in determining the porosity of the derived carbon. Different biomass sources, such as wood, coconut shells, rice husks, and corn cobs, have varying structures and compositions, leading to different pore structures in the resulting carbon material.

ii. *Carbonization temperature:* The temperature at which the biomass is carbonized (pyrolyzed) can influence the porosity. Higher temperatures tend to result in greater carbonization and the development of more micropores, while lower temperatures may favor the preservation of mesopores and macropores.

iii. *Activation:* Activation involves the chemical or physical treatment of carbonized biomass to create pores. Chemical activation typically involves treating the carbon with an activating agent like KOH or $ZnCl_2$, while physical activation is done through gasification with CO_2 or steam. The choice of activation method and conditions can be adjusted to control pore size and distribution.

iv. *Carbonization atmosphere:* The type of atmosphere (e.g., inert, oxidative, or reducing) during the carbonization process can affect the porosity of the resulting carbon. For instance, an inert atmosphere can promote the development of micropores, while an oxidative atmosphere may lead to larger pores.

v. *Precursor pre-treatment:* The porosity of the resultant carbon can be altered by pre-treating the biomass precursor, such as by washing, treating it with acid, or impregnating it with chemicals.

vi. *Post-treatment:* After carbonization, further treatments, such as chemical etching, can be applied to tailor the pore structure of the carbon material.

vii. *Template-directed synthesis:* Biomass-derived carbon can be synthesized using templates, where a sacrificial material with a specific shape and size is used to generate pores in the carbon. This allows precise control over the pore size and arrangement.

viii. *Mixing with other materials:* Mixing biomass-derived carbon with other materials like polymers, metal nanoparticles, or other carbon materials can also influence porosity and functionality.

ix. *Activation time and temperature:* If activation is used, the duration and temperature of activation can be adjusted to fine-tune the porosity of the carbon.

x. *Post-synthesis treatments:* After the carbon material is obtained, additional treatments such as acid washing, heat treatment, or chemical modification can be employed to modify the porosity.

The specific method used to tune the porosity of biomass-derived porous carbon will depend on the desired application and the characteristics of the biomass feedstock. Researchers and engineers often optimize these parameters to achieve the desired pore size distribution and surface area for a particular application, whether it's for water purification, energy storage in supercapacitors, or catalysis.

3.3 HETEROATOM-DOPED POROUS CARBON

Heteroatom-doped biomass-derived porous carbon materials have gained significant attention in recent years for various energy storage applications, including supercapacitors. These materials offer several advantages, such as high surface area, excellent electrical conductivity, and tunable pore structures. The heteroatoms (usually nitrogen, sulphur, or oxygen) introduced during the carbonization process can enhance the electrochemical performance of the carbon material by modifying its electronic properties and surface chemistry. Kim and coworkers presented an efficient and exothermic pyrolysis process involving a mixture of Mg/K/Mg·K-nitrate-urea-cellulose[17]. The result is nitrogen-doped porous carbon material with changeable pore properties after high-temperature carbonization and washing treatment, as shown in the SEM images of Figure 3.3.

The resulting porous carbon exhibits impressive properties in a two-electrode test, including a high specific capacitance of 279 F/g at 1 A/g in a 6 M KOH electrolyte, exceptional cycling stability of over 89% capacitance retention after 10,000 cycles, and robust rate capability. Liu et al. developed a method for creating dual-doped carbon microspheres based on alkali lignin, incorporating both nitrogen and phosphorus[18]. These microspheres, referred to as MLCM, were synthesized through a process involving pre-oxidation and carbonization of an ionic liquid solution containing lignin ([Mmim] DMP). They were then utilized as eco-friendly materials for supercapacitor electrodes and achieved 338.2 F/g specific capacitance at 0.8 A/g current density.

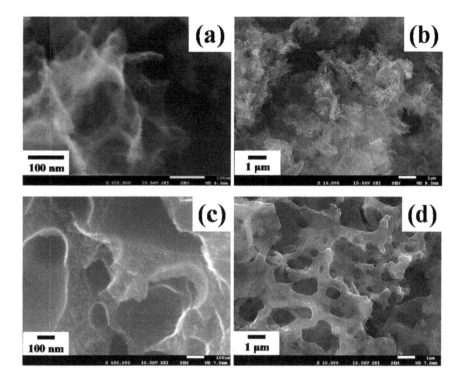

FIGURE 3.3   (a, b) SEM images after washing for sample Mg8K4Ur20Cot1g-900. (c, d) SEM images after washing for K10Ur20Cot1g-900[17].

3.3.1 Doping Mechanisms

In order to dope heteroatoms, non-carbon substances like nitrogen, sulphur, or phosphorus must be incorporated into the carbon lattice. The electrical structure of carbon can be changed by these heteroatoms, improving its electrochemical characteristics. Nitrogen doping, in particular, has received significant attention for its ability to improve capacitance and conductivity. Here are some of the key mechanisms involved in heteroatom doping in biomass-derived porous carbon:

i. *Precursor selection:* The choice of biomass precursor plays a crucial role in heteroatom doping. Biomass sources rich in nitrogen, sulphur, or oxygen-containing compounds are preferred because these elements will be incorporated into the final carbon structure.

ii. *Carbonization:* Carbonization is the process of heating the biomass precursor in an inert atmosphere (e.g., nitrogen or argon) at elevated temperatures (typically 400–900°C). During carbonization, volatile components are driven off, leaving behind a porous carbon structure.

iii. *Heteroatom incorporation:* Nitrogen, sulphur, and oxygen heteroatoms can be introduced through several mechanisms during carbonization:

- *Nitrogen doping:* Nitrogen is often introduced through the substitution of carbon atoms in the carbon lattice by nitrogen atoms, resulting in pyridinic, pyrrolic, and graphitic nitrogen.

- *Sulphur doping:* Sulphur can be incorporated by the presence of sulphur-containing compounds in the biomass precursor. During carbonization, sulphur can be retained in the carbon matrix.

- *Oxygen doping:* Oxygen can be incorporated through the oxidation of carbon atoms during carbonization or the presence of oxygen-containing functional groups in the biomass.

iv. **Chemical reactions:** During carbonization, chemical reactions involving the hetero-atoms can occur. For example, nitrogen-containing compounds may undergo pyrolysis and transformation to various nitrogen functionalities in the carbon structure.

v. **Templating effect:** The inherent porous structure of biomass-derived carbon can serve as a template for the formation of heteroatom-doped carbon. The heteroatoms can influence the pore size and distribution within the carbon material.

vi. **Surface functionalization:** Functional groups are frequently present on the surface of heteroatom-doped carbon materials, which can improve their reactivity and adsorption capabilities. Amino, carbonyl, hydroxyl, and thiol groups are a few examples of these functional groups.

vii. **Activation:** Post-treatment can further improve the development of porosity and heteroatom functionality in biomass-derived carbon through chemical or physical activation (e.g., activation with KOH or CO_2).

The complex process of heteroatom doping in biomass-derived porous carbon composites can be optimized by changing the biomass supply, the carbonization settings, and the post-treatment procedures. With the help of these techniques, carbon compounds can be created with tailored features for particular uses, such as increased catalytic activity or better electrochemical performance in energy storage devices.

3.3.2 Enhanced Electrochemical Performance

A possible method to improve the electrochemical performance of supercapacitors is heteroatom doping in porous carbon materials generated from biomass. It may result in better electron transfer kinetics, enhanced pseudocapacitance, and improved charge storage processes, among other advantages. Better electrolyte penetration and ion diffusion can be achieved by changing the wettability of the carbon surface with the addition of heteroatoms. The heteroatom, doping technique, and optimization procedure should be selected in accordance with the particular needs of the supercapacitor application. Here are some significant variables that are impacted by heteroatom doping in porous carbon generated from biomass for supercapacitors:

i. *Modification of surface chemistry:* Heteroatom doping modifies the porous carbon's surface chemistry, generating new active sites and enhancing the substance's wettability. The electrochemical activity can be improved by nitrogen doping by introducing pyridinic, pyrrolic, and graphitic nitrogen functionalities. Sulphur doping can produce functional groups containing sulphur, increasing pseudocapacitance. The introduction of phosphorus-containing groups through phosphorus doping can increase the material's conductivity and capacitance.

ii. *Defect formation:* The carbon lattice frequently experiences structural flaws brought on by heteroatom doping, which can promote ion diffusion and raise the active surface area. These flaws may serve as catalysts for redox reactions, improving the material's capacitive performance.

iii. *Improved conductivity:* Charge transport kinetics within the supercapacitor electrodes can be improved by heteroatom doping, which can increase the porous carbon material's electrical conductivity. Doping with nitrogen and phosphorus, in particular, can improve electron mobility.

iv. *Enhanced pseudocapacitance:* Some heteroatoms, such as sulphur and phosphorus, can contribute to pseudocapacitance due to their redox activity, which can significantly increase the overall capacitance of the material.

v. *Tailored doping levels and types:* The type and level of heteroatom doping should be carefully controlled to optimize the electrochemical performance for specific supercapacitor applications. The choice of biomass precursor and the doping method can influence the final properties of the carbon material.

vi. *Synergistic effects:* Combining different heteroatoms or co-doping with heteroatoms and other materials (e.g., metal oxides) can lead to synergistic effects, further improving the supercapacitor's performance.

3.4 HYBRID COMPOSITES

Hybrid composites of biomass-derived porous carbon represent an exciting frontier in materials science and sustainable technology. These advanced materials are forged at the intersection of renewable biomass resources and cutting-edge carbonization techniques, offering a multifaceted approach to addressing some of the most pressing challenges of our time. Hybrid composites combine porous carbon made from biomass with other materials, such as metal oxides, conductive polymers, or nanocarbons, to produce synergistic effects. These composites are made to maximize the benefits of each component, enhancing supercapacitors' performance. Due to their adaptability, these hybrid composites are beneficial in a range of applications. These composites are employed for efficient heavy metal and pollutant adsorption from wastewater in environmental remediation because of their outstanding porosity and chemical toughness. Additionally, they have applications in catalysis, a procedure in which metal nanoparticles are added to produce catalytic activity for a range of chemical transformations. The development of more efficient and ecologically friendly gas adsorption systems has also been made possible by the promise these materials have shown in the fields of gas storage and separation. Hybrid composites made of porous carbon generated from biomass have a wide range of potential applications in supercapacitors. They are ideal for a variety of energy storage applications because they have excellent cycle life, high energy and power densities, and quick charge-discharge speeds. Additionally, its sustainability is in line with the rising emphasis on green and renewable energy sources around the world. These hybrid composites have the potential to revolutionize energy storage technology, ushering in a new era of effective, environmentally friendly, and high-performance supercapacitors as research continues to push the boundaries of material design and synthesis.

3.4.1 Metal Oxide-Carbon Composites

Even though porous carbon materials on their own have good electrochemical characteristics, the addition of metal oxides enhances supercapacitors' overall performance. RuO_2, MnO_2, and Co_3O_4 are a few examples of metal oxides that are well-known for their high specific capacitance and pseudocapacitive behavior. They create a synergistic composite structure with porous carbon generated from biomass that combines the good conductivity and mechanical durability of carbon materials with the high capacitance of metal oxides. The following are the main benefits of combining metal oxides with porous carbon:

i. *High specific capacitance:* High specific capacitance is provided by metal oxide composites made with porous carbon generated from biomass, increasing the amount of energy they can store.

ii. *Enhanced cycling stability:* Metal oxides and porous carbon work together to increase the cycling stability of supercapacitors, resulting in long-term dependability.

iii. *Improved rate capability:* These composites are appropriate for applications needing quick charge and discharge because of their great rate capability.

iv. *Eco-friendly and sustainable:* When compared to conventional carbon sources, the use of carbon from biomass improves sustainability and lessens the impact on the environment.

Ganesh et al. produced the sweet potato-derived carbon framework using a cost-effective and straightforward low-temperature solution growth method[19]. This framework incorporates MnO_2 nanorods and is well-suited for large-scale commercial production. The value of specific capacitance for composite material increases from 468 F/g to 718 F/g in comparison with pure MnO_2 material. Moreover, the cycle stability is also improved from 78% to 89% after incorporating the MnO_2 with carbon when tested for 5000 cycles. Gouda and companions synthesized the two types of nanocomposites: one with cobalt oxide and the other with nickel oxide[20]. Nanoparticles of nickel oxide and cobalt oxide at various weight percentages (10%, 25%, 50%, and 75%) were incorporated into activated carbon nanocomposites. The activated carbon electrode exhibited the lowest specific capacitance at 105 F/g. However, the 25% nickel oxide-loaded cobalt oxide-activated carbon nanocomposite (25NiO@Co$_3$O$_4$-AC) and the 25% cobalt oxide-loaded nickel oxide-activated carbon nanocomposite (25Co$_3$O$_4$@NiO-AC) electrodes demonstrated the highest specific capacitances, reaching 800.9 F/g and 691.8 F/g, respectively, when tested at a current density of 1 A/g. Additionally, they achieved energy densities of 136.6 Wh/kg and 116.2 Wh/kg, respectively. These superior capacitance and energy density values can be attributed to the effective dispersion of nickel oxide and cobalt oxide within the prepared activated carbon matrix.

3.4.2 Conductive Polymer-Carbon Composites

Polymer composites with porous carbon are a cutting-edge and revolutionary class of materials that have immense potential for energy storage, particularly for supercapacitors. These composite materials create a hybrid material with remarkable electrochemical performance by fusing the special qualities of polymers and porous carbon structures. Porous carbon matrices made from biomass can include conductive polymers like polyaniline (PANI) or polypyrrole (PPy). The performance of these composites, which balance double-layer capacitance and pseudocapacitance, is improved overall. For instance, Mahato et al. described the production of a composite material consisting of activated carbon combined with semi-polycrystalline polyaniline (SPani-AC)[21]. This was achieved through the in-situ oxidative polymerization of aniline directly on the carbon surface, and the process took place in an aqueous HCl solution, with the temperature elevated to 60°C. The cyclic voltammetry curves display characteristics associated with surface-redox pseudocapacitance having a specific capacitance of 507 F/g when examined at a scan rate of 10 mV/s. Furthermore, the capacitive retention was determined to be 96% over the course of 4500 consecutive charge-discharge cycles. On the other hand, Chunfei et al. prepared the binary composite material by growing pseudocapacitive polypyrrole (PPy) microparticles

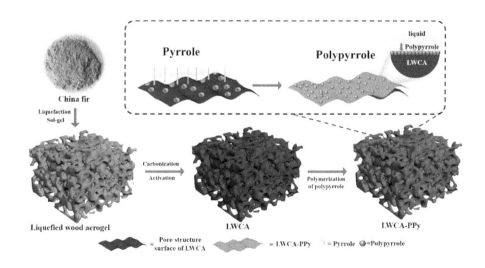

FIGURE 3.4   Schematic illustration for the synthesis of LWCA-PPy-x[22].

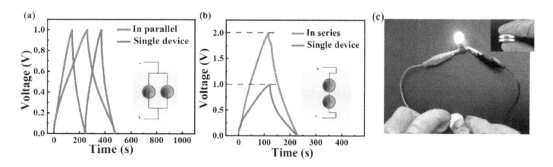

FIGURE 3.5 Practical utilization of coin cell made from LWCA-PPy-65. The GCD diagram of two coin cells connected in (a) parallel and (b) series. (c) Lightning of small bulb[22].

directly onto the surfaces of aerogels made from hierarchical porous liquefied wood carbon (LWCA), as shown in Figure 3.4[22].

The energy density and specific capacitance of the material are greatly improved by the addition of these pseudocapacitive PPy microparticles. The rapid mobility of electrolyte ions is made possible by the hierarchical structure of LWCA, which also plays a critical function in avoiding the aggregation of PPy microparticles. With a capacitance of 421.45 F/g and impressive cycle stability, the optimized hybrid capacitive material, LWCA-PPy-65 composite, exhibits great capacitive performance, maintaining 82.9% of its performance after 5000 cycles. When tested with a KOH electrolyte, the hybrid supercapacitor based on LWCA-PPy-65 exhibits a remarkable power density of 2012.8 W/kg and a high energy density of 52.0 Wh/kg and successfully lit a small bulb, as shown in Figure 3.5. Additionally, after 5000 cycles at a current density of 5 A/g, it maintains 92.81% of its capacity, demonstrating good cycling performance.

A large surface area of the porous carbon component is required in order to improve the capacitance of supercapacitors, which raises their ability to store energy. Additionally, the carbon's porosity enables effective ion transport and electrolyte penetration into the material, which results in quick charge and discharge rates. The composite is bound and structurally supported by the polymer matrix, which ensures its stability and toughness even in demanding working environments. To suit particular needs like mechanical strength, flexibility, or thermal stability, the polymers employed can be altered. By overcoming some of the major problems that supercapacitors encounter, such as low energy density and limited cycle life, these adaptive composites are revolutionizing the field of energy storage. With the advancement of this field of research, a new era of efficient and eco-friendly energy storage solutions is set to begin. With applications ranging from portable electronics to renewable energy systems and electric vehicles, polymer composites incorporating porous carbon are anticipated to greatly aid in the development of high-performance and long-lasting supercapacitors.

Porous carbon generated from biomass has developed into a malleable and resilient substance for supercapacitor electrodes. It is a practical choice for boosting the energy storage capacity of supercapacitors because of its sizable surface area, tunable porosity, compatibility with heteroatom doping, and hybrid composites. It is anticipated that porous carbon produced from biomass will be crucial in the creation of next-generation energy storage devices as this field of study progresses.

3.5 CONCLUSION

In conclusion, the advancement of supercapacitor technology is extremely promising for biomass-derived porous carbon materials. Supercapacitors are sought-after candidates for energy storage applications because of their remarkable properties, which include high surface area, heteroatom doping, and compatibility with hybrid composites. Porous carbon produced from biomass is expected to be crucial in the development of low-cost, high-performance supercapacitors as this field of study progresses.

NOTE

* Corresponding author

REFERENCES

[1] Abdul Ghani Olabi, Qaisar Abbas, Ahmed Al Makky, Mohammad Ali Abdelkareem, Supercapacitors as next generation energy storage devices: Properties and applications, Energy, **2022**, 248, 123617. DOI:10.1016/j.energy.2022.123617.

[2] Martin Winter, Ralph J. Brodd, What are batteries, fuel cells, and supercapacitors?, Chemical Reviews, **2004**, 104, 4245. DOI:10.1021/cr020730k.

[3] Sunaina Saini, Prakash Chand, Aman Joshi, Fabrication of ultrahigh supercapacitor device based on $ZnCo_2O_4@MnO_2$ with porous nanospheres decorated on flower-shaped structure, Journal of Energy Storage, **2023**, 71, 108209. DOI:10.1016/j.est.2023.108209.

[4] Stephanie L. Candelaria, Yuyan Shao, Wei Zhou, Xiaolin Li, Jie Xiao, Ji-Guang Zhang, Yong Wang, Jun Liu, Jinghong Li, Guozhong Cao, Nanostructured carbon for energy storage and conversion, Nano Energy, **2012,** 1, 195. DOI:10.1016/j.nanoen.2011.11.006.

[5] Marta Sevilla, Robert Mokaya, Energy storage applications of activated carbons: Supercapacitors and hydrogen storage, Energy & Environmental Science, **2014,** 7, 1250. DOI:10.1039/C3EE43525C.

[6] Bincy Lathakumary Vijayan, Izan Izwan Misnon, Chelladurai Karuppaiah, Gopinathan M. Anil Kumar, Shengyuan Yang, Chun-Chen Yang, Rajan Jose, Thin metal film on porous carbon as a medium for electrochemical energy storage, Journal of Power Sources, **2021,** 489, 229522. DOI:10.1016/j.jpowsour.2021.229522.

[7] Majid Shaker, Ali Asghar Sadeghi Ghazvini, Weiqi Cao, Reza Riahifar, Qi Ge, Biomass-derived porous carbons as supercapacitor electrodes—A review, New Carbon Materials, **2021,** 36, 546. DOI:10.1016/S1872-5805(21)60038-0.

[8] Sunaina Saini, Prakash Chand, Aman Joshi, Biomass derived carbon for supercapacitor applications: Review, Journal of Energy Storage, **2021,** 39, 102646. DOI:10.1016/j.est.2021.102646.

[9] T. Rajesh Kumar, Raja Arumugam Senthil, Zhigang Pan, Junqing Pan, Yanzhi Sun, A tubular-like porous carbon derived from waste American poplar fruit as advanced electrode material for high-performance supercapacitor, Journal of Energy Storage, **2020,** 32, 101903. DOI:10.1016/j.est.2020.101903.

[10] Junlei Xiao, Huiling Li, Hua Zhang, Shuijian He, Qian Zhang, Kunming Liu, Shaohua Jiang, Gaigai Duan, Kai Zhang, Nanocellulose and its derived composite electrodes toward supercapacitors: Fabrication, properties, and challenges, Journal of Bioresources and Bioproducts, **2022,** 7, 245. DOI:10.1016/j.jobab.2022.05.003.

[11] M. Salanne, B. Rotenberg, K. Naoi, K. Kaneko, P. L. Taberna, C. P. Grey, B. Dunn, P. Simon, Efficient storage mechanisms for building better supercapacitors, Nature Energy, **2016,** 1, 16070. DOI:10.1038/nenergy.2016.70.

[12] M. I. A. Abdel Maksoud, Ramy Amer Fahim, Ahmed Esmail Shalan, M. Abd Elkodous, S. O. Olojede, Ahmed I. Osman, Charlie Farrell, Ala'a H. Al-Muhtaseb, A. S. Awed, A. H. Ashour, David W. Rooney, Advanced materials and technologies for supercapacitors used in energy conversion and storage: A review, Environmental Chemistry Letters, **2021,** 19, 375. DOI:10.1007/s10311-020-01075-w.

[13] Mohammed Nabil Mahamad, Muhammad Abbas Ahmad Zaini, Zainul Akmar Zakaria, Preparation and characterization of activated carbon from pineapple waste biomass for dye removal, International Biodeterioration & Biodegradation, **2015,** 102, 274. DOI:10.1016/j.ibiod.2015.03.009.

[14] Prashant Dubey, Vishal Shrivastav, Priyanka H. Maheshwari, Shashank Sundriyal, Recent advances in biomass derived activated carbon electrodes for hybrid electrochemical capacitor applications: Challenges and opportunities, Carbon, **2020,** 170, 1. DOI:10.1016/j.carbon.2020.07.056.

[15] S. Sankar, Abu Talha Aqueel Ahmed, Akbar I. Inamdar, Hyunsik Im, Young Bin Im, Youngmin Lee, Deuk Young Kim, Sejoon Lee, Biomass-derived ultrathin mesoporous graphitic carbon nanoflakes as stable electrode material for high-performance supercapacitors, Materials & Design, **2019,** 169, 107688. DOI:10.1016/j.matdes.2019.107688.

[16] Xin-Sheng Li, Man-Man Xu, Yang Yang, Quan-Bo Huang, Xiao-Ying Wang, Jun-Li Ren, Xiao-Hui Wang, MnO_2@corncob carbon composite electrode and all-solid-state supercapacitor with improved electrochemical performance, **2019,** 12, 2379. DOI:10.3390/ma12152379.

[17] Cheong Kim, Chunyu Zhu, Yoshitaka Aoki, Hiroki Habazaki, Heteroatom-doped porous carbon with tunable pore structure and high specific surface area for high performance supercapacitors, Electrochimica Acta, **2019,** 314, 173. DOI:10.1016/j.electacta.2019.05.074.

[18] Chao Liu, Yi Hou, Youming Li, Huining Xiao, Heteroatom-doped porous carbon microspheres derived from ionic liquid-lignin solution for high performance supercapacitors, Journal of Colloid and Interface Science, **2022,** 614, 566. DOI:10.1016/j.jcis.2022.01.010.

[19] A. Ganesh, T. Sivakumar, G. Sankar, Biomass-derived porous carbon-incorporated MnO2 composites thin films for asymmetric supercapacitor: Synthesis and electrochemical performance, Journal of Materials Science: Materials in Electronics, **2022,** 33, 14772. DOI:10.1007/s10854-022-08397-1.

[20] Mostafa S. Gouda, Mona Shehab, Shacker Helmy, Moataz Soliman, Reda S. Salama, Nickel and cobalt oxides supported on activated carbon derived from willow catkin for efficient supercapacitor electrode, Journal of Energy Storage, **2023,** 61, 106806. DOI:10.1016/j.est.2023.106806.

[21] Neelima Mahato, T. V. M. Sreekanth, Kisoo Yoo, Jonghoon Kim, Semi-polycrystalline polyaniline-activated carbon composite for supercapacitor application, Molecules, **2023,** 28. DOI:10.3390/molecules28041520.

[22] Chunfei Lv, Xiaojun Ma, Ranran Guo, Dongna Li, Xuewen Hua, Tianyu Jiang, Hongpeng Li, Yang Liu, Polypyrrole-decorated hierarchical carbon aerogel from liquefied wood enabling high energy density and capacitance supercapacitor, Energy, **2023,** 270, 126830. DOI:10.1016/j.energy.2023.126830.

CHAPTER 4

Biomass-Derived Porous Carbon for Removal of Organic Pollutants from Water

Xinhua Qi,* Xiaoning Liu, Xiaoping Wang, and Haiqing Zhang

ABSTRACT

With the increasing types and contents of organic pollutants in water, the harm to the ecological environment is extremely serious. At present, there is an urgent need for effective methods to remove these pollutants. According to the investigation results, biomass-derived porous carbon materials have the potential to replace existing traditional materials because of their low price, high pore content, and strong plasticity. Moreover, it has been proven that biomass-derived porous carbon is a promising environmental material that can be used to efficiently remove organic pollutants from water in many ways. Therefore, the purpose of this chapter is to describe the mechanism and related research progress of the removal of organic pollutants by biomass-derived porous carbon through (i) adsorption, (ii) photocatalysis, and (iii) chemical oxidation pathway. This is helpful to better understand and promote the related research on the removal of organic pollutants by biomass-derived porous carbon.

4.1 INTRODUCTION

With the acceleration of urbanization and the rapid development of industry and agriculture, environmental pollution is becoming more and more serious, showing the diversification of pollution components and ways. Coincidentally, the speed of environmental pollution far exceeds its own remediation capacity and the development speed of environmental system and remediation technology, resulting in a serious environmental pollution

DOI: 10.1201/9781003520566-4

problem. Organic pollutants are common pollutants in water environment. In recent years, organic pollutants such as dyes, phenols, antibiotics, and perfluorinated compounds in water have attracted much attention because of their complex molecular structure, strong toxicity, and difficulty of degradation[1,2]. Countries all over the world have established strict emission standards for organic pollutants. Therefore, it is urgent to develop and improve a variety of economical, efficient, and environmentally friendly environmental materials and organic pollutant removal technologies[3-5].

Biomass refers to all biological organisms produced under photosynthesis, such as plants, animals, microorganisms, and so on, in which biomass energy is also an indispensable energy for human survival[6,7]. It is the fourth-largest energy after coal, oil and natural gas. Biomass is a renewable resource because of its low price, non-toxic, and low pollution. Biomass continues to regenerate at a rate of about 170 billion tons every year, which is equivalent to 15 to 20 times the current annual oil output if converted into energy[8]. The research and application of biomass energy utilization technology has become a hot issue of great concern. The porous carbon materials prepared from biomass have the characteristics of low cost, large specific surface area, high stability, and so on. The application performance of biomass-based materials can be improved by a variety of modification methods, and they have a strong affinity for the molecules of environmental organic pollutants[9]. Researchers have tried to use biomass-based porous materials to remove organic pollutants in a variety of ways[10,11]. This chapter will introduce the application of biomass-based porous materials in the removal of organic pollutants from water in three ways: adsorption, photocatalysis, and chemical oxidation.

4.2 ADSORPTION OF ORGANIC POLLUTANTS

4.2.1 Adsorption Mechanism of Biomass-Derived Porous Carbon for Organic Pollutants

In general, the adsorption of organic pollutants by biomass-derived porous carbon is the result of the compound action of multiple mechanisms[12,13]. Among them, the pore filling effect, hydrophobic interaction, π-π bond electron donor-acceptor interaction, charge interaction, and hydrogen bonding interaction are the main mechanisms affecting the adsorption of biomass-derived porous carbon and organic pollutants. The main force types of organic pollutants adsorbed by biochar are shown in Figure 4.1.

4.2.1.1 Pore Filling Effect

Pore filling effect is the main way of interaction between adsorption materials and pollutant molecules in the process of adsorption of organic pollutants by biomass-derived porous carbon[15]. In biochar with low solute concentration or low volatile matter content, the adsorption of organic pollutant molecules may be mainly determined by pore filling mechanism and follow the nonlinear Langmuir adsorption isotherm. Interestingly, this mechanism is usually accompanied by pore blockage in biomass-derived porous carbon. The pore network of biomass carbon is composed of macropores of > 50 nm, mesopores of 2–50 nm, and micropores of < 2 nm, in

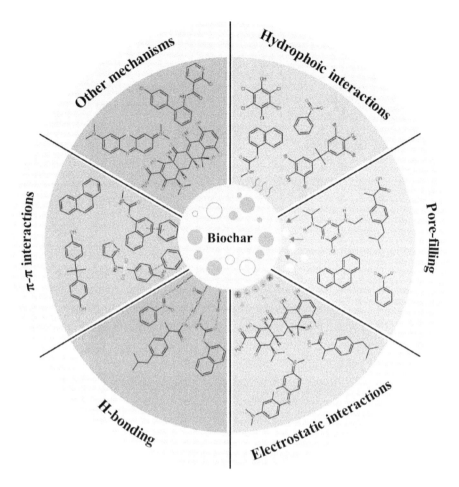

FIGURE 4.1 Summary of proposed mechanisms for organic contaminants adsorption onto biochar[14].

which micropores and mesopores of 2–20 nm are more favorable for the adsorption of biomass-derived porous carbon and organic pollutants[16,17]. In addition, when the molecular size of the adsorbate is close to the pore size of the biochar, the overlapping adsorption potential relative to the pore wall, that is, the pore filling mechanism, can enhance the adsorption of pollutant molecules[18], as shown in Figure 4.2. In the process of biomass-derived porous carbon preparation, a large amount of organic matter in biomass is pyrolyzed and carbonized to produce a carbon skeleton with a large number of pores[19]. Therefore, biomass-derived porous carbon is considered to be an ideal material for adsorbing organic pollutants.

4.2.1.2 Hydrophobic Interaction

The hydrophobicity of organic pollutant molecules and biomass-derived porous carbon materials is an important factor affecting the adsorption effect[19]. In general, the biomass-derived porous carbon with low oxidation activity on the material surface is hydrophobic,

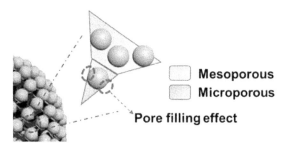

FIGURE 4.2   Pore filling effect of organic pollutants adsorbed by biomass-derived porous carbon[18].

and it can hold hydrophobic organic pollutant molecules through hydrophobicity and partition adsorption mechanism[15,20]. The adsorption of hydrophobic organic pollutants by biomass carbon mainly occurs on the material surface, so the number of organic pollutant molecules adsorbed by biomass-derived porous carbon is related to the surface properties of biomass carbon materials. The molecular structure and physicochemical properties of organic pollutants also affect the hydrophobic interaction between organic pollutants and biomass-derived porous carbon materials, for example, when there are hydrophobic functional groups in the molecular structure of soluble organic pollutants, such as methyl (-CH$_3$), nitro (-NO$_2$), alkyl (-R), ester (-COO-), ether bond (-O-) and halogen atom (-X) [14,21]. They can also be attached to the surface of biomass-derived porous carbon materials by hydrophobic interaction. In addition, the organic pollutants fixed by biomass-derived porous carbon through hydrophobic interaction will give priority to occupy the active sites on the material surface, thus reducing the occupation of active sites by other competitive substances (such as electrons, hydrophilic molecular organics)[2,4,22]. This is helpful to increase the adsorption capacity and selectivity of biomass-derived porous carbon to hydrophobic organic pollutants. However, biomass carbon with more hydrophilic groups can enhance its adsorption of hydrophilic organic pollutants to some extent. Therefore, in practical application, according to the hydrophilic and hydrophobic characteristics of the target organic pollutants, the surface characteristics of biomass-derived porous carbon materials should be adjusted through the preparation process, so as to achieve large capacity and high selectivity for biomass carbon adsorption of target organic pollutants[23]. Generally speaking, the biomass-derived porous carbon prepared by high temperature pyrolysis has higher hydrophobicity than that prepared by low temperature pyrolysis. NaOH activation can also enhance the hydrophobicity of biomass-derived porous carbon.

4.2.1.3   π-π Bond Electron Donor-Acceptor Interaction

The π-π interaction is formed by the conjugation of the single bond C bond of the composite material with the benzene ring of the pollutant[24,25]. The functional groups such as carboxylic acid and nitro group in the molecular structure of the target organic pollutant can act as π-electron acceptors, while the aromatic components on the biomass-derived porous carbon materials can act as π-electron donors, and the π-π interactions can immobilize the organic pollutant in the biomass-derived porous carbon materials[26,27]. Carboxylic acid, nitro and ketone groups on the surface of biomass carbon act as electron acceptors and interact with aromatic molecules to form π-π electron donor receptors, thus enhancing the adsorption of aromatic molecules[28]. Different types of hydroxyl and amino groups in biomass-derived porous carbon materials can also be used as π-electron donor sites. The amount of organic pollutant molecules adsorbed by biomass-derived porous carbon materials increase with the increase of the number of oxygen-containing functional groups, mainly due to π-π electron donor-acceptor interaction[29]. π-π interaction is a common way for biomass-derived porous carbon materials to adsorb organic pollutants[30]. In general, the importance of π-π interaction has been emphasized in most of the studies on the adsorption of organic pollutants by biomass-derived porous carbon materials.

4.2.1.4 Electrostatic Interaction

Electrostatic interaction refers to when the biomass-derived porous carbon materials and the target organic matter molecule charge is opposite, they can be attracted to each other through the electrostatic interaction of attraction of a mode of action, which is a common mode of action of biomass-derived porous carbon materials adsorption of organic pollutants molecules. The magnitude of the electrostatic attraction depends on the distance between the two atoms and the electric charge of each atom. pH in the reaction system is the most important factor affecting electrostatic interaction[2,31]. Many studies have shown that the intensity of electrostatic interaction may change with the shielding effect of electrolyte solution. That is, the electrostatic interaction between biomass-derived porous carbon materials and target organic molecules is affected by pH and ionic strength in the reaction system, in which pH can affect the surface charge characteristics of biomass-derived porous carbon materials[32]. The zero point charge pHpzc of biochar material is the point at which the net charge on the solid surface of the material is equal to zero and the potential difference caused by free charge between the two phases (solid/liquid) is also zero when the concentration of potential ions in the reaction system is determined to be a certain value. When pH < pHpzc, due to the protonation of functional groups on the surface of biomass carbon with positive charge at low pH, anionic organic pollutants can be adsorbed by charge action. On the other hand, when pH > pHpzc, the surface of biomass-derived porous carbon materials with charge can adsorb cationic organic pollutants by electrostatic reaction[33]. When the electrostatic interaction between ionic organic pollutants and biomass-derived porous carbon materials is repulsive, the removal rate of organic pollutants increases with the increase of ionic strength of biomass-derived porous carbon materials solution, and when the electrostatic interaction is attractive, the removal rate of pollutants decreases with the increase of ionic strength of the solution, which may be due to the competition of negatively charged ions in biomass-derived porous carbon materials.

4.2.1.5 Hydrogen Bonding Interaction

Hydrogen bond is a special intermolecular or intramolecular interaction between hydrogen and other atoms such as oxygen, fluorine, nitrogen and so on[34]. Biomass-derived porous carbon material is rich in polar groups, which enables biomass-derived porous carbon materials and organic pollutants containing electronegative elements to interact with each other through hydrogen bonding[35]. Hydrogen bonding is the main mechanism for biomass-derived porous carbon adsorption materials to hold polar organic compounds. The functional groups (hydroxyl group, carboxyl group, aldehyde group, etc.) contained in biomass-derived porous carbon adsorption materials and organic pollutants contribute to the formation of hydrogen bonds[36].

4.2.1.6 Other Mechanisms

In addition to the above adsorption mechanisms, some potential mechanisms (such as nucleophilic addition, cation bridging, cation/anion exchange, acid-base interaction) cannot be ignored[1,27,37]. Due to the specificity of the physicochemical properties of biochar and organic pollutants, the adsorption mechanism of different biomass-derived porous carbon adsorption materials to organic pollutants may be significantly different in various aspects. Moreover, the adsorption of organic pollutants by biomass-derived porous carbon materials is a complex process of multiple mechanisms. Therefore, the study of the adsorption of organic pollutants on the surface of biomass-derived porous carbon materials should be considered from the perspective of multiple mechanisms[38].

4.2.2 Adsorption of Different Types of Organic Pollutants by Biomass-Derived Porous Carbon Materials

4.2.2.1 Organic Dyes

Organic dyes are the most common organic pollutants in clothing, paper and other manufacturing industries[39]. It can threaten the ecological environment through the following pathways: (i) The majority of dyes are toxic (such as benzidine dyes, some azo dyes, etc. (ii) Dyes released into the environment enter living organisms through the food chain, posing a threat to the health of humans or other organisms. (iii) The visual pollution of water colour. Coloured water not only affects the visual senses but also prevents sunlight from penetrating the water, weakens photosynthesis of aquatic organisms, and affects the growth of aquatic organisms. When the dye is decomposed in the water environment, it will consume a lot of oxygen, resulting in anoxia, odour and corruption, which is not conducive to the growth of aquatic animals and plants. Therefore, how to effectively remove organic matter and dyes in the water environment has received extensive attention from countries around the world.

In order to develop effective technologies for removing organic dyes, researchers have made great efforts[40,41]. Adsorption, advanced oxidation, photocatalysis, biodegradation and other methods have been explored and used to remove organic dyes from aqueous environments. Adsorption is considered as an ideal method to remove dyes and other pollutants because of its cost-effectiveness, simplicity, and wide range of applications, etc. Biomass is one of the most abundant renewable carbon sources in the world, and carbon materials prepared from biomass have been widely used in the field of dye removal. If the form and function of biochar are designed on this basis, organic dyes in water can be removed more effectively by improving the preparation and modification methods of biomass carbon materials. As shown in Figure 4.3, the schematic diagram of adsorption of organic dyes by 3D porous biomass-based porous carbon materials is shown.

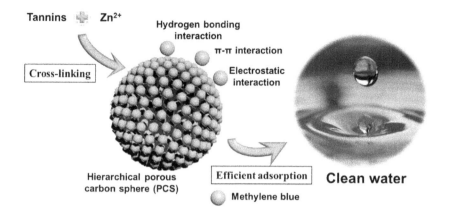

FIGURE 4.3   Schematic diagram of adsorption of organic dyes by 3D biomass-derived porous carbon adsorption materials[18].

4.2.2.2 Organic Phenolic Compounds

Organic phenolic compounds are common chemical reagents, widely used in the production of drugs, disinfectants, preservatives and other products[42]. In the use and synthesis of organic phenolic compounds, the production of phenol-containing industrial wastewater is inevitable. At present, organic phenols have become one of the common pollutants in water. Most of the phenolic compounds molecules contain complex aromatic structures, which are highly toxic and difficult to degrade. Their release into the ecological environment will pose a great threat to water, soil and human health. The main performance is:

i) Toxic to the Human Body   Phenolic compounds are prototypical poisons, which have toxic effects on all living things. Phenol can be chemically reacted through contact with human skin and mucous membranes to form insoluble protein, making cells lose vitality. The high concentration of phenol solution will make protein coagulate. Phenol can also penetrate deep, causing deep tissue damage, necrosis, until systemic poisoning. Long-term consumption of phenol-contaminated water can cause dizziness, anemia, and various neurological conditions.

ii) Hazards to Water and Aquatic Organisms   Water polluted by phenol-containing sewage will have serious adverse consequences. Due to the high oxygen consumption of phenol-containing wastewater, the oxygen balance in the water body will be disrupted. When water contains phenol to 0.002–0.015 mg/L, chlorination will produce chlorophenol odor and cannot be used as drinking water. When the water contains 0.1–0.2 mg/L, the fish have a phenolic taste, and the high concentration causes a large number of fish deaths.

iii) Hazards to Crops   Direct irrigation with untreated phenol-containing wastewater (100–750 mg/L) will cause crop death and yield reduction.

Therefore, in the water pollution control laws of countries around the world, phenolic wastewater is regarded as one of the wastewater that must be treated, and strict discharge standards have been set. Adsorption, electrochemical oxidation, Fenton reaction, photo-oxidation, ozone oxidation, extraction, membrane separation and other technologies have been studied and should be applied to the treatment of phenolic organic wastewater. Among them, due to the operational flexibility and reusability of adsorbents, adsorption is considered to be an effective and economical method to remove phenolic compounds from water bodies[43]. Biomass-derived porous carbon materials have abundant pore channels and surface functional groups, and the π-π bond electron donor acceptor interaction between biomass charcoal and organic phenolic compounds is usually the main mechanism of adsorption and removal of such pollutants.

4.2.2.3 Antibiotic

Antibiotics, as a chemical substance that can treat diseases, can be synthesized by artificial chemical processes or produced or metabolized by microorganisms[44]. The use of antibiotics to treat diseases has good therapeutic effects, fast effects, and minimal side effects, making it widely used not only in human disease treatment but also in aquaculture, animal husbandry, poultry farming, and plant cultivation. Excessive use of antibiotics leads to a large amount of residues in the Earth's ecological environment. Due to the fact that about 30% to 90% of antibiotics cannot be absorbed by any animal, there are many metabolic antibiotic residues in both water, land, and organic animals and plants. Antibiotics that have not been completely metabolized and transformed in human and animal bodies are distributed throughout their activities and metabolism, greatly affecting the survival and reproduction of other terrestrial and aquatic animals and plants. They not only disrupt the ecological environment balance in some areas but also expand environmental pollution. With the increasing demand for antibiotics, the types of antibiotics are gradually becoming diverse. A variety of antibiotics have serious impacts on the water and soil environment through biological metabolism or disposal, not only affecting the growth status of animals and plants but also causing significant harm to human health[45].

A large number of studies have confirmed that biochar materials can effectively remove antibiotics from water through adsorption. The adsorption of antibiotics on biomass-derived porous carbon materials is based on π-π interactions, hydrogen bonding interactions, electrostatic interactions, pore filling mechanisms, or the synergistic effects of multiple adsorption mechanisms[46].

4.2.2.4 Perfluorinated and Polyfluoroalkyl Substances (PFASs)

Perfluorinated and polyfluoroalkyl substances (PFASs) are a large class of new artificial aliphatic organic compounds, in which all or part of the hydrogen atoms on the carbon chain are replaced by fluorine atoms[47]. They have strong stability and are widely used in production activities such as medical equipment, pesticides, and textiles. Carbon fluorine bonds (C-F) are the strongest covalent bonds in organic chemistry, endowing PFASs with extremely strong stability, including thermal stability, chemical stability, and biological stability. Due to its excellent physical and chemical properties, it has been widely used in various industrial and consumer products in the past few decades. It is estimated that over 4700 PFASs are produced and used in human daily life[48]. Due to the lack of research on environmental toxicology and legal regulation of PFASs in the past, they have been widely distributed in various corners of the Earth and pose a potential threat to humanity that cannot be ignored. Currently, they have also received attention from researchers from various countries. Due to the lack of understanding in the early stage, many ecological environment media are polluted by PFASs. The hazards of PFASs mainly include environmental toxicity, animal toxicity, and human toxicity. If not disposed of in a timely manner, it may lead to serious consequences

Currently, the adsorption and removal of PFASs using biochar materials is a research hotspot. Previous studies have shown that the adsorption and treatment of PFASs using biochar materials mainly involve electrostatic interactions, hydrogen bonding, hydrophobic interactions, van der Waals forces, and π-π interactions[49]. Among them, electrostatic and hydrophobic effects are considered to be the main forces for biomass-derived porous carbon adsorbents to hold PFASs molecules (Figure 4.4).

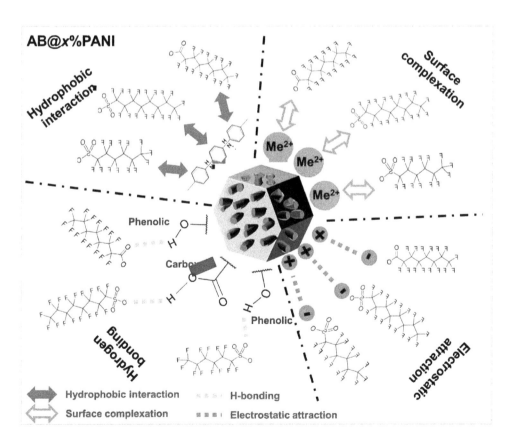

FIGURE 4.4   Mechanism of adsorption of perfluorinated compounds by a metal modified biomass carbon[50].

4.3 PHOTOCATALYTIC DEGRADATION OF ORGANIC POLLUTANTS

With the development of modern industry, a large amount of organic matter in various industrial wastewater, including animal and plant fibers, dyes, sugars, fats, organic acids, and organic raw materials, has caused serious harm to the environment and the ecosystem. With the discharge of organic pollutants, the various chemical components in the water will change, which will eventually lead to water pollution and affect the growth of aquatic organisms. Therefore, it is urgent to develop an environmentally friendly treatment technology to degrade organic pollutants in water.

Biomass-derived porous carbon material, as a new type of environmentally friendly functional material, has been widely used in pollutant removal due to its high efficiency of adsorption and photocatalytic degradation. Photocatalysis has the advantages of mild reaction conditions, high utilization rate of light absorption, high degradation efficiency, and non-toxic degradation products[51]. Due to the unique surface adsorption characteristics of the photocatalyst, it can improve the photocatalytic degradation performance, almost completely remove organic pollutants without secondary pollution[52], and has been widely studied in the field of environmental remediation and energy production, making photocatalyst the most promising technology in the degradation of organic pollutants. There are abundant persistent free radicals in biomass-derived porous carbon. When biomass-derived porous carbon is irradiated by visible or ultraviolet light, it will absorb light energy of specific wavelengths and produce reactive oxygen species (ROS). ROS plays an important role in the photocatalytic oxidation and degradation reaction process, thus improving the degradation and removal of organic pollutant[53].

4.3.1 Pure Biomass-Derived Porous Carbon Photocatalyst

Biomass-derived porous carbon is obtained by high temperature pyrolysis in an inert gas protected atmosphere (hypoxia) and is a good substitute for carbonaceous materials. It has the characteristics of high carbon content, good chemical stability and hydrophilicity, large specific surface area, high cation exchange capacity, rich surface functional groups and pore structure, etc., and is widely used in organic pollutant degradation as an adsorbent and efficient photocatalyst[54].

Gasim et al. summarized the surface active sites and functions of biomass-derived porous carbon and introduced the application of biomass-derived porous carbon in pollutant treatment[55]. The application prospect of biomass-derived porous carbon was analyzed from the aspects of surface characteristics, organic pollutant degradation mechanism, and active site participation in catalytic reaction. Wang et al. found that under solar irradiation, biomass-derived porous carbon can generate ROS through biomass-derived porous carbon base and dissolved organic matter[56]. Research by Bhavani et al. shows that biomass-derived porous carbon, as a low-cost environmentally friendly functional material, has high photo-response, effective carrier separation efficiency, and electron-hole recombination reduction and can be used for environmental organic pollutant degradation, energy generation, and waste treatment[57]. The photocatalytic reaction mechanism of biomass-derived porous carbon is shown in Figure 4.5. Therefore, based on the broad application prospects of biomass-derived porous carbon in environmental pollution control, the design of biomass-derived porous carbon based photocatalysts and organic pollutant degradation have been widely studied.

4.3.2 Metal/Non-Metal-Supported Biomass-Derived Porous Carbon Photocatalyst

The catalytic rate of photocatalyst is mainly dependent on the physical and chemical properties of the material, such as microstructure characteristics, specific surface area, particle size, light harvesting capacity, transfer and separation of photogenerated electron-hole pairs, and band gap. The co-catalyst synthesized with biomass-derived porous carbon as the carrier and modified by metal/non-metal components is also a common strategy to improve the photocatalytic performance, which is a good method to generate a large number of active free radicals, and can promote the separation of photogenerated charge and photocatalytic degradation efficiency[58]. Metal/non-metal doping can effectively adjust the energy band arrangement and regulate the microstructure and photoelectric properties.

Chen et al. prepared TiO_2/Fe-Cu-HBS (hollow biocarbon spheres) composite photocatalyst through mutually supportive growth method to achieve the degradation of tetracycline hydrochloride (TC), Rhodamine B (RhB), and carbamazepine (CBZ)[59]. In this process, Fe and Cu doping in biopolymer starch by mechanical activation technology can break the primary crystal structure and increase the specific surface area. According to research reports, the introduction of metals such as Fe and Cu can effectively regulate the electrical properties of materials, and their oxygen-containing compounds also have strong visible light absorption capacity. Cu^{2+}/Cu^+ and Fe^{3+}/Fe^{2+} have strong redox capacity, which can improve the transport and separation efficiency of photogenerated charge[60]. In particular, after Fe and Cu were incorporated into starch and carbonized, Fe-Cu-BC hollow spherical structure was obtained. Meanwhile, the surface electron density of starch-derived porous carbon increased, which was conducive to carrier transport and separation, and a large number of active adsorption sites increased, thus improving the photocatalytic degradation performance. Jing et al. prepared Zn/Fe bimetallic modified spartina biomass-derived porous carbon heterostructure ZnO-Fe@SC by magnetized

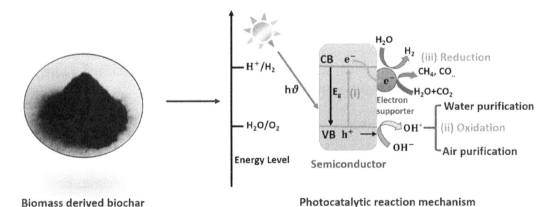

Biomass derived biochar   **Photocatalytic reaction mechanism**

FIGURE 4.5   Biomass-derived porous carbon and its photocatalytic reaction mechanism[57].

carbonization, which has excellent catalytic performance for the degradation of malachite green[61]. Photophysical characterization shows that the band gap of ZnO-Fe@SC is smaller than that of pure ZnO, which may be due to surface defects arising from bimetal (Zn-Fe) modification of biomass-derived porous carbon and the formation of Zn-O-C bonds in ZnO-Fe@SC. Under photoexcitation, electrons in the valence band of ZnO-Fe@SC composite catalyst are activated and transferred to the conduction band, which enhances the photogenerated electron transition ability and improves the photocatalytic activity.

Peng et al. prepared S, N co-doped La_2S_3 modified biomass-derived porous carbon (La_2S_3/SN-biochar) composite material by one-pot vulcanization method, which has excellent adsorption-photocatalytic degradation activity for tetracycline hydrochloride (TCH)[62]. When the non-metal is incorporated into the matrix of biomass-derived porous carbon, it can be used as an electron catcher so that the electrons can be effectively separated, and at the same time provide a large number of active sites for the adsorption of organic pollutants. In addition, the band arrangement positions of La_2S_3 and SN-biochar indicate the successful construction of the Z-scheme junction, which is conducive to the rapid transfer and separation of photogenic carriers, and enhances the visible light absorption capacity and utilization. Zheng et al. prepared $ZnFe_2O_4$/B, N-co-doped biochar (ZnFe/BN-biochar) by microwave-assisted pyrolysis technology[63]. Based on the abundant carboxyl, hydroxyl, and other functional groups on the surface, chelate with Zn^{2+} and Fe^{3+}. Moreover, these functional groups, such as O-H, C-N, C=O, B-O, Zn-O, and Fe-O, can form a large number of hydrogen bonds and effectively react with TCH molecules through hydrogen bonds and electrostatic interactions. In addition, B and N doping results in high specific surface area and abundant active sites in the composite, which improves the adsorption and photon capture abilities of TCH molecules.

4.3.3 Biomass-Derived Porous Carbon/Semiconductor Composite Photocatalyst

Due to the limited surface adsorption sites of biomass-derived porous carbon and the underdeveloped porous structure, the adsorption performance of biomass-derived porous carbon is affected. At the same time, the accessibility of photocatalyst to pollutant molecules is affected, and the removal rate of pollutants is reduced. Therefore, surface modification of biomass-derived porous carbon has attracted the attention of researchers. Biomass-derived porous carbon is considered to be a good carrier to improve the degradation performance of pollutants, and the introduction of other photocatalysts in the biomass-derived porous carbon framework significantly improves the degradation rate of organic pollutants. The construction of heterogeneous composites is considered an effective strategy to improve photocatalytic performance. When biomass-derived porous carbon and semiconductor materials are composed of composite photocatalysts, they can regulate the microstructure characteristics, electrochemical properties, and active free radicals of the composite materials during the photocatalytic oxidation process, which usually has the effects of reducing the band gap (Eg), increasing the specific surface area, improving carrier transfer and separation, and promoting the formation of active free radicals.

4.3.3.1 Reduce the Band Gap

It has been reported that the band gap of biomass-derived porous carbon/semiconductor composite photocatalyst is reduced compared to the raw material. The main reason is that the metal-O-C bond formed between the semiconductor and biomass-derived porous carbon effectively builds a bridge for the carrier transmission.

Djellabi et al. prepared a highly hydrophilic lignocellulosic biomass/TiO_2 nanocomposite photocatalyst for the reduction of Cr^{6+} ions by a simple sol-gel method[64]. As shown in Figure 4.6(a), the connection between Ti-O-C transfers photogenerated electrons from the TiO_2 conduction band to the surface of biomass-derived porous carbon, promoting the separation and efficient transport of charge carriers, and improving the photocatalytic performance. Hou et al. synthesized N-doped biomass-derived porous carbon (NCC) and prepared BiOI/NCC composite photocatalyst using NCC as a self-sacrificing template[65]. The photocatalytic performance of the composites showed that the degradation efficiency of RhB was significantly higher than that of pure BiOI. As shown in Figure 4.6(b, c), it can be seen from the UV-Vis that the spectrum lines after surface treatment with HCl and

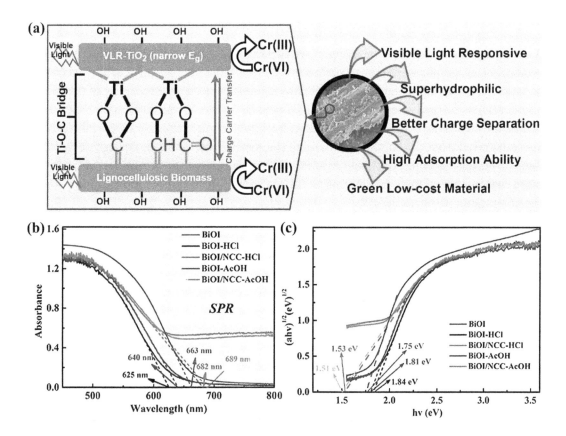

FIGURE 4.6 (a) Ti-O-C charge transfer mechanism[64]. (b, c) UV-Vis spectra and calculated band gaps[65].

AcOH appear blue shift, because the acid treatment process makes BiOI become ultra-thin structure and extends the band gap[66]. However, after recombination with NCC, the spectral lines are obviously redshifted and the band gap is narrowed. The reason for this change is mainly the formation of Bi-O-C bond, which forms a good heterogeneous interface between NCC and BiOI and plays an electron-receptor bridging role in the process of electron transfer, thus promoting the effective separation of charge carriers. A similar situation was reported by Leichtweis et al.[67]. After the synthesis of $CuFe_2O_4$ and biomass-derived porous carbon composite, the band gap is reduced by Fe-O-C and Cu-O-C bond bridging, and the photocatalytic degradation performance of RhB is optimized.

According to a large number of studies and reports, the band gap of biomass-derived porous carbon supported semiconductor photocatalyst will decrease, but according to Zhu et al.[68], this trend only exists before the critical load of biomass-derived porous carbon. Once this load is exceeded, the surface of the composite photocatalyst will be affected by phase transition and photon effect, thus reducing the photocatalytic degradation performance. Therefore, it is necessary to analyze the photocatalytic mechanism according to the photoelectrochemical characteristics of the synthesized biomass-derived porous carbon/semiconductor composite photocatalyst.

4.3.3.2 Promote Charge Transfer and Separation

As we all know, biomass-derived porous carbon is a good oxidation-reducing agent; because of its graphene-like structure, some biomass-derived porous carbon has semiconductor properties, which can significantly enhance the catalytic activity of composite photocatalysts formed after biomass-derived porous carbon is combined with other semiconductor materials. When light is irradiated, photoexcitons are generated, and electrons are transferred from the conduction band of the semiconductor to the surface of biomass-derived porous carbon, effectively inhibiting the recombination of carriers, resulting in more electron-hole pairs participating in the photocatalytic redox reaction and improving the utilization rate of carriers[69].

Sun et al. used bamboo cellulose as raw material for calcined biomass-derived porous carbon and constructed a ternary $Ni_{0.1}Co_{0.9}Fe_2O_4$/g-C_3N_4/biochar hybrid photocatalyst[70]. Compared with pure g-C_3N_4 and $Ni_{0.1}Co_{0.9}Fe_2O_4$ under visible light irradiation, CN/NCF/BC-1:1:1 composite photocatalyst with the best mass ratio showed outstanding photocatalytic degradation activity for methylene blue (MB), and the degradation rate of MB reached 96.7% within 2 hours. By electrochemical characterization, it is shown that the prepared target catalyst has smaller band gap energy, the best carrier separation efficiency, enhanced light absorption capacity, and lower charge transport resistance. As shown in Figure 4.7(a, b), the strongest photocurrent response and the smallest charge transfer impedance of CN/NCF/BC-1:1:1 composite photocatalyst prove that the composite has the best carrier separation efficiency. Zhang et al. prepared biomass-derived porous carbon (MBC) from wood meal raw materials and used it as a skeleton, synthesized MBC@Cu_2O composite photocatalyst to treat and degrade wastewater containing dyes, realized the synergistic effect of the biomass-derived porous carbon skeleton and Cu_2O heterostructure, and enabled the composite photocatalyst to efficiently degrade methyl orange (MO)[71]. According to the electrochemical characterization analysis in Figure 4.7(c, d), MBC@Cu$_2$O significantly inhibits carrier recombination and prolongs charge lifetime, thus showing significantly enhanced photocatalytic degradation performance.

4.3.3.3 Increase Specific Surface Area

The surface structure of photocatalyst can significantly optimize its photocatalytic performance. Therefore, many researches focus on increasing the specific surface area and active adsorption site, and preparing biomass-derived porous carbon/semiconductor composite photocatalyst is an effective strategy to increase the specific surface area and active site. However, biomass-derived porous carbon forms many micropores and large specific surface area in the process of high-temperature pyrolysis and carbonization, and particles may agglomerate or aggregate in the micropores of biomass-derived porous carbon when the semiconductor material is loaded on its surface, resulting in the specific surface area of

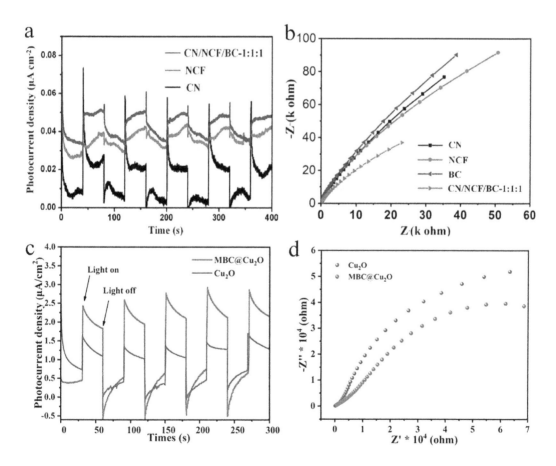

FIGURE 4.7 (a) Transient photocurrent response curves of synthetic photocatalysts. (b) EIS spectrum of the prepared photocatalysts[70]. (c) Transient photocurrent response curves. (d) EIS spectra of Cu$_2$O and MBC@Cu$_2$O[71].

the composite being smaller than that of biomass-derived porous carbon, but it has some advantages and significance for photocatalytic reactions.

He et al. prepared biomass-derived porous carbon/ZnO photocatalyst to treat organic pollutant MB in water and obtained the best photocatalytic degradation composite material for MB by adjusting the molar ratio of biomass-derived porous carbon and ZnO[72]. The specific surface area of the prepared biomass-derived porous carbon is as high as 2634.83 m^2/g, which provides a good space for the loading and uniform dispersion of ZnO, and the abundant micropores are conducive to improving the adsorption capacity of biomass-derived porous carbon. Although the specific surface area of the composite is lower than that of pure biomass-derived porous carbon, it is nearly 27 times that of pure ZnO, and its average pore diameter and pore volume distribution show the highest trend. Therefore, the unique surface characteristics of the composite photocatalyst with the best molar ratio promote the improvement of the photocatalytic performance.

Kang et al. synthesized honeycomb CdS/sulphur modified biomass-derived porous carbon composite (CdS/S-BC), which significantly enhanced the photocatalytic performance of RhB adsorption[73]. S-BC improves the specific surface area and pore volume of biomass-derived porous carbon, making it a good catalyst carrier. It also shows that annealing BC at high temperature in H_2S atmosphere is beneficial to increase the specific surface area and pore volume. Then, a series of composite materials, x-CdS/S-BC, were prepared by adjusting the molar amount of CdS (x represents the molar amount of CdS). Among them, 1-CdS/S-BC has the strongest photocatalytic degradation performance for RhB compared with other synthetic materials. Due to the optimal loading capacity, the substrate accessibility is improved, the surface agglomeration and photocorrosion are reduced, and the specific surface area is increased, so that 1-CdS/S-BC can significantly improve the degradation and removal ability of RhB.

4.3.3.4 Promote the Formation of Active Free Radicals

Active free radicals are essential for oxidation in photocatalytic reaction systems. Wang et al. synthesized ACB-Bi_2WO_6 and ACB-Bi_2MOO_6 composite photocatalysts using active biomass-derived porous carbon (ACB) as the carrier, and tested their adsorption-photocatalytic degradation properties for RhB, tetracycline (TC), and norfloxacin (NOR)[74]. The highest occupied molecular orbital (HOMO) energy level of biomass-derived porous carbon is lower than the valence band energy level of semiconductor, while the lowest unoccupied molecular orbital (LUMO) energy level of ACB is more negative than the conduction band energy level of semiconductor, and heterojunctions with such distribution characteristics belong to typical type II heterojunctions[75]. Therefore, it can be seen that the direction of electron migration is from the conduction of ACB to Bi_2WO_6. In addition, biomass-derived porous carbon has good electrical conductivity, abundant surface oxygen-containing functional groups, and persistent free radicals, and electrons have a special π-π stack structure[3,45]. As a good cocatalyst, biomass-derived porous carbon is conducive to stimulating more active free radicals to participate in the photocatalytic reaction process, thus producing more •OH and •O_2^-, optimizing the photocatalytic degradation ability.

Singlet oxygen ($^1O_2$) is a strong oxidizing agent that has a short life span, is not easily quenched by antioxidants, and has been widely studied in the field of environmental remediation. Typically, excited photosensitizers, which transfer energy to dissolved oxygen in solution, can produce $^1O_2$. Photosensitivity is divided into two types. Type I photosensitivity generates $^1O_2$ through the transfer of functional groups and carriers on the surface of the material. Type II photosensitivity is to excite photosensitizers such as dyes into a triplet state and then, through energy transfer, drive dissolved oxygen molecules to produce $^1O_2$. As an oxidizing agent, biomass-derived porous carbon has abundant surface functional groups and good surface properties, which can inhibit the quenching of $^1O_2$, stimulate many oxidants (persulfate and H_2O_2, etc.), drive $^1O_2$ to participate in the catalytic reaction, and degrade organic pollutants. Yang et al. effectively degraded sulfaethazine by preparing nitrogen-doped biomass-derived porous carbon[76]. The experimental results show that $^1O_2$ is the main active oxygen involved in photocatalytic reaction, which is mainly produced by electron transfer and energy transfer. The DFT calculation shows that pyridine N doping is also beneficial to excite the photosensitive material into a triplet state and sensitize O_2 to $^1O_2$ through energy transfer.

In summary, the researchers developed a variety of surface functionalization strategies for biomass-derived porous carbon, and prepared target photocatalysts with excellent photocatalytic performance and cycle stability for degrading various pollutants such as dyes, pharmaceutical products, and phenolic compounds. Hybrid photocatalysts have broad application prospects in the removal of environmental pollutants and environmental remediation and provide a new way to prepare various high-performance biomass-derived porous carbon modified hybrid photocatalysts.

4.4 CHEMICAL OXIDATION METHODS

Advanced oxidation processes (AOPs) are techniques for breaking down and mineralizing harmful organic pollutants predominantly in liquid situations by generating large amounts of reactive radicals via physical or chemical pathways. The core function of AOPs is their ability to produce reactive oxygen species (ROS), including free radicals and non-free radicals. A free radical is usually correlated with the species that contains one or more unpaired electrons, such as hydroxyl radicals ($\bullet OH$), persulfate radicals ($\bullet SO_4^-$), superoxide radicals ($\bullet O_2^-$), peroxyl radicals ($RO_2\bullet$), and hydroperoxyl ($HO_2\bullet$) radicals. And those free radicals can be easily converted from non-radicals (e.g., hydrogen peroxide (H_2O_2), hypochlorous acid (HOCl), and ozone (O_3), as well as singlet oxygen ($^1O_2$)[30]. AOPs include chemical oxidation (e.g., Fenton reaction, ozonation, and persulfate oxidation), photocatalysis, electrocatalysis, sonocatalysis, radiolysis, and supercritical water oxidation processes.

Among various advanced oxidation processes (AOPs), chemical oxidation approaches are considered the most promising techniques for the degradation of organic pollutants. Chemical oxidation methods not only can degrade organic contaminants quickly and completely, but they are also inexpensive and cause less harm. The valuable functional carbon-based materials conversed from biomass have also attracted attention as effective activating materials for chemical oxidants.

4.4.1 Native Biomass-Derived Porous Carbon as Catalysts

Biomass can be converted to porous carbon by direct hydrothermal and pyrolysis processes, and the final production can be adapted to different chemical oxidation reactions. Native biomass-derived porous carbon materials have the properties of large surface area, high conductivity, high stability, low cost, and abundant oxygen-containing functional groups, which are beneficial for their application as catalysts for chemical oxides. However, their structures are highly dependent on the types of biomass and calcination processes. The degree of graphitization, surface area, pore structure, defect sites, and oxygen-containing functional groups of native biomass-derived porous carbon materials all affect their activity as catalysts for oxidants during organic pollution degradation.

The degree of graphitization of biomass-derived porous carbon material has been proven to play a decisive role in the activation of oxidants, and the degree of graphitization can be adjusted by controlling the material calcination temperature[77]. The higher the degree of sp^2 hybridization, the higher the degree of graphitization and the higher the catalytic activity, which contributes to the electron transfer between oxidants and biochar. For example, Seok-Young Oh et al. contrasted the catalytic properties of biochar in phenol ozonation with those of other carbonaceous components, including graphite and granular activated carbon, and demonstrated that graphitic structures (C-π) could potentially be the cause for strengthening carbonaceous material-ozone systems[78].

Native biomass-derived porous carbon catalysts with high specific surface area and pore structure are beneficial to adsorb reactants and make them contact the active sites on the surface and inside of the catalyst to promote the oxidation reaction[79,80]. The various porous structures of carbon-based materials are conducive to the adsorption of different kinds of organic pollutants, which makes them easily contact the active sites inside the catalysts.

The defect sites of biomass-derived porous carbon material (edge sites and oxygen vacancy sites) can also reduce the adsorption energy of oxidants on the native biomass-derived porous carbon catalysts, which is to the benefit of the direct transfer of electrons from organic pollutants or carbon to oxidants[81,82].

Additionally, the oxygen-containing functional groups (e.g., (-C=O, -OH, and -COOH) on the surface of carbon materials affect the redox properties, adsorption capacity, and catalytic performance[83,84]. The carbonyl group on the surface of carbon material has lone pair electrons, which can interact with oxidants through hydrogen bond. The electrons are transferred to oxidants through hydrogen bond to form singlet oxygen. For instance, Xin et al. used waste leather as precursor to prepare porous carbon materials (LPC) and explored the catalytic performance of diethyl phthalate removal in $Fe(III)/H_2O_2$ system. As shown in Figure 4.8, the result of oxygen-containing functional group masking experiment showed that carbonyl group in LPC serves as electron supplier to facilitate direct Fe(III) reduction[85].

4.4.2 Heteroatom-Doped Biomass-Derived Porous Carbon as Catalysts

In order to improve the activation performance of biomass-derived porous carbon as catalysts in chemical oxidation, heteroatoms can be introduced into the carbon-based materials to achieve higher activity. The introduced heteroatoms have different atomic radii, atomic

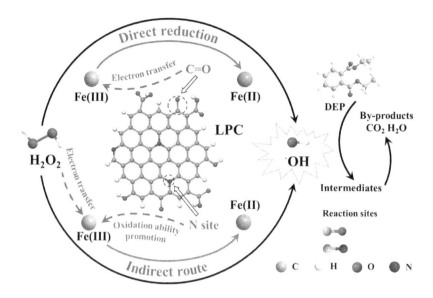

FIGURE 4.8   The proposed mechanism of the LPC/Fe(III)/H_2O_2 system[85].

orbitals, electron densities, and electronegativity with carbon. Therefore, the doping of heteroatoms can adjust the surface properties of carbon materials by breaking the network structure of the original inert carbon skeleton and inducing the unbalanced distribution of electrons. Consequently, more active sites (e.g., pyridinic N, pyrrolic N, graphitic N, and C=O groups) generate on the surface and inside the carbon materials, and their catalytic capacity is enhanced at the same time.

It is widely recognized that incorporation of heteroatoms (such as boron (B), nitrogen (N), and sulphur (S)) into the carbon-based materials could significantly change the surface structure (e.g., enhance π-π generation, increase pore structures, improve the surface area, and change intermolecular distance) and modify surface chemical properties (e.g., increase or decrease the surface functional groups, modify hydrophilic and hydrophobic properties, and improve dispersibility) as well. For example, Hung et al. prepared boron-doped biochar by using pristine brown algae and boric acid as catalysts for the activation of peroxymonosulfate (PMS). As C-N-B heterostructures were formed, both C and B atoms were significantly activated in a synergistic way, with the electron-deficient atoms (B) assisting the binding of the electron-rich atoms (N) and PMS to generate reactive complexes that eventually improved the degradation of polycyclic aromatic hydrocarbons[86].

Nitrogen atoms have a similar atomic radius to carbon atoms (0.75A vs 0.77A), so the load can reach a high level. Besides, due to the large difference in electronegativity between N and C (3.07 vs 2.55), the local electronic structure can be adjusted so that the adjacent C atoms have a high positive charge density[87–89]. In addition, in the conjugated carbon network, nitrogen atoms also have the capacity to properly affect the spin density and charge distribution of surrounding carbon atoms. They can also disrupt bonds and create lattice distortion, which increases the chemical potential of the catalytic reaction. In the application of oxidant activation, N-doping can simultaneously improve the adsorbtion ability and catalytic capacity of biochar. Electron-rich N atoms can form Lewis base sites in graphite carbon to coordinate the redox process and promote electron transfer by activating adjacent sp^2 carbon atoms to improve catalytic capacity[90]. Hence, nitrogen doping is an easy and effective way of improving catalytic activity by introducing a variety of nitrogen atoms in various configurations (graphite nitrogen, pyridine nitrogen, and pyrrole nitrogen). Xin et al. synthesized herb residue nitrogen-doped biochar (N-BC) and used it to activate peroxymonosulfate (PMS). Results showed that pyridinic-N played a pivotal role in the $^1O_2$-dominated pathway. The decomposition of PMS into $^1O_2$ was assisted by pyridinic N, which has the maximum electron density and demonstrated strong attraction to PMS molecules. Furthermore, pyridinic N could also activate the π electron of sp^2 C atoms, which would facilitate PMS activation and the production of •OH and $•SO_4^-$[91].

Doping sulphur atoms would lead to increase in surface area and a growing number of surface functional groups, along with boron and nitrogen. S-groups in biomass-derived porous carbon have a dual character, and they might be related to the synthesis procedure. On the one hand, S-groups in N/S co-doped carbon materials would accelerate the formation of N-groups. In addition, the presence of thiophene S increased the local positive charges of biomass-derived porous carbon, which encouraged electron transportation and improved peroxydisulfate (PDS) activation capacity[92]. On the other hand, S-doping made biomass-derived porous carbon negatively charged. Its electrostatic repulsion towards peroxydisulfate (PDS) and 2,4-dichlorophenol (DCP) hindered their interface contact thus leading to the decreased of degradation efficiency[93].

4.4.3 Transition Metal Composite Biomass-Derived Porous Carbon as Catalysts

The combination of transition metal and biomass-derived porous carbon materials could not only inherit some properties of conventional carbon materials, such as the electrostatic interaction, surface functional groups, pore structures, and π-π structures but also have the following effects. The rich pore structure of carbon material can help the transition metal disperse evenly so that it does not aggregate into larger particles because of magnetism. Pure transition metal has high charge transfer efficiency between different valence states. But they have the shortcoming of poor dispersibility on the surface. As a result, during the catalysis process, metal may leach out in the form of ions, which may lead to secondary pollution[82]. The coupling of transition metal and oxygen-containing functional groups on carbon material can not only inhibit metal ion leaching but also improve the activity of catalyst and maintain its good stability[94]. Transition metal composite biomass-derived porous carbon catalysts have been proven to be one of the most efficient catalysts. A variety of transition metals (e.g., Mn, Cu, Ni, Fe, Co, and Mg) have been proven to synergetically activate chemical oxide with biomass-derived carbon catalysts.

He et al. fabricated MnO_x-loaded biochar derived from waste tea leaves (Mn-nWT) for heterogeneous catalytic ozonation to degrade 2,3,5-trimethylpyrazine (TMP) and found that loading metal oxides onto biochar synergistically promoted the heterogeneous catalytic ozonation process by enhancing the physical and chemical properties. Oxygen vacancies, surface hydroxyl groups, and multivalent Mn sites on the catalysts are examples of

Lewis acid sites that could adsorb ozone and lengthen its O-O bond, promoting ozone dissociation and the production of reactive oxygen species[95].

Zhang et al. used gallic acid as the carbon precursor to synthesize Co-cross-linked ordered mesoporous carbon catalysts (OMC-Co-Tx) and employed them as peroxymonosulfate (PMS) activators for removal of sulfamethoxazole (SMX) from aqueous solutions. In addition to raising the removal ratio, the addition of Co species promoted the development of ordered mesoporous structures, which improved the mass transfer and diffusion of reaction species as well as the increase in the specific surface area of catalysts[96].

Bimetallic biochar composites are also used in chemical oxidation systems apart from individual transition metal-loaded biochar catalysts. For instance, Xin et al. synthesized biochar-modified $CuFeO_2$ ($CuFeO_2$/BC) by low temperature hydrothermal method and evaluated its catalytic performance in Fenton-like system for degradation of tetracycline (TC). In addition to reducing the agglomeration of $CuFeO_2$ particles, the addition of biochar significantly increased the BET surface area from 16.8 to 37.3 m^2/g, allowing for greater catalytic activity for the removal of tetracycline (TC). Figure 4.9 showed that the synergistic interaction of the Fe^{3+}/Fe^{2+} and Cu^{2+}/Cu^+ redox cycles would be the contributing factor to the high catalytic efficiency[97].

4.4.4 Miscellaneous Biomass-Derived Porous Carbon as Catalysts

The combination of biomass-derived porous carbon with other materials has also been explored. Compared with using them separately, the combination of two types of substances can better increase the catalytic performance for chemical oxidation.

Yang et al. formed the layered double hydroxides (LDH) on rice straw biochar (RSBC) via a simple two-step strategy: the adsorption of metal ions and the hydrothermal process. The stabilized RSBC-LDH exhibited outstanding catalytic performance for peroxymonosulfate (PMS) activation to efficiently degrade diethyl phthalate (DEP) and rhodamine B

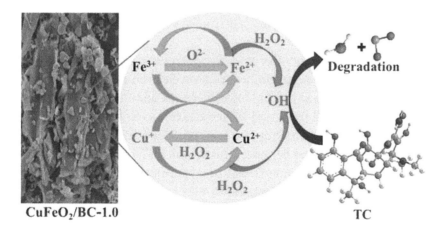

FIGURE 4.9 Proposed mechanism schematic of H_2O_2 activation on the surface of $CuFeO_2$/BC-1.0 for TC degradation[97].

(RhB) (more than 90% within 240 min). The Co^{2+} in layered double hydroxides (LDH) was the main active site in the peroxymonosulfate (PMS) activation process, and $\cdot OH$, $\cdot SO_4^-$, and $^1O_2$ were the main reactive oxygen species in this system[98].

4.4.5 Catalytic Oxidation Mechanisms with Biomass-Derived Porous Carbon

In the whole process of activated oxidants, organic pollutants are usually oxidized at the beginning due to the low oxidation potential of the non-radical pathway, while radicals can further mineralize organic pollutants and degrade intermediates into CO_2 and H_2O. Therefore, the mineralization of organic pollutants in water environment requires the interaction of radicals and non-radicals.

4.4.5.1 Radical Way

The radical pathway depends on the generation of hydroxyl radicals ($\cdot OH$), sulfate radicals ($\cdot SO_4^-$), and superoxide ion radicals ($\cdot O_2^-$). Hydroxyl radicals ($\cdot OH$) are nonselective reactive oxygen species that can react with the majority of organic pollutants. However, the diffusion of hydroxyl radicals ($\cdot OH$) in solution was restricted due to their short half-lives. Compared with hydroxyl radicals ($\cdot OH$), sulfate radicals ($\cdot SO_4^-$) have a higher standard redox potential ($E_0 = 2.5–3.1$ V) and a longer half-life (30–40 µs). Therefore, it can contact with antibiotics in the aqueous solution more sufficiently, and the corresponding degree of mineralization is higher. Typically, the catalytic ozonation process results in the production of superoxide ion radicals ($\cdot O_2^-$). It usually reacts first to form various other active species (e.g. $\cdot SO_4^-$, $\cdot OH$, and $^1O_2$) rather than immediately oxidizing organic pollutants[84].

4.4.5.2 Non-radical Way

Generally speaking, non-radical pathways are divided into direct electron transfer and singlet oxygen ($^1O_2$). In the direct electron transfer process (Figure 4.10), biomass-derived porous carbon catalysts only serve as a channel for the quick transfer of electrons from organic contaminants to oxidants and selectively oxidize organic pollutants with electron-rich moieties. As for the generation of singlet oxygen ($^1O_2$), C=O groups, structural defects, and electronegative N in catalysts were all reactive sites for decomposition of chemical oxide to produce $^1O_2$[99].

Non-radical way has relatively weak oxidative potential compared with free radical way, so it is more selective for the electrophilic attack on electron-rich substances such as dyestuffs, phenols, and antibiotics with big molecule[101]. In addition, common inorganic ions and natural organic matter in the actual wastewater can scour free radicals, which will not only reduce its oxidation capacity but also may produce harmful by-products. However, the non-free radical pathway has better selectivity and mild oxidation potential and is less affected by inorganic ions, so the non-free radical pathway is more suitable for the complex water environment in practical application[102].

4.5 CONCLUSION AND PROSPECT

To sum up, biomass-derived porous carbon materials are an ideal material for the removal of organic pollutants in water, and they can efficiently remove organic pollutants by

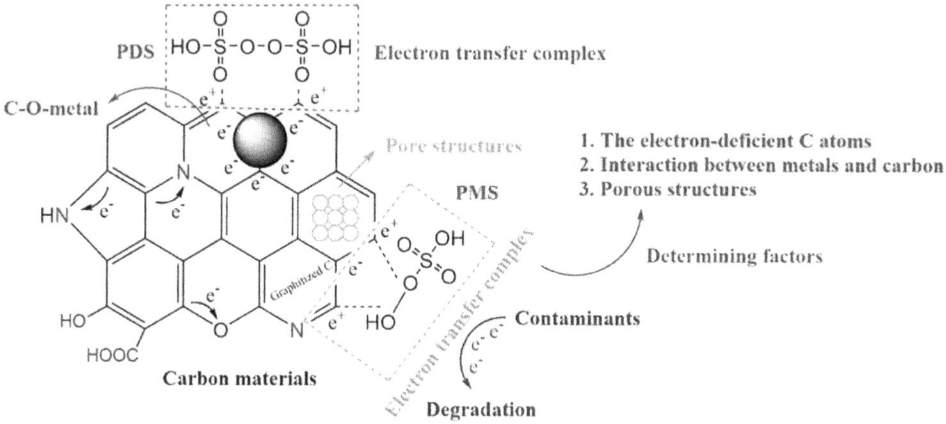

FIGURE 4.10   Generation mechanisms of electron transfer complex in carbon-persulfate-based AOPs[100].

adsorption, photocatalysis, chemical oxidation, and other ways. Follow-up research should continue to pay attention to the application of biomass porous carbon materials in the removal of organic pollutants from the ecological environment. However, there are still many bottlenecks to be explored in the current research.

i. Biomass materials have many advantages, and most of the studies are still limited to small-scale experiments in the laboratory. In the actual environment with larger scale and more complex application environment, what is the effect of biomass materials? What is the economic benefit of removing organic pollutants from porous biomass materials? Will the interaction between it and organic pollutants be disturbed by complex environmental factors? These problems need to be further studied and explored.

ii. The research on the action mechanism of biomass-derived porous carbon materials in many dimensions, such as water body, soil, atmosphere, and so on, in the process of removing organic pollutants needs to be further improved.

iii. At present, a variety of modification processes have appeared in many studies to improve the pollutant removal performance of biomass materials. Will this modified material, especially some organic and metal modified materials, do harm to the ecological environment? What is the relevant toxicology? In addition, how to recycle the used materials is also an important issue to be considered. The commercialization of biomass-derived porous carbon materials and the establishment of biochar quality standards are of great significance for future research.

NOTE

* Corresponding author

REFERENCES

[1] Mohammad Boshir Ahmed, John L. Zhou, Huu Hao Ngo, Wenshan Guo, Md Abu Hasan Johir, Kireesan Sornalingam, Single and competitive sorption properties and mechanism of functionalized biochar for removing sulfonamide antibiotics from water, Chemical Engineering Journal, **2017**, 311, 348. DOI:10.1016/j.cej.2016.11.106.

[2] Yingjie Dai, Naixin Zhang, Chuanming Xing, Qingxia Cui, Qiya Sun, The adsorption, regeneration and engineering applications of biochar for removal organic pollutants: A review, Chemosphere, **2019**, 223, 12. DOI:10.1016/j.chemosphere.2019.01.161.

[3] M. J. Ahmed, B. H. Hameed, Insight into the co-pyrolysis of different blended feedstocks to biochar for the adsorption of organic and inorganic pollutants: A review, Journal of Cleaner Production, **2020**, 265, 121762. DOI:10.1016/j.jclepro.2020.121762.

[4] Bingbing Qiu, Qianni Shao, Jicheng Shi, Chenhao Yang, Huaqiang Chu, Application of biochar for the adsorption of organic pollutants from wastewater: Modification strategies, mechanisms and challenges, Separation and Purification Technology, **2022,** 300, 121925. DOI:10.1016/j.seppur.2022.121925.

[5] Yuyan Liu, Jiawei Chen, Effect of ageing on biochar properties and pollutant management, Chemosphere, **2022,** 292, 133427. DOI:10.1016/j.chemosphere.2021.133427.

[6] Fasih Ullah Haider, Xiukang Wang, Usman Zulfiqar, Muhammad Farooq, Saddam Hussain, Tariq Mehmood, Muhammad Naveed, Yuelin Li, Cai Liqun, Qudsia Saeed, Ishtiaq Ahmad, Adnan Mustafa, Biochar application for remediation of organic toxic pollutants in contaminated soils; An update, Ecotoxicology and Environmental Safety, **2022,** 248, 114322. DOI:10.1016/j.ecoenv.2022.114322.

[7] Yi-Chen Ethan Li, Sustainable biomass materials for biomedical applications, ACS Biomaterials Science & Engineering, **2019,** 5, 2079. DOI:10.1021/acsbiomaterials.8b01634.

[8] Yufeng Ma, Yanan Xiao, Yaoli Zhao, Yu Bei, Lihong Hu, Yonghong Zhou, Puyou Jia, Biomass based polyols and biomass based polyurethane materials as a route towards sustainability, Reactive and Functional Polymers, **2022,** 175, 105285. DOI:10.1016/j.reactfunctpolym.2022.105285.

[9] Nitesh Kasera, Praveen Kolar, Steven G. Hall, Nitrogen-doped biochars as adsorbents for mitigation of heavy metals and organics from water: A review, Biochar, **2022,** 4, 17. DOI:10.1007/s42773-022-00145-2.

[10] Lei Huang, Zhixuan Luo, Xuexia Huang, Yian Wang, Jia Yan, Wei Liu, Yufang Guo, Samuel Raj Babu Arulmani, Minhua Shao, Hongguo Zhang, Applications of biomass-based materials to remove fluoride from wastewater: A review, Chemosphere, **2022,** 301, 134679. DOI:10.1016/j.chemosphere.2022.134679.

[11] Nhamo Chaukura, Bhekie B. Mamba, Shivani B. Mishra, Porous materials for the sorption of emerging organic pollutants from aqueous systems: The case for conjugated microporous polymers, Journal of Water Process Engineering, **2017,** 16, 223. DOI:10.1016/j.jwpe.2017.02.001.

[12] Amina Othmani, Juliana John, Harishkumar Rajendran, Abdeldjallil Mansouri, Mika Sillanpää, Padmanaban Velayudhaperumal Chellam, Biochar and activated carbon derivatives of lignocellulosic fibers towards adsorptive removal of pollutants from aqueous systems: Critical study and future insight, Separation and Purification Technology, **2021,** 274, 119062. DOI:10.1016/j.seppur.2021.119062.

[13] Mukarram Zubair, Hamidi Abdul Aziz, Ihsanullah Ihsanullah, Mohd Azmier Ahmad, Mamdouh A. Al-Harthi, Biochar supported CuFe layered double hydroxide composite as a sustainable adsorbent for efficient removal of anionic azo dye from water, Environmental Technology & Innovation, **2021,** 23, 101614. DOI:10.1016/j.eti.2021.101614.

[14] Zirui Luo, Bin Yao, Xiao Yang, Lingqing Wang, Zhangyi Xu, Xiulan Yan, Lin Tian, Hao Zhou, Yaoyu Zhou, Novel insights into the adsorption of organic contaminants by biochar: A review, Chemosphere, **2022,** 287, 132113. DOI:10.1016/j.chemosphere.2021.132113.

[15] Panagiotis Regkouzas, Evan Diamadopoulos, Adsorption of selected organic micro-pollutants on sewage sludge biochar, Chemosphere, **2019,** 224, 840. DOI:10.1016/j.chemosphere.2019. 02.165.

[16] Mandu Inyang, Eric Dickenson, The potential role of biochar in the removal of organic and microbial contaminants from potable and reuse water: A review, Chemosphere, **2015,** 134, 232. DOI:10.1016/j.chemosphere.2015.03.072.

[17] Abhishek Mandal, Neera Singh, T. J. Purakayastha, Characterization of pesticide sorption behaviour of slow pyrolysis biochars as low cost adsorbent for atrazine and imidacloprid removal, Science of the Total Environment, **2017,** 577, 376. DOI:10.1016/j.scitotenv.2016.10.204.

[18] Wenping Cao, Danni Li, Sisuo Zhang, Jing Ren, Xiaoning Liu, Xinhua Qi, Synthesis of hier-archical porous carbon sphere via crosslinking of tannic acid with Zn^{2+} for efficient adsorption of methylene blue, Arabian Journal of Chemistry, **2023,** 16, 105122. DOI:10.1016/j.arabjc.2023.105122.

[19] Dinesh Mohan, Ankur Sarswat, Yong Sik Ok, Charles U. Pittman, Organic and inorganic con-taminants removal from water with biochar, a renewable, low cost and sustainable adsor-bent—A critical review, Bioresource Technology, **2014,** 160, 191. DOI:10.1016/j.biortech.2014.01.120.

[20] Xinlei Liu, Rong Ji, Yu Shi, Fang Wang, Wei Chen, Release of polycyclic aromatic hydrocar-bons from biochar fine particles in simulated lung fluids: Implications for bioavailability and risks of airborne aromatics, Science of the Total Environment, **2019,** 655, 1159. DOI:10.1016/j.scitotenv.2018.11.294.

[21] K. S. A. Sohaimi, N. Ngadi, H. Mat, I. M. Inuwa, Syieluing Wong, Synthesis, characterization and application of textile sludge biochars for oil removal, Journal of Environmental Chemical Engineering, **2017,** 5, 1415. DOI:10.1016/j.jece.2017.02.002.

[22] Ning Cheng, Bing Wang, Pan Wu, Xinqing Lee, Ying Xing, Miao Chen, Bin Gao, Adsorption of emerging contaminants from water and wastewater by modified biochar: A review, Environmental Pollution, **2021,** 273, 116448. DOI:10.1016/j.envpol.2021.116448.

[23] Emilio Rosales, Jessica Meijide, Marta Pazos, María Angeles Sanromán, Challenges and recent advances in biochar as low-cost biosorbent: From batch assays to continuous-flow systems, Bioresource Technology, **2017,** 246, 176. DOI:10.1016/j.biortech.2017.06.084.

[24] Qianzhen Fang, Shujing Ye, Hailan Yang, Kaihua Yang, Junwu Zhou, Yue Gao, Qinyi Lin, Xiaofei Tan, Zhongzhu Yang, Application of layered double hydroxide-biochar composites in wastewater treatment: Recent trends, modification strategies, and outlook, Journal of Hazardous Materials, **2021,** 420, 126569. DOI:10.1016/j.jhazmat.2021.126569.

[25] Yuan-Bin Zheng, Jian-Xin Jia, Wei Shi, Jia-Jie Long, A sustainable one-step pretreatment of cotton gray fabric with 18-crown-6 as phase transfer in supercritical carbon dioxide, The Journal of Supercritical Fluids, **2021,** 175, 105269. DOI:10.1016/j.supflu.2021.105269.

[26] Hao Zheng, Qian Zhang, Guocheng Liu, Xianxiang Luo, Fengmin Li, Yipeng Zhang, Zhenyu Wang, Characteristics and mechanisms of chlorpyrifos and chlorpyrifos-methyl adsorption onto biochars: Influence of deashing and Low Molecular Weight Organic Acid (LMWOA) aging and co-existence, Science of The Total Environment, **2019,** 657, 953. DOI:10.1016/j.scitotenv.2018.12.018.

[27] Jian Shen, Gordon Huang, Chunjiang An, Shan Zhao, Scott Rosendahl, Immobilization of tetrabromobisphenol A by pinecone-derived biochars at solid-liquid interface: Synchrotron-assisted analysis and role of inorganic fertilizer ions, Chemical Engineering Journal, **2017,** 321, 346. DOI:10.1016/j.cej.2017.03.138.

[28] Yuhong Fu, Yafei Shen, Zhendong Zhang, Xinlei Ge, Mindong Chen, Activated bio-chars derived from rice husk via one- and two-step KOH-catalyzed pyrolysis for phenol adsorp-tion, Science of the Total Environment, **2019,** 646, 1567. DOI:10.1016/j.scitotenv.2018.07.423.

[29] M. J. Ahmed, B. H. Hameed, Adsorption behavior of salicylic acid on biochar as derived from the thermal pyrolysis of barley straws, Journal of Cleaner Production, **2018,** 195, 1162. DOI:10.1016/j.jclepro.2018.05.257.

[30] Xinbo Zhang, Yongchao Zhang, Huu Hao Ngo, Wenshan Guo, Haitao Wen, Dan Zhang, Chaocan Li, Li Qi, Characterization and sulfonamide antibiotics adsorption capacity of spent coffee grounds based biochar and hydrochar, Science of the Total Environment, **2020,** 716, 137015. DOI:10.1016/j.scitotenv.2020.137015.

[31] Bo Wang, Yansong Jiang, Fayun Li, Dengyue Yang, Preparation of biochar by simultaneous carbonization, magnetization and activation for norfloxacin removal in water, Bioresource Technology, **2017,** 233, 159. DOI:10.1016/j.biortech.2017.02.103.

[32] Xiaofei Tan, Yunguo Liu, Guangming Zeng, Xin Wang, Xinjiang Hu, Yanling Gu, Zhongzhu Yang, Application of biochar for the removal of pollutants from aqueous solutions, Chemosphere, **2015,** 125, 70. DOI:10.1016/j.chemosphere.2014.12.058.

[33] Mohammad Boshir Ahmed, John L. Zhou, Huu Hao Ngo, Wenshan Guo, Adsorptive removal of antibiotics from water and wastewater: Progress and challenges, Science of the Total Environment, **2015,** 532, 112. DOI:10.1016/j.scitotenv.2015.05.130.

[34] Akila G. Karunanayake, Olivia A. Todd, Morgan L. Crowley, Lindsey B. Ricchetti, Charles U. Pittman, Renel Anderson, Todd E. Mlsna, Rapid removal of salicylic acid, 4-nitroaniline, benzoic acid and phthalic acid from wastewater using magnetized fast pyrolysis biochar from waste Douglas fir, Chemical Engineering Journal, **2017,** 319, 75. DOI:10.1016/j.cej.2017.02.116.

[35] Mahtab Ahmad, Sang Soo Lee, Xiaomin Dou, Dinesh Mohan, Jwa-Kyung Sung, Jae E. Yang, Yong Sik Ok, Effects of pyrolysis temperature on soybean stover- and peanut shell-derived biochar properties and TCE adsorption in water, Bioresource Technology, **2012,** 118, 536. DOI:10.1016/j.biortech.2012.05.042.

[36] Fei Lian, Baoshan Xing, Black Carbon (Biochar) in water/soil environments: Molecular structure, sorption, stability, and potential risk, Environmental Science & Technology, **2017,** 51, 13517. DOI:10.1021/acs.est.7b02528.

[37] Cuiping Chen, Wenjun Zhou, Daohui Lin, Sorption characteristics of N-nitrosodimethylamine onto biochar from aqueous solution, Bioresource Technology, **2015,** 179, 359. DOI:10.1016/j.biortech.2014.12.059.

[38] Chen Zhang, Cui Lai, Guangming Zeng, Danlian Huang, Chunping Yang, Yang Wang, Yaoyu Zhou, Min Cheng, Efficacy of carbonaceous nanocomposites for sorbing ionizable antibiotic sulfamethazine from aqueous solution, Water Research, **2016,** 95, 103. DOI:10.1016/j.watres.2016.03.014.

[39] Hongbo Li, Xiaoling Dong, Evandro B. da Silva, Letuzia M. de Oliveira, Yanshan Chen, Lena Q. Ma, Mechanisms of metal sorption by biochars: Biochar characteristics and modifications, Chemosphere, **2017,** 178, 466. DOI:10.1016/j.chemosphere.2017.03.072.

[40] Xue Li, Jiwei Luo, Hui Deng, Peng Huang, Chengjun Ge, Huamei Yu, Wen Xu, Effect of cassava waste biochar on sorption and release behavior of atrazine in soil, Science of the Total Environment, **2018,** 644, 1617. DOI:10.1016/j.scitotenv.2018.07.239.

[41] Hasan Saygılı, Fuat Güzel, High surface area mesoporous activated carbon from tomato processing solid waste by zinc chloride activation: Process optimization, characterization and dyes adsorption, Journal of Cleaner Production, **2016,** 113, 995. DOI:10.1016/j.jclepro.2015.12.055.

[42] Fatima Elayadi, Wafaa Boumya, Mounia Achak, Younes Chhiti, Fatima Ezzahrae M'hamdi Alaoui, Noureddine Barka, Chakib El Adlouni, Experimental and modeling studies of the removal of phenolic compounds from olive mill wastewater by adsorption on sugarcane bagasse, Environmental Challenges, **2021,** 4, 100184. DOI:10.1016/j.envc.2021.100184.

[43] Songlei Lv, Chunxi Li, Jianguo Mi, Hong Meng, A functional activated carbon for efficient adsorption of phenol derived from pyrolysis of rice husk, KOH-activation and EDTA-4Na-modification, Applied Surface Science, **2020,** 510, 145425. DOI:10.1016/j.apsusc.2020.145425.

[44] Yujie Ben, Caixia Fu, Min Hu, Lei Liu, Ming Hung Wong, Chunmiao Zheng, Human health risk assessment of antibiotic resistance associated with antibiotic residues in the environment: A review, Environmental Research, **2019,** 169, 483. DOI:10.1016/j.envres.2018.11.040.

[45] Muhammad Zaheer Afzal, Xue-Fei Sun, Jun Liu, Chao Song, Shu-Guang Wang, Asif Javed, Enhancement of ciprofloxacin sorption on chitosan/biochar hydrogel beads, Science of the Total Environment, **2018,** 639, 560. DOI:10.1016/j.scitotenv.2018.05.129.

[46] Isabel T. Carvalho, Lúcia Santos, Antibiotics in the aquatic environments: A review of the European scenario, Environment International, **2016,** 94, 736. DOI:10.1016/j.envint.2016.06.025.

[47] Dushanthi M. Wanninayake, Comparison of currently available PFAS remediation technologies in water: A review, Journal of Environmental Management, **2021,** 283, 111977. DOI:10.1016/j.jenvman.2021.111977.

[48] José L. Domingo, Martí Nadal, Human exposure to per- and polyfluoroalkyl substances (PFAS) through drinking water: A review of the recent scientific literature, Environmental Research, **2019,** 177, 108648. DOI:10.1016/j.envres.2019.108648.

[49] S. Sd Elanchezhiyan, Subbaiah Muthu Prabhu, Jonghun Han, Young Mo Kim, Yeomin Yoon, Chang Min Park, Synthesis and characterization of novel magnetic $Zr-MnFe_2O_4$@rGO nanohybrid for efficient removal of PFOA and PFOS from aqueous solutions, Applied Surface Science, **2020,** 528, 146579. DOI:10.1016/j.apsusc.2020.146579.

[50] Yeonji Yea, Gyuri Kim, Dengjun Wang, Sewoon Kim, Yeomin Yoon, S. Sd Elanchezhiyan, Chang Min Park, Selective sequestration of perfluorinated compounds using polyaniline decorated activated biochar, Chemical Engineering Journal, **2022,** 430, 132837. DOI:10.1016/j.cej.2021.132837.

[51] Oussama Baaloudj, Noureddine Nasrallah, Rachida Bouallouche, Hamza Kenfoud, Lotfi Khezami, Aymen Amin Assadi, High efficient cefixime removal from water by the sillenite $Bi_{12}TiO_{20:}$ Photocatalytic mechanism and degradation pathway, Journal of Cleaner Production, **2022,** 330, 129934. DOI:10.1016/j.jclepro.2021.129934.

[52] Ya Yang, Hulin Zhang, Sangmin Lee, Dongseob Kim, Woonbong Hwang, Zhong Lin Wang, Hybrid energy cell for degradation of methyl orange by self-powered electrocatalytic oxidation, Nano Letters, **2013,** 13, 803. DOI:10.1021/nl3046188.

[53] Goutham Rangarajan, Ramin Farnood, Role of persistent free radicals and lewis acid sites in visible-light-driven wet peroxide activation by solid acid biochar catalysts—A mechanistic study, Journal of Hazardous Materials, **2022,** 438, 129514. DOI:10.1016/j.jhazmat.2022.129514.

[54] Shunli Yu, Juan Zhou, Yanmei Ren, Zhiwang Yang, Ming Zhong, Xiaoqiang Feng, Bitao Su, Ziqiang Lei, Excellent adsorptive-photocatalytic performance of zinc oxide and biomass derived N, O-contained biochar nanocomposites for dyes and antibiotic removal, Chemical Engineering Journal, **2023,** 451. DOI:10.1016/j.cej.2022.138959.

[55] Mohamed Faisal Gasim, Zheng-Yi Choong, Pooi-Ling Koo, Siew-Chun Low, Mohamed-Hussein Abdurahman, Yeek-Chia Ho, Mardawani Mohamad, I. Wayan Koko Suryawan, Jun-Wei Lim, Wen-Da Oh, Application of biochar as functional material for remediation of organic pollutants in water: An overview, Catalysts, **2022,** 12, 210. DOI:10.3390/catal12020210.

[56] Rong-Zhong Wang, Dan-Lian Huang, Yun-Guo Liu, Chen Zhang, Cui Lai, Xin Wang, Guang-Ming Zeng, Xiao-Min Gong, Abing Duan, Qing Zhang, Piao Xu, Recent advances in biochar-based catalysts: Properties, applications and mechanisms for pollution remediation, Chemical Engineering Journal, **2019,** 371, 380. DOI:10.1016/j.cej.2019.04.071.

[57] Palagiri Bhavani, Murid Hussain, Young-Kwon Park, Recent advancements on the sustainable biochar based semiconducting materials for photocatalytic applications: A state of the art review, Journal of Cleaner Production, **2022,** 330, 129899. DOI:10.1016/j.jclepro.2021.129899.

[58] Yuwen Zhou, Shiyi Qin, Shivpal Verma, Taner Sar, Surendra Sarsaiya, Balasubramani Ravindran, Tao Liu, Raveendran Sindhu, Anil Kumar Patel, Parameswaran Binod, Sunita Varjani, Reeta Rani Singhnia, Zengqiang Zhang, Mukesh Kumar Awasthi, Production and beneficial impact of biochar for environmental application: A comprehensive review, Bioresource Technology, **2021,** 337, 125451. DOI:10.1016/j.biortech.2021.125451.

[59] Jian Chen, Wuxiang Zhang, Xiaoqiang Li, Renyu Huang, Qingling Liu, Yanjuan Zhang, Tao Gan, Zuqiang Huang, Huayu Hu, Mutually supportive growth strategy to engineer a hollow biochar sphere-supported TiO_2 composite with improved interfacial compatibility for efficient visible light-driven photocatalysis, Journal of Environmental Chemical Engineering, **2023**, 11, 110327. DOI:10.1016/j.jece.2023.110327.

[60] Yixiao Wu, Guifen Feng, Renyu Huang, Beiling Liang, Tao Gan, Huayu Hu, Yanjuan Zhang, Zhenfei Feng, Zuqiang Huang, Simultaneous growth strategy for constructing a Cu–Fe/carboxylate-decorated carbon composite with improved interface compatibility and charge transfer to boost the visible photocatalytic degradation of tetracycline, Chemical Engineering Journal, **2022**, 448, 137608. DOI:10.1016/j.cej.2022.137608.

[61] Hua Jing, Lili Ji, Zilong Li, Zhen Wang, Ran Li, Kaixuan Ju, Zn/Fe bimetallic modified Spartina alterniflora-derived biochar heterostructure with superior catalytic performance for the degradation of malachite green, Biochar, **2023**, 5, 29. DOI:10.1007/s42773-023-00227-9.

[62] Hao Peng, Heju Wang, Liping Wang, Congying Huang, Xiaogang Zheng, Jing Wen, Efficient adsorption-photocatalytic removal of tetracycline hydrochloride over La_2S_3 modified biochar with S,N-codoping, Journal of Water Process Engineering, **2022**, 49, 103038. DOI:10.1016/j.jwpe.2022.103038.

[63] Hao Peng, Yanxiang Li, Jing Wen, Xiaogang Zheng, Synthesis of $ZnFe_2O_4$/B, N-codoped biochar via microwave-assisted pyrolysis for enhancing adsorption-photocatalytic elimination of tetracycline hydrochloride, Industrial Crops and Products, **2021**, 172, 114066. DOI:10.1016/j.indcrop.2021.114066.

[64] Ridha Djellabi, Bo Yang, Ke Xiao, Yan Gong, Di Cao, Hafiz Muhammad Adeel Sharif, Xu Zhao, Caizhen Zhu, Junmin Zhang, Unravelling the mechanistic role of TiOC bonding bridge at titania/lignocellulosic biomass interface for Cr(VI) photoreduction under visible light, Journal of Colloid and Interface Science, **2019**, 553, 409. DOI:10.1016/j.jcis.2019.06.052.

[65] Jianhua Hou, Ting Jiang, Rui Wei, Faryal Idrees, Detlef Bahnemann, Ultrathin-layer structure of BiOI microspheres decorated on N-doped biochar with efficient photocatalytic activity, Frontiers in Chemistry, **2019**, 7. DOI:10.3389/fchem.2019.00378.

[66] Zaiyong Jiang, Xizhuang Liang, Yuanyuan Liu, Tao Jing, Zeyan Wang, Xiaoyang Zhang, Xiaoyan Qin, Ying Dai, Baibiao Huang, Enhancing visible light photocatalytic degradation performance and bactericidal activity of BiOI via ultrathin-layer structure, Applied Catalysis B: Environmental, **2017**, 211, 252. DOI:10.1016/j.apcatb.2017.03.072.

[67] Jandira Leichtweis, Siara Silvestri, Nicoly Welter, Yasmin Vieira, Paloma I. Zaragoza-Sánchez, Alma C. Chávez-Mejía, Elvis Carissimi, Wastewater containing emerging contaminants treated by residues from the brewing industry based on biochar as a new CuFe2O4/biochar photocatalyst, Process Safety and Environmental Protection, **2021**, 150, 497. DOI:10.1016/j.psep.2021.04.041.

[68] Zhi Zhu, Wenqian Fan, Zhi Liu, Yang Yu, Hongjun Dong, Pengwei Huo, Yongsheng Yan, Fabrication of the metal-free biochar-based graphitic carbon nitride for improved 2-Mercaptobenzothiazole degradation activity, Journal of Photochemistry and Photobiology A: Chemistry, **2018**, 358, 284. DOI:10.1016/j.jphotochem.2018.03.027.

[69] Xibao Li, Weiwei Wang, Fan Dong, Zhiqiang Zhang, Lu Han, Xudong Luo, Juntong Huang, Zhijun Feng, Zhi Chen, Guohua Jia, Tierui Zhang, Recent advances in noncontact external-field-assisted photocatalysis: From fundamentals to applications, ACS Catalysis, **2021**, 11, 4739. DOI:10.1021/acscatal.0c05354.

[70] Jing Sun, Xuemei Lin, Jie Xie, Yongzheng Zhang, Qi Wang, Zongrong Ying, Facile synthesis of novel ternary g-C_3N_4/ferrite/biochar hybrid photocatalyst for efficient degradation of methylene blue under visible-light irradiation, Colloids and Surfaces A: Physicochemical and Engineering Aspects, **2020**, 606, 125556. DOI:10.1016/j.colsurfa.2020.125556.

[71] Ying zhang, XiaoJuan Li, Junfeng Chen, Yanan Wang, Zhuoying Cheng, Xueqi Chen, Xing Gao, Minghui Guo, Porous spherical Cu_2O supported by wood-based biochar skeleton for the adsorption-photocatalytic degradation of methyl orange, Applied Surface Science, **2023**, 611, 155744. DOI:10.1016/j.apsusc.2022.155744.

[72] Ya He, Yafei Wang, Jin Hu, Kaijun Wang, Youwen Zhai, Yuze Chen, Yunbiao Duan, Yutian Wang, Weijun Zhang, Photocatalytic property correlated with microstructural evolution of the biochar/ZnO composites, Journal of Materials Research and Technology, **2021**, 11, 1308. DOI:10.1016/j.jmrt.2021.01.077.

[73] Fuyan Kang, Cai Shi, Weici Li, Malin Eqi, Zeshun Liu, Xiaogang Zheng, Zhanhua Huang, Honeycomb like CdS/Sulphur-modified biochar composites with enhanced adsorption-photocatalytic capacity for effective removal of rhodamine B, Journal of Environmental Chemical Engineering, **2022**, 10, 106942. DOI:10.1016/j.jece.2021.106942.

[74] Tongtong Wang, Amit Kumar, Xin Wang, Di Zhang, Yi Zheng, Guogang Wang, Qingliang Cui, Jinjun Cai, Jiyong Zheng, Construction of activated biochar/Bi_2WO_6 and /Bi_2MoO_6 composites to enhance adsorption and photocatalysis performance for efficient application in the removal of pollutants and disinfection, Environmental Science and Pollution Research, **2023**, 30, 30493. DOI:10.1007/s11356-022-24049-7.

[75] Dinghan Liu, Jianfeng Huang, Xingwang Tao, Dan Wang, One-step synthesis of C–Bi_2WO_6 crystallites with improved photo-catalytic activities under visible light irradiation, RSC Advances, **2015**, 5, 66464. DOI:10.1039/C5RA11381D.

[76] Jingjing Yang, Wen Zhu, Qian Yao, Guining Lu, Chen Yang, Zhi Dang, Photochemical reactivity of nitrogen-doped biochars under simulated sunlight irradiation: Generation of singlet oxygen, Journal of Hazardous Materials, **2021**, 410, 124547. DOI:10.1016/j.jhazmat.2020.124547.

[77] Yaowen Gao, Zhenhuan Chen, Yue Zhu, Tong Li, Chun Hu, New insights into the generation of singlet oxygen in the metal-free peroxymonosulfate activation process: Important role of electron-deficient carbon atoms, Environmental Science & Technology, **2020**, 54, 1232. DOI:10.1021/acs.est.9b05856.

[78] Seok-Young Oh, Thi-Hai Anh Nguyen, Ozonation of phenol in the presence of biochar and carbonaceous materials: The effect of surface functional groups and graphitic structure on the formation of reactive oxygen species, Journal of Environmental Chemical Engineering, **2022**, 10, 107386. DOI:10.1016/j.jece.2022.107386.

[79] Chencheng Qin, Hou Wang, Xingzhong Yuan, Ting Xiong, Jingjing Zhang, Jin Zhang, Understanding structure-performance correlation of biochar materials in environmental remediation and electrochemical devices, Chemical Engineering Journal, **2020**, 382, 122977. DOI:10.1016/j.cej.2019.122977.

[80] Junhao Chen, Xiaolu Yu, Cheng Li, Xin Tang, Ying Sun, Removal of tetracycline via the synergistic effect of biochar adsorption and enhanced activation of persulfate, Chemical Engineering Journal, **2020**, 382, 122916. DOI:10.1016/j.cej.2019.122916.

[81] Jun Wang, Xiaoguang Duan, Jian Gao, Yi Shen, Xiaohui Feng, Zijun Yu, Xiaoyao Tan, Shaomin Liu, Shaobin Wang, Roles of structure defect, oxygen groups and heteroatom doping on carbon in nonradical oxidation of water contaminants, Water Research, **2020**, 185, 116244.

[82] Yang Yu, Ning Li, Xukai Lu, Beibei Yan, Guanyi Chen, Yanshan Wang, Xiaoguang Duan, Zhanjun Cheng, Shaobin Wang, Co/N co-doped carbonized wood sponge with 3D porous framework for efficient peroxymonosulfate activation: Performance and internal mechanism, Journal of Hazardous Materials, **2022**, 421, 126735. DOI:10.1016/j.jhazmat.2021.126735.

[83] Kangmeng Zhu, Xisong Wang, Mengzi Geng, Dong Chen, Heng Lin, Hui Zhang, Catalytic oxidation of clofibric acid by peroxydisulfate activated with wood-based biochar: Effect of

biochar pyrolysis temperature, performance and mechanism, Chemical Engineering Journal, **2019,** 374, 1253.

[84] Wuqi Huang, Sa Xiao, Hua Zhong, Ming Yan, Xin Yang, Activation of persulfates by carbonaceous materials: A review, Chemical Engineering Journal, **2021,** 418, 129297. DOI:10.1016/j.cej.2021.129297.

[85] Xin Lv, Chenying Zhou, Zhichao Shen, Yuchen Zhang, Chuanshu He, Ye Du, Zhaokun Xiong, Rongfu Huang, Peng Zhou, Bo Lai, Waste leather derived porous carbon boosted Fenton oxidation towards removal of diethyl phthalate: Mechanism and long-lasting performance, Journal of Hazardous Materials, **2023,** 458, 132040. DOI:10.1016/j.jhazmat.2023.132040.

[86] Changmao Hung, Chiuwen Chen, Chinpao Huang, Jiawei Cheng, Chengdi Dong, Algae-derived metal-free boron-doped biochar as an efficient bioremediation pretreatment for persistent organic pollutants in marine sediments, Journal of Cleaner Production, **2022,** 336, 130448. DOI:10.1016/j.jclepro.2022.130448.

[87] Ya Liu, Chunmao Chen, Xiaoguang Duan, Shaobin Wang, Yuxian Wang, Carbocatalytic ozonation toward advanced water purification, Journal of Materials Chemistry A, **2021,** 9, 18994.

[88] Dahu Ding, Shengjiong Yang, Xiaoyong Qian, Liwei Chen, Tianming Cai, Nitrogen-doping positively whilst sulfur-doping negatively affect the catalytic activity of biochar for the degradation of organic contaminant, Applied Catalysis B: Environmental, **2020,** 263, 118348. DOI:10.1016/j.apcatb.2019.118348.

[89] Wei Ren, Gang Nie, Peng Zhou, Hui Zhang, Xiaoguang Duan, Shaobin Wang, The intrinsic nature of persulfate activation and N-doping in carbocatalysis, Environmental Science & Technology, **2020,** 54, 6438. DOI:10.1021/acs.est.0c01161.

[90] Yongfeng Zhu, Wenbo Wang, Huifang Zhang, Xiushen Ye, Zhijian Wu, Aiqin Wang, Fast and high-capacity adsorption of Rb^+ and Cs^+ onto recyclable magnetic porous spheres, Chemical Engineering Journal, **2017,** 327, 982. DOI:10.1016/j.cej.2017.06.169.

[91] Xin Li, Jiayue Wang, Lu Xia, Rujun Cheng, Jianqiu Chen, Jingge Shang, Peroxymonosulfate activation by nitrogen-doped herb residue biochar for the degradation of tetracycline, Journal of Environmental Management, **2023,** 328, 117028. DOI:10.1016/j.jenvman.2022.117028.

[92] Yuyang Zhou, Xiaochun Wu, Jin Zhang, Zhenxing Wang, In situ formation of tannic (TA)-aminopropyltriethoxysilane (APTES) nanospheres on inner and outer surface of polypropylene membrane toward enhanced dye removal capacity, Chemical Engineering Journal, **2022,** 433, 133843. DOI:10.1016/j.cej.2021.133843.

[93] Jiangfang Yu, Lin Tang, Ya Pang, Yaoyu Zhou, Haopeng Feng, Xiaoya Ren, Jing Tang, Jiajia Wang, Lifei Deng, Binbin Shao, Non-radical oxidation by N,S,P co-doped biochar for persulfate activation: Different roles of exogenous P/S doping, and electron transfer path, Journal of Cleaner Production, **2022,** 374, 133995. DOI:10.1016/j.jclepro.2022.133995.

[94] Xiaoguang Duan, Hongqi Sun, Shaobin Wang, Metal-free carbocatalysis in advanced oxidation reactions, Accounts of Chemical Research, **2018,** 51, 678. DOI:10.1021/acs.accounts.7b00535.

[95] Yinning He, Yi Chen, Jinzhe Li, Da Wang, Shuang Song, Feilong Dong, Zhiqiao He, Efficient degradation of 2,3,5-trimethylpyrazine by catalytic ozonation over MnO_x supported on biochar derived from waste tea leaves, Chemical Engineering Journal, **2023,** 464, 142525. DOI:10.1016/j.cej.2023.142525.

[96] Haiqing Zhang, Richard Lee Smith, Haixin Guo, Xinhua Qi, Cobalt cross-linked ordered mesoporous carbon as peroxymonosulfate activator for sulfamethoxazole degradation, Chemical Engineering Journal, **2023,** 472, 145060. DOI:10.1016/j.cej.2023.145060.

[97] Shuaishuai Xin, Guocheng Liu, Xiaohan Ma, Jiaxin Gong, Bingrui Ma, Qinghua Yan, Qinghua Chen, Dong Ma, Guangshan Zhang, Mengchun Gao, Yanjun Xin, High efficiency heterogeneous Fenton-like catalyst biochar modified $CuFeO_2$ for the degradation of

tetracycline: Economical synthesis, catalytic performance and mechanism, Applied Catalysis B: Environmental, **2021,** 280, 119386. DOI:10.1016/j.apcatb.2020.119386.

[98] Qiang Yang, Peixin Cui, Cun Liu, Guodong Fang, Meiying Huang, Qiuyue Wang, Yiyi Zhou, Hongbo Hou, Yujun Wang, In situ stabilization of the adsorbed Co^{2+} and Ni^{2+} in rice straw biochar based on LDH and its reutilization in the activation of peroxymonosulfate, Journal of Hazardous Materials, **2021,** 416, 126215. DOI:10.1016/j.jhazmat.2021.126215.

[99] Yaobin Ding, Xueru Wang, Libin Fu, Xueqin Peng, Cong Pan, Qihang Mao, Chengjun Wang, Jingchun Yan, Nonradicals induced degradation of organic pollutants by peroxydisulfate (PDS) and peroxymonosulfate (PMS): Recent advances and perspective, Science of the total Environment, **2021,** 765, 142794. DOI:10.1016/j.scitotenv.2020.142794.

[100] Haoyu Luo, Hengyi Fu, Hua Yin, Qintie Lin, Carbon materials in persulfate-based advanced oxidation processes: The roles and construction of active sites, Journal of Hazardous Materials, **2022,** 426. DOI:10.1016/j.jhazmat.2021.128044.

[101] Yidi Chen, Xiaoguang Duan, Chaofan Zhang, Shaobin Wang, Nanqi Ren, Shih-Hsin Ho, Graphitic biochar catalysts from anaerobic digestion sludge for nonradical degradation of micropollutants and disinfection, Chemical Engineering Journal, **2020,** 384, 123244. DOI:10.1016/j.cej.2019.123244.

[102] Ning Jiang, Haodan Xu, Lihong Wang, Jin Jiang, Tao Zhang, Nonradical oxidation of pollutants with single-atom-Fe(III)-activated persulfate: Fe(V) being the possible intermediate oxidant, Environmental Science & Technology, **2020,** 54, 14057. DOI:10.1021/acs.est.0c04867.

CHAPTER 5

Biomass-Derived Porous Carbon for CO₂ Capture

Wenhui Jia, Shuangjun Li,[†] Junyao Wang,
Shuai Deng, Huiyan Zhang, and Xiangzhou Yuan*

ABSTRACT

CO_2 capture is considered a promising and potential approach for mitigating climate change, especially in the context of carbon neutrality. To alleviate critical environmental issues caused by excessive CO_2 emissions as well as biomass mismanagement, upcycling biomass into porous carbon materials for high-performance CO_2 capture has been extensively investigated as a sustainable and practical route. This chapter mainly addresses synthesis methodologies of biomass-derived porous carbons (BDPCs) and then comprehensively evaluates their CO_2 capture performance based on CO_2 adsorption capacity, CO_2 selectivity, and cycling stability. As one typical data-driven approach, machine learning (ML) has emerged as one powerful tool for accurately predicting BDPC-based CO_2 capture performance and effectively providing inverse design guidelines for synthesizing high-performance BDPCs. In addition, from the perspective of environmental benefits, the life cycle assessment (LCA) of BDPC-based CO_2 capture is fully performed to provide valuable information for researchers and policymakers. With concerted efforts, upcycling biomass and its waste into porous carbon materials for high-performance CO_2 capture has advantages of mitigating climate change caused by CO_2 emissions and environmental pollution caused by biomass mismanagement, which is beneficial to achieving several of UN sustainable development goals (UN SDGs) including Goals 13, 9, and 15.

5.1 INTRODUCTION

5.1.1 Necessities of CO₂ Capture

In recent years, greenhouse gas emissions, especially CO_2 emissions, have been increasing year by year, causing significant adverse impacts on our entire ecosystem. For example, according

DOI: 10.1201/9781003520566-5

to a report by the International Energy Agency, global CO_2 emissions from energy combustion in 2022 already exceed 360 billion short tons[1]. By 2023, the atmospheric CO_2 concentration has exceeded 419 ppm, and more importantly, CO_2 growth rate shows an increasing trend with an average of 2.5 ppm per year[2]. According to the World Meteorological Organization (WMO) estimations, with fossil fuels still dominating the global energy demand, the concentration of CO_2 in the atmosphere will continue to increase and may even reach a staggering 550 ppm by 2050[3,4]. Meanwhile, the WMO warned that the growing concentration of greenhouse gases will further drive global temperatures to rise, and subsequent climate change is having increasingly adverse impacts on today's socioeconomic development[5]. However, according to the Intergovernmental Panel on Climate Change (IPCC), using pre-industrial temperatures as a baseline, the global temperature has already risen by 1.5 °C, causing significant issues for the global climate system[6]. In addition, the United Nations stated in 2021 that there are still numerous formidable challenges to achieving the goals of the Paris Agreement and the 2030 Agenda for Sustainable Development[7].

In this context, carbon capture, utilization, and storage (CCUS), a non-electric and carbon-negative technology, has emerged as a promising and practical technical approach for reducing CO_2 emissions, which is essential for mitigating climate change and achieving carbon neutrality. It is worthy of note that CO_2 utilization will not be addressed in this chapter. To the best of our knowledge, existing CCUS technologies remove a limited amount of CO_2 from the atmosphere and large-point sources (i.e., power plant and cement plant). Therefore, further advancing development and deployment of CCUS technologies is essential to achieve carbon neutrality.

5.1.2 Methods of Carbon Capture

As displayed in Figure 5.1(a), three main types of carbon capture technologies, including pre-combustion capture, post-combustion capture, and oxy-fuel combustion, are used for capturing CO_2 from power plant and large point resources[8]. Pre-combustion capture is the process of converting the carbon content in fuel into CO_2 prior to combustion, thus enabling the capture of CO_2. Post-combustion capture refers to the process of capturing CO_2 from the gas mixture after fuel combustion, by separating it from the flue gas. Oxy-fuel combustion refers to complete combustion of the fuel in an atmosphere with highly pure oxygen. Among the mentioned approaches (summarized in Figure 5.1(b)), post-combustion CO_2 capture technologies (i.e., absorption and adsorption) are widely regarded as one of the most cost-effective methods for practical CO_2 capture. However, despite the successful commercialization of the absorption process, challenges persist concerning the relatively low CO_2 concentration in post-combustion gases, as well as issues such as solvent losses, corrosion at critical points, and the high costs associated with CO_2 capture technology[9,10].

Currently, CO_2 emissions from distributed carbon sources cannot be ignored compared to centralized carbon sources; therefore, CO_2 capture from the atmosphere is another effective measure to mitigate climate change[11]. Against this background, Lancker first proposed direct air capture (DAC) in 1999[12]. DAC has received much attention in the subsequent two decades as an emerging CO_2 mitigation method due to the fact that it captures CO_2 directly from the air and that its operation is not constrained by location or time. It suggests that DAC is not dependent on a specific source of CO_2 or constrained by the timing of CO_2 emissions, making it more flexible to be deployed[13]. However, one of the main challenges with the current DAC technology is the low atmospheric CO_2 concentration (~420 ppm), which places high demands on adsorption performance of the adsorbent and energy requirements of the regeneration process. The degradation of adsorbent performance is another major challenge for DAC, due to the presence of co-existing water

(a)

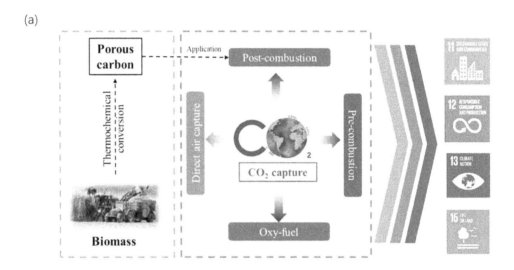

(b)

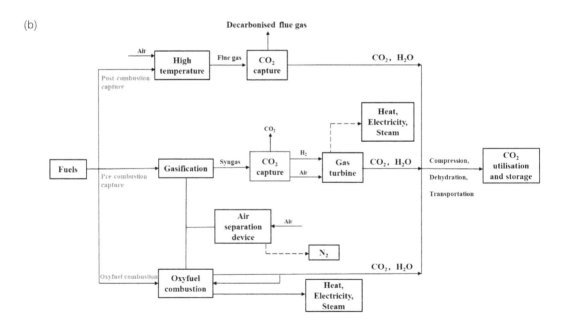

FIGURE 5.1 (a) Three major applications for CO_2 capture. (b) Flow chart of post-combustion capture, pre-combustion capture, and oxygen-enriched combustion technologies.

vapor and degradation of cyclic chemical stability[13]. Although DAC technology is still at an early stage of development, further ground-breaking research, and economic support to realize large-scale DAC applications are necessary in today's carbon-neutral context.

In addition to DAC, the development of low-cost, high-CO_2-permeable membranes has been widely studied as an emerging CO_2 capture technology. Compared to other CO_2 capture technologies, membrane separation routes are more energy-efficient and environmentally friendly, as they do not require additional chemicals and are driven mainly by concentration or pressure differences. However, due to the generally low thermal stability of membranes, pre-cooling of the flue gas is required when used in post-combustion CO_2 capture process, and this has limited the large-scale commercial application of membrane separation technology[14].

It is worth noting that solid carbon-based CO_2 adsorption, a promising CO_2 capture route, has also been widely investigated due to its good cost-effectiveness, well-developed porous structure, and excellent cycling stability. For adsorbent regeneration, energy consumption is one key indicator of its environmental impacts and practical feasibility. Because high energy consumption implies high global warming potential led by carbon-based adsorbent regeneration. Therefore, in the context of carbon neutrality, the desired solid carbon materials for CO_2 adsorption must have the advantages of cost-effectiveness, high CO_2 capture capacity and selectivity, low energy consumption for adsorbent regeneration, fast adsorption-desorption kinetics, and excellent cyclic stability, which are displayed in Figure 5.2.

Biomass, as a sustainable and non-fossil organic material from agriculture, forestry, and urban areas, has shown tremendous potential in the preparation of CO_2 adsorbents due to its abundant and accessible characteristics as well as being a renewable carbon resource. Therefore, in recent years, researchers have extensively studied the development

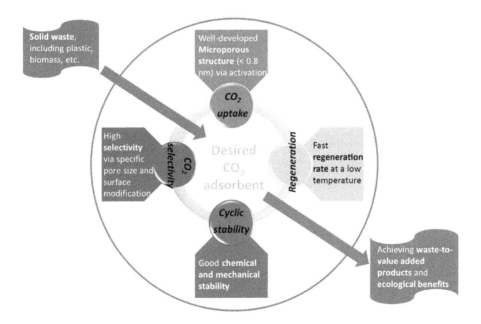

FIGURE 5.2   Major criteria of the desired biomass-based CO_2 adsorbents[15].

of biomass-derived porous carbons (BDPCs) for CO_2 adsorption. In 2012, Olivares-Marin and Maroto-Valer[16] first reported technologies for solid CO_2 adsorbents derived from solid waste materials and addressed their practical application in CO_2 capture. Dissanayake et al.[17] evaluated the potential for CO_2 adsorption using virgin and engineered biochar and discussed the factors influencing CO_2 adsorption performance and issues in the preparation of biomass-based CO_2 adsorbents. The development of BDPCs for CO_2 capture alleviates environmental impacts including climate change (or global warming) caused by large amount of CO_2 emission and environmental pollution caused by biomass mismanagement. Therefore, biomass, as an effective carbon precursor for synthesizing porous carbon, has sparked research interest in preparation of BDPCs for CO_2 capture.

The preparation of high-performance BDPCs for CO_2 capture commonly includes carbonization, activation, and surface modification. The purpose of carbonization is to valorize biomass into biochar and/or hydrochar using thermochemical conversion (e.g., pyrolysis, hydrothermal carbonization, and gasification). After performing carbonization, the resulting biochar/hydrochar commonly displays poor textural properties, including small surface areas and pore volume, which are summarized in Table 5.1. The poor textural properties are not available for effectively capturing CO_2, because only the microporous structure in BDPCs is beneficial to increasing their CO_2 capture performance[18–20]. It suggests that further development of porous structure in biochar/hydrochar is an essential step for synthesizing high-performance CO_2 adsorbents. Therefore, activation, mainly physical and chemical activations, is widely performed as a second step to further develop microporous structure in biochar/hydrochar.

The activating agents commonly used in physical activation include CO_2, steam, ammonia, and oxygen. Among them, CO_2 is widely used in physical activation due to its low reactivity at high temperatures and its relatively clean properties. Chemical activating agents mainly include KOH, NaOH, H_3PO_4, H_2O_2, and $ZnCl_2$, among others. KOH activation has been recognized as one of the most effective agents for developing microporous carbon materials with excellent CO_2 capture performance. Physical activation is relatively green but widely operates in the high temperature range (i.e., even up to 900 °C). Compared to physical activation, chemical activation has the advantages of lower energy consumption, lower reaction temperature, and shorter reaction time. However, the activating agents used

TABLE 5.1 Solid Waste-Derived Products from Various Preparation Methods[15]

	Solid Waste	Biochar/Hydrochar	Porous Carbon
Carbon content (%)	15–80	40–90	80–95
Preparation method	Drying pretreatment	Pyrolysis or hydrothermal process	Pyrolysis or hydrothermal process followed by physical or chemical activation
Effective functional groups	Few	Few	Numerous
Surface area (m^2/g)	NA	≤800	≥1000
Total pore volume (cm^3/g)	NA	≤0.3	≥0.5

in chemical activation have adverse effects on the environment, for example, water pollution caused by heavy metals occurs when using $ZnCl_2$ as the chemical agent. In addition, washing treatment is required to remove the extra chemical agents from activated carbon materials, which might lead to secondary pollution. Moreover, equipment corrosion and environmental contamination generated during chemical activation should be carefully treated in order to make chemical activation greener and more sustainable.

5.2 PREPARATION OF BIOMASS-DERIVED POROUS CARBON

Upcycling biomass into porous carbon for CO_2 adsorption has been extensively studied, which is mainly summarized in Table 5.2. As mentioned in Chapter 2, carbonization, activation, and surface modification of biomass are commonly recognized as the main technical routes to obtain BDPC with well-developed pore structure and effective active site or functional groups for high-performance CO_2 capture. Here we mainly address the conventional synthesis method first and then introduce ML-aided guidelines for high-performance CO_2 adsorbents. The key performance indicators for evaluating CO_2 adsorbents are discussed to comprehensively understand how to design high-performance CO_2 adsorbents from biomass and its waste. Finally, the environmental impacts of BDPCs for CO_2 capture are assessed from a life-cycle perspective, significantly providing critical guidelines to researchers and policymakers.

5.2.1 Synthesis of BDPCs for Carbon Capture

In addition to carbonization and activation discussed in Chapter 2, surface modification is one more step to introduce effective surface functional groups or provide abundant active sites for adsorbing carbon dioxide molecules[30–32], which is widely used to enhance the CO_2 adsorption capacity as well as CO_2 selectivity. The main methods of surface modification include heteroatom-doping (i.e., nitrogen, sulphur, and oxygen) and metal-doping treatments.

As a typical heteroatom-doping treatment, nitrogen (N)-doping is considered one of the most promising and effective approaches for achieving surface functionalization of carbon materials[33,34]. The commonly used reagents for nitrogen-doping treatment include $NaNH_2$, CH_4N_2O, and ammonia. And ammonia is considered the main nitrogen source for synthesizing N-doping porous carbons[35–37]. Amination, as one N-doping treatment, involves the use of ammonia as an activator and nitrogen source for N-doping treatment under high-temperature conditions, -NH_2, -H, and -NH radicals are formed, and further react with the carbon and ultimately form functional groups such as—NH_2, pyridinic, pyrrolic—on the surface of the carbon[38,39]. Equation (1) represents a possible route to prepare N-doped porous carbon using biomass as a precursor[40]. In addition, compared to other nitrogen functional groups, pyrrolic nitrogen is the most effective in enhancing the CO_2 adsorption performance of carbon materials.

$$\text{Biomass (C)} + NH_3 \rightarrow C^* + H^* + NH^* + NH_2^* \rightarrow H_2 + \text{C-NH} + \text{C-NH}_2 \tag{1}$$

The introduction of oxygen (O)-containing functional groups on the surface of carbon materials can also significantly improve the CO_2 adsorption capacity of carbon materials.

TABLE 5.2　Research and Development on Biomass-Derived Porous Carbons (BDPCs) for CO_2 Capture

Feedstock	Synthesis Approach			Textural Properties		N Content (%)	S Content (%)	CO₂ Adsorption Capacity		CO₂/N₂ Selectivity		Ref.
	Carbonization	Activation	Surface Modification	Specific Surface Area (m²/g)	Total Pore Volume (cm³/g)			25 °C & 1 Bar	0 °C & 1 Bar	ISAT Method	Henry's Law	
Cornstalks	Pyrolysis at 400 °C for 2 h	K₂CO₃ activation at 600 °C for 2 h	Melamine modification	1136.31	0.39	7.39	–	3.11	–	–	–	[21]
Macadamia nutshell	Carbonization at 500 °C	KOH activation at 700 °C	Melamine modification	1604	0.66	2.62	–	4.35	6.61	20(10/90)	–	[22]
Waste distiller's grains	Carbonization at 500 °C for 2 h	KOH activation at 550 °C for 2 h	Thiourea modification	1800	0.98	6.54	0.20	4.10	7.02	32 (15/85)	–	[23]
Common Oak leaves	–	KOH activation at 700 °C for 1 h	–	1842	0.88	–	–	5.44	6.17	74.36 (15/85)	132	[24]
Corn kernels	Pre-carbonization at 400 °C for 2 h	KOH activation at 900 °C for 1 h	–	1726	1.01	0.33	–	3.29	7.86	19 (15/85)	–	[25]
Walnut shell	–	KOH activation at 800 °C for 2 h	Thiourea modification	1412	0.77	4.89	0.71	5.51	7.26	25.59(15/85)	–	[26]
Hazelnut shell	Carbonization at 500 °C for 2 h	KOH activation at 550 °C	Thiourea modification	1600	0.67	3.01	0.42	4.30	6.43	17(10/90)	–	[27]
Original licorice	Hydrothermal treatment at 270 °C for 2 h	KOH activation at 600 °C for 2 h	Urea modification	1305	0.733	7.98	–	3.89	6.43	21.3 (15/85)	–	[28]
Lotus leaves	Carbonization at 500 °C for 1 h	KOH activation at 550 °C	Melamine modification	1487	0.69	4.09	–	3.87	5.44	20(10/90)	-	[29]

Noted: KOH = potassium hydroxide; K₂CO₃ = potassium carbonate; ISAT = initial surface absorption test; CO₂ = carbon dioxide; Specific surface area is calculated by BET; Pore volume is calculated by N₂ adsorption data; N and S contents are determined by XPS.

Yuan et al.[41] prepared CO_2 adsorbents using urea as a nitrogen source and KOH as a chemical activator by one-pot synthesis at 700 °C. In the study, it was found that the oxygen content in the carbon material was further increased, and the increased oxygen atoms could provide Lewis basic sites for CO_2 adsorption, which are beneficial to improving the CO_2 adsorption capacity and selectivity. In principle, oxygen atoms combine with surface carbon atoms to form acidic or basic functional groups. Moreover, oxygen-containing acidic functional groups can decompose at 800 °C[42]. Currently, sulphuric acid, nitric acid, and hydrogen peroxide are widely used to introduce oxygen-rich functional groups onto the surface of porous carbon[30]. Metal-doping treatment has also attracted much attention in studies related to the performance improvement of CO_2 adsorption. Metal-doping treatment mainly involves the doping of metal oxides, metal hydroxides, and metal oxyhydroxides. Metal-doping treatment not only further enables porous carbon to produce more surface-active sites but also makes acidic CO_2 molecules easily combine with basic metal oxides, hydroxides, and oxyhydroxides. Suitable metal-doping treatment has been proven to be effective in enhancing the surface properties and CO_2 adsorption capacity of carbon materials[43,44]. Liu et al.[45] developed a series of alkaline metal-doped carbons by introducing metal ions (Li, Na, and K) into the carbon skeleton and found that Li-doped porous carbons reached a CO_2 uptake of 4.1 mmol/g with a CO_2/N_2 selectivity of 47 at 25 °C.

In addition to single-atom doping treatment, dual-heteroatom doping treatment is also a current research hotspot for synthesizing high-performance CO_2 adsorbents. For example, CN_2H_4S is used as both N and S sources in the diatomic doping treatment to synthesize high-efficiency CO_2 adsorbents with more surface active sites.[34] Furthermore, some lignin-derived porous carbons have high functional groups due to their own elemental availability. For example, pueraria-derived porous carbon contains 4.16 wt% N[46], and celery-derived porous carbon contains 4.62 wt% N and 2.56 wt% S[47].

However, due to the need to utilize chemical solvents for treatment during the surface activation described above, the treated carbon materials need to be further washed with hydrochloric acid and/or distilled water to remove possible residual chemical solvents therein, but this causes waste contamination, and the issue of CO_2 emission during the reaction process should be carefully considered.

5.2.2 Data-Driven Synthesis Optimization

Currently, no practical guideline is available for synthesizing high-performance BDPC-based CO_2 capture. The main approach to the experimental screening of porous carbon formulations is still a direct approach that relies on the experience and intuition of the researcher and by performing constant trial and error methods[48], and the other method is to perform simulation screening using ab initio computational methods to synthesize target compounds with desired properties[49]. These are time- and labor-consuming approaches. More importantly, contributions of main factors (i.e., textural properties, functional groups, CO_2 adsorption conditions) are still unclear. Therefore, developing more efficient approaches is urgent for accelerating synthesis of porous carbon materials, optimizing CO_2 adsorption, and providing valuable insights from overall process, which is finally promoting BDPC-based CO_2 capture.

ML, one of the most widespread data-driven approaches, is a computer program that leans from historical data to understand a system and optimize future tasks. ML combines direct and silico methods as an approach with great potential to better help us analyze and optimize the preparation method of porous carbon materials with desirable properties. Recently, different ML methods have been developed and applied in CO_2 capture[50–53]. Currently, the use of ML in porous carbon preparation is still mainly focused on predicting the target properties of porous carbon while ensuring the validity of the simulation on the basis of a certain level of accuracy[48]. Nowadays, some ML models have been able to predict the CO_2 adsorption capacity of porous carbon relatively accurately based on its structural and textural properties and to further determine the relationship between CO_2 adsorption capacity and the corresponding adsorption conditions[54,52]. In addition, recent studies have confirmed that ML can be effectively used to analyze the preparation of BDPCs for CO_2 adsorption, as shown in Figure 5.3.

The data-driven ML approach has led to an increasing research interest in the analysis of BDPC preparation due to its ultra-high computational power, analytical capability, and multidisciplinary interactive applications. Data collection, formatting, and pre-processing are the first and most important steps in the preparation of BDPCs with specific CO_2 capture properties using ML analysis. Moreover, before using ML for analysis and prediction, it is necessary to fill in and/or discard data from the collected dataset for final accurate prediction.

In addition, it is necessary to preprocess the collected dataset before using ML for analysis and prediction. Correcting errors and filling in the gaps is achieved by performing operations on the data set such as filling in missing date and discarding outliers in order to ultimately make accurate predictions. Missing data in a dataset is usually filled in by methods such as replacing missing data with median, plurality, or randomly selected data. However, data with obvious anomalies can be discarded directly. The preprocessed data is usually divided into two main categories: training datasets (70–90%) and test datasets (10–30%), which are used to train and test the developed ML models, respectively.

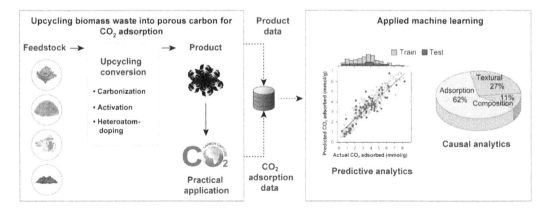

FIGURE 5.3 Application of ML to the preparation of engineered biochar from biomass for carbon dioxide (CO_2) capture[32].

Tree-based ML algorithms that can be adapted to classification and regression scenarios have been widely used to predict CO_2 capture performance using BPDCs[55–57] and include various methods such as decision trees, gradient boosting (GB), and random forests (RF). The decision tree is a tree structure in which each internal node represents a judgement on an attribute, each branch represents the output of a judgement, and each leaf node represents a classification result. The GB method is trained using negative gradient information about the model's loss function, and the weak predictive model produced at each step is trained and integrated into the final strong model. RF, on the other hand, uses decision trees as the basic unit to train and predict samples by integrating a large number of decision trees.

There is a difference in the prediction accuracy of different ML models, with the RF and Gradient Enhanced Decision Tree (GBDT) models having higher prediction accuracy. For example, Ma et al.[58] predicted the CO_2 adsorption capacity of porous carbon by applying ML to 1594 CO_2 adsorption datasets. The results show that the R^2 of the RF model exceeds 0.97 on both training and test data, with good prediction performance. Zhu et al.[59] analyzed 6244 CO_2 adsorption datasets of 155 porous carbons by ML to predict the CO_2 adsorption capacity of porous carbon materials. The results show that under certain adsorption conditions, the RF model exhibits good accuracy and predictive performance ($R^2 > 0.9$) in the test dataset. In addition, Yuan et al.[60] designed various tree-based models to predict the CO_2 adsorption capacity of BDPC by applying ML to 527 data points collected from peer-reviewed publications, among which the GBDT model showed the best prediction performance on both training and test data with R^2 of 0.98 and 0.84, respectively.

The ML method has been widely used for simulation prediction of CO_2 adsorption by BDPCs and has shown excellent performance in predicting the performance of adsorbent materials. However, the current studies are still mainly focused on the prediction of CO_2 adsorption performance of biochar[54], and the use of ML technology to directly guide the design of biochar with excellent CO_2 adsorption capacity is still insufficient. Secondly, although research on gas separation and sequestration using solid waste-derived porous carbon has been carried out for nearly a decade, the lack of a systematic and comprehensive database containing information on all experimentally synthesized porous carbons and their detailed physicochemical properties remains a major challenge in guiding the synthesis of high-performance porous carbons using ML techniques. And the lack of reliable data also limits the development of highly data-driven ML techniques to some extent, as data sets collected in the early stages are susceptible to various factors before being applied to ML techniques[61]. It is worth noting that development of ML models for BDPCs-based CO_2 capture is the basic and plain application of ML algorithms, the more interesting and valuable thing is to apply ML modeling approaches to efficiently design specific BDPCs-based CO_2 adsorbents and adsorption parameters.

5.3 CO_2 ADSORPTION PROPERTIES OF BIOMASS-DERIVED POROUS CARBON

Ideal porous carbon materials for CO_2 capture should possess several advantageous features such as high CO_2 adsorption capacity, high CO_2 selectivity, low energy consumption for adsorbent regeneration, rapid adsorption-desorption kinetics, and high cyclic stability. These characteristics have also been widely applied in the evaluation of CO_2 adsorption performance of BDPCs, which are also important data sources for data-driven synthesis optimization.

5.3.1 CO_2 Adsorption Capacity and Selectivity

CO_2 adsorption capacity includes dynamic adsorption capacity and isothermal adsorption capacity. According to the report from the National Energy Technology Laboratory (NETL) of the United States, adsorbent suitable for CO_2 capture should have a CO_2 adsorption capacity of 3 mmol/g at a temperature of 25 °C and a pressure of 1 bar with a flue gas carbon dioxide concentration of 15–20%[62]. Currently, most BDPCs can be effectively used for CO_2 capture after carbonization, activation, and surface modification.

CO_2 isothermal adsorption capacity is mainly obtained by experimental measurements or experimental simulations under equilibrium conditions, and the main adsorption analyzers used so far include ASAP2020, 2420, 2460 (Micromeritics), Autosorb-Iq (Quantachrome), and 3H-2000PS2 (Beshide). The isothermal adsorption capacity is mainly influenced by factors such as the chemical nature of the adsorbent, the surface structure, and the operating conditions (temperature and pressure of CO_2). Dynamic adsorption capacity is the capacity of CO_2 adsorbent to adsorb CO_2 from a gas mixture containing CO_2 at a certain temperature, pressure, and gas flow rate, which is influenced by the nature of the adsorbent material itself (e.g., pore structure and chemical composition) and operating parameters (e.g., gas flow rate, and composition). Currently, dynamic CO_2 adsorption data are mainly obtained by thermogravimetric analysis and vertical/horizontal fixed-bed reactors.

Compared with the isothermal adsorption capacity obtained under equilibrium conditions, the dynamic adsorption capacity obtained under non-equilibrium conditions accurately evaluates the CO_2 capture performance in real environmental applications, as it considers the effects of dynamically varying CO_2 concentrations and pressures, gas flow rates, and contact times. In addition, the dynamic adsorption capacity can be optimally tuned to the adsorption process by adjusting various operating parameters (e.g., flow rate, temperature, and pressure) to obtain high CO_2 adsorption capacity.

Isothermal and dynamic CO_2 adsorption capacities are important indicators for evaluating the CO_2 capture performance of BDPCs. However, when assessing the suitability of CO_2 adsorbents based on commercial cyclic processes, it is also necessary to conduct a comprehensive analysis of the working capacity during the complete CO_2 adsorption-desorption cycle and a detailed swing analysis during the CO_2 capture[63]. In the swing analysis of CO_2 capture performance, the main factors affecting the working capacity of CO_2 adsorbent are operating temperature and pressure. In recent years, the influence of operating temperature and pressure on the CO_2 capture performance of BDPCs has been widely studied. Coromina et al.[64] prepared activated carbon by hydrothermal carbonization using Jujun grass and Camellia japonica and further determined its CO_2 absorption capacity. The results showed that the BDPC obtained at an activation temperature of 600–800 °C and with a KOH/hydrothermal carbon of 2 exhibited excellent CO_2 uptake capacity at 25 °C, reaching 1.5 mmol/g and 5.0 mmol/g at 0.15 and 1 bar, respectively. Mallesh et al.[65] also conducted research on CO_2 adsorption and desorption of PC at temperatures of 70 °C and 140 °C. The results showed that BDPC exhibits high working capacity over a wide temperature range.

And the influence of operating pressure can be addressed through pressure variation processes, which can also save time. In conclusion, working capacity is an important indicator for screening high-performance CO_2 adsorbents, and preliminary evaluation of CO_2 capture performance of BDPCs can be conducted through swing analysis.

Surface modification of BDPCs is considered a promising route to enhance CO_2 selectivity. According to the Lewis basicity theory, oxygen and nitrogen functional groups are regarded as effective basic sites for CO_2 adsorption, significantly enhancing the CO_2 adsorption capacity and the CO_2 selectivity over N_2[66,41]. In principle, selective adsorption of CO_2 from a binary mixture of carbon dioxide and nitrogen gas can only occur when the pore diameter of the adsorbent is larger than the kinetic diameter of CO_2 and smaller than the kinetic diameter of N_2.

The selectivity of adsorbents for CO_2 over other gases (e.g., N_2) in flue gases is calculated primarily by the ideal adsorption solution theory (IAST) method (Equation (2)), and an alternative method is the Henry's law method (Equation (3)). The IAST method treats the adsorbent as an ideal solution and determines the adsorption capacity of the adsorbent under specific adsorption conditions (temperature, pressure) by measuring the CO_2 adsorption results, thus predicting the adsorption selectivity of the adsorbent for CO_2[67]. When using the experimental ideal adsorption solution theory pure gas isotherms for the prediction of adsorption behavior of binary mixtures, the single-component isotherms must be fitted with a suitable model, and the final selectivity depends mainly on the mechanism of the adsorption and the extent of the experimental data obtained. Among other things, adsorption isotherms need to be obtained at equilibrium conditions and certain temperatures and pressures.

$$S(CO_2/N_2)=x(CO_2)/x(N_2)\cdot y(CO_2)/y(N_2) \qquad (2)$$

$$S(CO_2/N_2)=K_H(CO_2)/K_H(N_2) \qquad (3)$$

The target gas selectivity should be calculated from its Henry's constant (K_H) as in Equation (3). However, the ratio of the gas Henry's constants can only reflect the true selectivity of the gas mixture at low pressure and low gas loading conditions[3,68]. Calculation of the selectivity of CO_2 over N_2 gas can be performed directly using the Ideal Adsorption Solution Theory and Henry's Law methods by using the obtained volumetric single-gas adsorption isotherms without using any special equipment for gas mixture measurements.

5.3.2 Cyclic CO_2 Capture

The actual BDPC-based CO_2 capture must go through an adsorption-desorption cycle operation. Therefore, it is necessary to select and optimize a suitable operating condition to obtain the best CO_2 capture performance (i.e., maximum working capacity). The cyclic performance of BDPCs-based CO_2 capture is commonly evaluated by four main approaches, including temperature swing adsorption (TSA), pressure swing adsorption (PSA), vacuum swing adsorption (VSA), and electric swing adsorption (ESA).

PSA utilizes pressure differences to achieve CO_2 separation, i.e., CO_2 gas adsorption under high pressure conditions followed by CO_2 gas desorption under low pressure conditions[69]. Unlike PSA, VSA performs CO_2 desorption under vacuum conditions and CO_2 adsorption under atmospheric pressure conditions[70]. Moreover, VPSA, which is a combination of VSA and PSA, has also received extensive attention. VPSA performs CO_2 adsorption at higher than atmospheric pressure and CO_2 desorption under vacuum. Due to its time-saving regeneration under vacuum conditions, VPSA has been considered one promising and practical route for industrial-scale CO_2 capture. It is worth noting that compressors and vacuum pumps required in VPSA are energy-intensive equipment, suggesting that further development of the VPSA process is necessary to achieve the lowest possible energy consumption. TSA technology is another promising route for CO_2 capture because BDPC-based CO_2 capture is mainly controlled by physical adsorption and has a high CO_2 adsorption capacity under ambient conditions. Compared with PSA/VSA, TSA does not require compression or vacuum treatment, implying its low energy requirement for running cyclic CO_2 adsorption-desorption[63,71]. In addition to pressure and temperature-driven processes, electric swing adsorption (ESA) is also an effective CO_2 capture approach that utilizes low-voltage electric current for rapid heating to establish the Joule effect and has the advantages of short heating time and high regeneration efficiency. ESA achieves higher CO_2 uptake, purity, and recovery within the same period of time compared to traditional TSA[72].

Both performance comparison and optimization of BDPC-based CO_2 capture cycle configurations need to be carried out with suitable cycle performance metrics. CO_2 purity, recovery rate, productivity, and energy consumption are usually the main indicators for evaluating the performance of carbon dioxide cycle capture[73]. The evaluation of the CO_2 adsorption performance of BDPC only involves the metrics of CO_2 adsorption capacity, selectivity, and recovery[74]. It can be found that for BDPCs currently applied to CO_2 adsorption, they are rarely considered from the perspective of energy consumption. At the same time, due to the lack of a unified energy consumption calculation method, the current CO_2 adsorption energy efficiency assessment specifications need to be further improved, and this aspect also needs to be further explored.

5.3.3 Other Parameters

When evaluating the CO_2 adsorption performance of adsorbents, factors such as moisture resistance, cost-effectiveness, and iso-exhaustion Qst should also be considered. Cost-effectiveness is an inherent advantage of BDPCs since it is a way of value-added utilization of biomass. The Qst value is related to the energy consumption for regeneration of the CO_2 adsorbent, with a higher Qst value representing higher energy consumption. Meanwhile, the magnitude of the Qst value depends on the configuration of the porous carbon, and the Qst value increases after doping with functional groups. Additionally, moisture resistance is considered an important indicator for evaluating the performance of CO_2 adsorbents.

The CO_2 adsorption performance of porous carbon materials is usually seriously affected when the moisture content in the gas mixture is high. Firstly, the water in the gas will enter the pores of the porous carbon to make the CO_2 adsorption efficiency decrease. On the one hand, the water molecules present in the pores will reduce the effective pore volume for CO_2 adsorption, and on the other hand, the water molecules will compete with CO_2 for adsorption sites, which will lead to the reduction of CO_2 adsorption sites. Secondly, under high humidity conditions, water molecules may interact with the adsorbent, thereby causing damage to the CO_2 adsorbent's structure. In conclusion, the moisture present in the gas mixture significantly reduces the CO_2 adsorption efficiency and selectivity for CO_2. Therefore, high-performance CO_2 adsorbents should have a certain degree of humidity resistance to ensure the adsorption efficiency and adsorption selectivity of CO_2 without the need for pre-treatment of the gas such as drying and purification[75]. The CO_2 adsorption performance of porous carbon should be comprehensively considered in terms of various performance indexes in order to efficiently screen CO_2 adsorbents with optimal CO_2 adsorption performance.

5.3.4 Environmental Impact Assessment

With increasing environmental concerns, researchers are focusing more and more on evaluating the environmental impacts of BDPCs-based CO_2 capture. Life-cycle assessment (LCA) is recognized as an effective tool for assessing the environmental performance of porous carbon during its preparation and application. Based on ISO 14040 and ISO 14044, process-based life cycle assessment (LCA) consists of four distinct phases: goal and scope definition, which describes the functional unit (FU) and system boundaries, life cycle inventory analysis (LCI), life cycle impact assessment (LCIA), and final interpretation[76]. The FU is the key parameter that enables the quantitative assessment of the performance of the production system during the LCA study, and it must be selected with due consideration of the scope and objectives of the LCA study in order to allow comparisons between different products and processes. In current studies, production output (e.g., BDPC quality) or waste input quality are often used as FUs in LCA studies for BDPC. Moreover, there may be more than one type of FUs to reflect the multifunctionality of the full process of a BDPC system. For instance, Puettmann et al.[77] defined three FU in the study, including 1000 kg of marketable biochar, fixed carbon as a proportion of biochar, and 1000 kg of forest residues.

Figure 5.4 illustrates the system boundary for typical BDPC preparation process. A full life-cycle process is usually divided into three stages, including preparation of feedstock, production of BDPCs, and final application of BDPC. It is important to note that in terms of BDPC application expansion, BDPC also has great potential as a soil conditioner and long-term carbon sequestration material due to its well-developed pore structure and rich carbon content.

Theoretically, global warming potential (GWP) is one of the primary assessment indicators for the environmental impacts of production and preparation process. According to Hersh and Mirkouei[78], the GWP of organic waste-derived biochar based on pine sawdust can be brought to 4.1 kg $CO_{2\text{-eq}}$/kg. The GWP of BDPC production has also been reported in some other studies, while the biochar was prepared from other types of biomass and solid waste, such as woody biomass (7.6–8.6 kg $CO_{2\text{-eq}}$/kg)[22], eucalyptus wastes (5.6–8.6 kg $CO_{2\text{-eq}}$/kg)[79], and olive waste cakes (11.096 kg $CO_{2\text{-eq}}$/kg)[76].

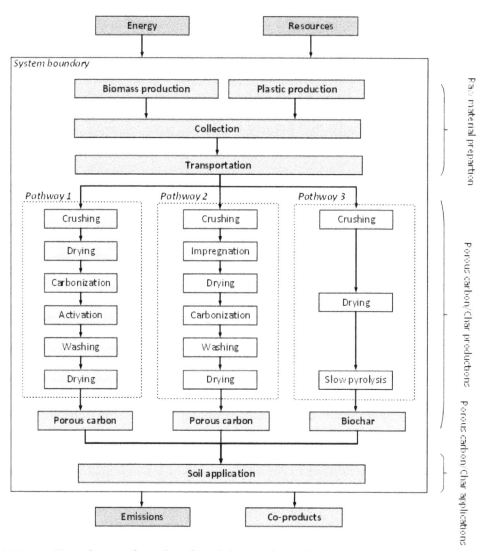

FIGURE 5.4   Typical system boundary for solid waste derived porous carbon life-cycle assessment.

Moreover, a circular research mode that performs a life cycle assessment of BDPC used for CO_2 capture has been studied, verifying that a negative GHG emission of -5.8 kg $CO_{2\text{-eq}}$/kg can be obtained from a cradle-to-grave perspective[50]. It can be concluded that if LCA studies on porous carbon are conducted with the system boundaries selected to consider the biomass production and the application of biochar in CO_2 capture, soil amelioration, or some other environmentally friendly applications, BDPC is expected to achieve a negative GWP.

In addition, LCA should also be conducted with due consideration of some other environmental impacts, such as resources depletion-water (WU), eutrophication (EP), ozone depletion (ODP), acidification (AP), particulate matter (RI), abiotic depletion potential (ADP), photochemical ozone formation (POFP), primary energy demand (PED), ionizing radiation-human health effects (IRP), and ecotoxicity-freshwater (ET), during thermochemical processes and activation.

Overall, BDPC-based CO_2 capture is quite an emerging technical approach, but there are still many challenges in the current LCA for BDPC-based CO_2 capture system. The process of preparation and utilization of BDPCs involves a variety of product outputs and applications; thus, different FU and system boundary selections make it difficult to compare the results of different LCA studies. In addition, more than half of the current inventory data on the preparation aspects of BDPC are directly derived from laboratory-scale experimental data. LCA results based on laboratory-scale experimental data do not accurately represent the real environmental impacts of the industrial-scale BDPC-based CO_2 capture. Furthermore, many investigations have not measured the emissions of pollutants from detailed synthesis process or have faced difficulties in monitoring, which may lead to an underestimation of the environmental risks of BDPC. Overall, there should be a standard framework to support LCA of BDPC to make results reliable and comparable.

5.4 CONCLUSION AND PERSPECTIVES

Excessive CO_2 emissions, as one of the main causes of climate change, have many negative impacts on the environment and entire ecosystem, such as the continued increase in global temperatures, the melting of glaciers, the decrease in sea ice, and the frequency of natural disasters such as forest wildfires. Considering that biomass is carbon-rich but carbon-neutral resources, BDPC-based CO_2 capture is one promising and practical route for simultaneously mitigating climate change caused by CO_2 emissions and environmental pollution caused by biomass mismanagement.

Upcycling biomass into porous carbon materials for high-performance CO_2 adsorption, carbonization, activation, and surface modification are widely used to effectively develop the good microporosity and efficiently increase surface functional groups. The CO_2 adsorption performance is evaluated from different perspectives, including isotherms, kinetics, and cyclic performances. The common methodologies for developing high-performance BDPCs-based CO_2 adsorbents are intuition-based, one factor at a time, and the experimental design is mainly via response surface methodologies, which are very time- and labor-consuming approaches. As a typical data-driven approach, ML is applied for developing BDPCs with high-performance CO_2 capture and investigating underlying relationships among synthesis parameters, textural properties of BDPCs, and CO_2 capture conditions. Moreover, a life-cycle assessment is performed on BDPC-based CO_2 capture to comprehensively display the environmental impacts, suggesting that CO_2 capture using BDPCs is expected to be carbon neutral or even carbon negative.

First, the cyclic CO_2 adsorption-desorption performance of BDPCs should be comprehensively evaluated in terms of CO_2 adsorption capacity, CO_2 selectivity relative to other gases, cycle stability, purity, recovery, and energy consumption, but a systematic approach to accurately assess the energy consumption is still lacking, and thus further research on the assessment of the energy consumption in the BDPCs-based adsorption-desorption cycle is necessary.

Second, the use of ML-based techniques has shown great promise in the preparation of porous carbon with desirable properties, but a limited number of studies have been conducted on the direct preparation of porous carbon using ML techniques. In addition, by analyzing the laboratory-scale CO_2 adsorption data of BDPCs using ML, it was found that the CO_2 adsorption performance of BDPC was mainly affected by both intrinsic (texture properties) and extrinsic (adsorption conditions) conditions. Therefore, it is necessary to accelerate the establishment of a complete database containing information related to all experimentally prepared BDPCs and their detailed physicochemical properties.

Third, a comprehensive assessment of the environmental impacts of CO_2 adsorption by BDPCs from a life-cycle perspective has received widespread attention. However, LCA studies of BDPCs using industrial-scale data are still missing in most studies, and the results of LCA based on laboratory-scale data do not accurately reflect the environmental impacts of CO_2 adsorption using BDPC on an industrial scale. In addition, the different FUs and system boundaries in the studies also create difficulties in comparing the results of different studies. At the same time, when we conduct a comprehensive assessment of the environmental impact of BDPC, we should carefully consider the environmental pollution caused by the emission of CO_2, toxic substances, and chemical reagents during the synthesis of BDPC.

In conclusion, effective action is urgently needed to accelerate the achievement of the goals of carbon neutrality. The sustainable and efficient value addition of biomass into porous carbon materials for CO_2 adsorption can not only effectively mitigate the climate problem mainly caused by CO_2 emissions but also solve the problem of potential biomass energy waste. Therefore, the utilization of BDPCs for CO_2 capture is a promising technology in a dual-carbon context. Its future research direction should focus on the preparation of BDPC with high adsorption capacity, high CO_2 selectivity, high cyclic stability, and high moisture resistance and can be used with CO_2 adsorption on an industrial scale, as well as further refinement of the methodology for evaluating the CO_2 adsorption performance and environmental performance of BDPC.

NOTES

† Wenhui Jia and Shuangjun Li contributed equally as the first authors.
* Corresponding author

REFERENCES

[1] Lichao Ge, Can Zhao, Simo Chen, Qian Li, Tianhong Zhou, Han Jiang, Xi Li, Yang Wang, Chang Xu, An analysis of the carbonization process and volatile-release characteristics of coal-based activated carbon, Energy, **2022**, 257. DOI:10.1016/j.energy.2022.124779.

[2] Monitoring Laboratory, Trends in atmospheric carbon dioxide, **2016**. Available from https://gml.noaa.gov/ccgg/trends/gl_trend.html.

[3] M. Oschatz, M. Antonietti, A search for selectivity to enable CO_2 capture with porous adsorbents, Energy and Environmental Science, **2018**, 11, 57. DOI:10.1039/c7ee02110k.

[4] Qiang Wang, Jizhong Luo, Ziyi Zhong, Armando Borgna, CO_2 capture by solid adsorbents and their applications: Current status and new trends, Energy and Environmental Science, **2011**, 4, 42. DOI:10.1039/c0ee00064g.

[5] World Meteorological Organization, Statement on the state of the global climate in 2018, **2019**. Available from www.cma.gov.cn/en2014/news/News/201903/t20190330_519202.html.

[6] Intergovernmental Panel on Climate Change, Special report global warming of 1.5 °C, **2018**. Available from www.ipcc.ch/site/assets/uploads/sites/2/2019/05/SR15_Chapter2_Low_Res.pdf.

[7] United Nations Economic Commission for Europe, Carbon capture, use and storage, **2021**. Available from https://unece.org/.

[8] Avanthi Deshani Igalavithana, Seung Wan Choi, Jin Shang, Aamir Hanif, Pavani Dulanja Dissanayake, Daniel C. W. Tsang, Jung-Hwan Kwon, Ki Bong Lee, Yong Sik Ok, Carbon dioxide capture in biochar produced from pine sawdust and paper mill sludge: Effect of porous structure and surface chemistry, Science of the Total Environment, **2020**, 739. DOI:10.1016/j.scitotenv.2020.139845.

[9] Niall MacDowell, Nick Florin, Antoine Buchard, Jason Hallett, Amparo Galindo, George Jackson, Claire S. Adjiman, Charlotte K. Williams, Nilay Shah, Paul Fennell, An overview of CO2 capture technologies, Energy & Environmental Science, **2010**, 3, 1645. DOI:10.1039/c004106h.

[10] Peter Markewitz, Wilhelm Kuckshinrichs, Walter Leitner, Jochen Linssen, Petra Zapp, Richard Bongartz, Andrea Schreiber, Thomas E. Mueller, Worldwide innovations in the development of carbon capture technologies and the utilization of CO_2, Energy & Environmental Science, **2012**, 5, 7281. DOI:10.1039/c2ee03403d.

[11] María Erans, Eloy S. Sanz-Pérez, Dawid P. Hanak, Zeynep Clulow, David M. Reiner, Greg A. Mutch, Direct air capture: Process technology, techno-economic and socio-political challenges, Energy & Environmental Science, **2022**, 15, 1360. DOI:10.1039/d1ee03523a.

[12] Klaus S. Lackner, Hans-Joachim Ziock, Patrick Gerard Grimes, Carbon dioxide extraction from air: Is it an option?,**1999**. Available from www.osti.gov/servlets/purl/770509.

[13] L. Jiang, W. Liu, R. Q. Wang, A. Gonzalez-Diaz, M. F. Rojas-Michaga, S. Michailos, M. Pourkashanian, X. J. Zhang, C. Font-Palma, Sorption direct air capture with CO_2 utilization, Progress in Energy and Combustion Science, **2023**, 95. DOI:10.1016/j.pecs.2022.101069.

[14] Xiaoxing Wang, Chunshan Song, Carbon capture from flue gas and the atmosphere: A perspective, Frontiers in Energy Research, **2020**, 8. DOI:10.3389/fenrg.2020.560849.

[15] Xiangzhou Yuan, Junyao Wang, Shuai Deng, Manu Suvarna, Xiaonan Wang, Wei Zhang, Sara Triana Hamilton, Ammar Alahmed, Aqil Jamal, Ah-Hyung Alissa Park, Xiaotao Bi, Yong Sik Ok, Recent advancements in sustainable upcycling of solid waste into porous carbons for carbon dioxide capture, Renewable and Sustainable Energy Reviews, **2022**, 162. DOI:10.1016/j.rser.2022.112413.

[16] M. Olivares-Marin, M. Maroto-Valer, Development of adsorbents for CO_2 capture from waste materials: A review, Greenhouse Gases-Science and Technology, **2012**, 2, 20. DOI:10.1002/ghg.45.

[17] Pavani Dulanja Dissanayake, Siming You, Avanthi Deshani Igalavithana, Yinfeng Xia, Amit Bhatnagar, Souradeep Gupta, Harn Wei Kua, Sumin Kim, Jung-Hwan Kwon, Daniel C. W. Tsang, Yong Sik Ok, Biochar-based adsorbents for carbon dioxide capture: A critical review, Renewable & Sustainable Energy Reviews, **2020**, 119. DOI:10.1016/j.rser.2019.109582.

[18] Jiaxin Li, Beata Michalkiewicz, Jiakang Min, Changde Ma, Xuecheng Chen, Jiang Gong, Ewa Mijowska, Tao Tang, Selective preparation of biomass-derived porous carbon with controllable pore sizes toward highly efficient CO_2 capture, Chemical Engineering Journal, **2019**, 360, 250. DOI:10.1016/j.cej.2018.11.204.

[19] Ganesh K. Parshetti, Shamik Chowdhury, Rajasekhar Balasubramanian, Biomass derived low-cost microporous adsorbents for efficient CO_2 capture, Fuel, **2015**, 148, 246. DOI:10.1016/j.fuel.2015.01.032.

[20] Jaroslaw Serafin, Urszula Narkiewicz, Antoni W. Morawski, Rafal J. Wrobel, Beata Michalkiewicz, Highly microporous activated carbons from biomass for CO_2 capture and effective micropores at different conditions, Journal of CO_2 Utilization, **2017**, 18, 73. DOI:10.1016/j.jcou.2017.01.006.

[21] Xiaofang Yuan, Jianfei Xiao, Murat Yılmaz, Tian C. Zhang, Shaojun Yuan, N, P Co-doped porous biochar derived from cornstalk for high performance CO_2 adsorption and electrochemical energy storage, Separation and Purification Technology, **2022**, 299. DOI:10.1016/j.seppur.2022.121719.

[22] Jiali Bai, Jiamei Huang, Qiyun Yu, Muslum Demir, Murat Kilic, Bilge Nazli Altay, Xin Hu, Linlin Wang, N-doped porous carbon derived from macadamia nut shell for CO_2 adsorption, Fuel Processing Technology, **2023**, 249. DOI:10.1016/j.fuproc.2023.107854.

[23] Lan Luo, Chunliang Yang, Fei Liu, Tianxiang Zhao, Heteroatom-N,S co-doped porous carbons derived from waste biomass as bifunctional materials for enhanced CO_2 adsorption and conversion, Separation and Purification Technology, **2023,** 320. DOI:10.1016/j.seppur.2023.124090.

[24] Jarosław Serafin, Orlando F. Cruz, Promising activated carbons derived from common oak leaves and their application in CO_2 storage, Journal of Environmental Chemical Engineering, **2022,** 10. DOI:10.1016/j.jece.2022.107642.

[25] Runping Wu, Qing Ye, Kai Wu, Lanyang Wang, Hongxing Dai, Highly efficient CO_2 adsorption of corn kernel-derived porous carbon with abundant oxygen functional groups, Journal of CO_2 Utilization, **2021,** 51. DOI:10.1016/j.jcou.2021.101620.

[26] Meng Cao, Yu Shu, Qiuhong Bai, Cong Li, Bang Chen, Yehua Shen, Hiroshi Uyama, Design of biomass-based N, S co-doped porous carbon via a straightforward post-treatment strategy for enhanced CO_2 capture performance, Science of the Total Environment, **2023,** 884. DOI:10.1016/j.scitotenv.2023.163750.

[27] Changdan Ma, Tingyan Lu, Jiawei Shao, Jiamei Huang, Xin Hu, Linlin Wang, Biomass derived nitrogen and sulfur co-doped porous carbons for efficient CO_2 adsorption, Separation and Purification Technology, **2022,** 281. DOI:10.1016/j.seppur.2021.119899.

[28] Yabin Zhou, Peng Tan, Ziqian He, Cheng Zhang, Qingyan Fang, Gang Chen, CO_2 adsorption performance of nitrogen-doped porous carbon derived from licorice residue by hydrothermal treatment, Fuel, **2022,** 311. DOI:10.1016/j.fuel.2021.122507.

[29] Qian Li, Shenfang Liu, Linlin Wang, Fangyuan Chen, Jiawei Shao, Xin Hu, Efficient nitrogen doped porous carbonaceous CO_2 adsorbents based on lotus leaf, Journal of Environmental Sciences, **2021,** 103, 268. DOI:10.1016/j.jes.2020.11.008.

[30] Gurwinder Singh, Kripal S. Lakhi, Sanchita Sil, Sheshanath V. Bhosale, InYoung Kim, Khalid Albahily, Ajayan Vinu, Biomass derived porous carbon for CO_2 capture, Carbon, **2019,** 148, 164. DOI:10.1016/j.carbon.2019.03.050.

[31] Xiangzhou Yuan, Pavani Dulanja Dissanayake, Bin Gao, Wu-Jun Liu, Ki Bong Lee, Yong Sik Ok, Review on upgrading organic waste to value-added carbon materials for energy and environmental applications, Journal of Environmental Management, **2021,** 296. DOI:10.1016/j.jenvman.2021.113128.

[32] Xiangzhou Yuan, Manu Suvarna, Sean Low, Pavani Dulanja Dissanayake, Ki Bong Lee, Jie Li, Xiaonan Wang, Yong Sik Ok, Applied machine learning for prediction of CO_2 adsorption on biomass waste-derived porous carbons, Environmental Science & Technology, **2021,** 55, 11925. DOI:10.1021/acs.est.1c01849.

[33] Adetola E. Ogungbenro, Dang V. Quang, Khalid A. Al-Ali, Lourdes F. Vega, Mohammad R. M. Abu-Zahra, Physical synthesis and characterization of activated carbon from date seeds for CO_2 capture, Journal of Environmental Chemical Engineering, **2018,** 6, 4245. DOI:10.1016/j.jece.2018.06.030.

[34] Zhixiu Yang, Guojie Zhang, Ying Xu, Peiyu Zhao, One step N-doping and activation of biomass carbon at low temperature through $NaNH_2$. An effective approach to CO_2 adsorbents, Journal of CO_2 Utilization, **2019,** 33, 320. DOI:10.1016/j.jcou.2019.06.021.

[35] Zhen Geng, Qiangfeng Xiao, Hong Lv, Bing Li, Haobin Wu, Yunfeng Lu, Cunman Zhang, One-step synthesis of microporous carbon monoliths derived from biomass with high nitrogen doping content for highly selective CO_2 capture, Scientific Reports, **2016,** 6. DOI:10.1038/srep30049.

[36] Liping Guo, Jie Yang, Gengshen Hu, Xin Hu, Linlin Wang, Yifan Dong, Herbert DaCosta, Maohong Fan, Role of hydrogen peroxide preoxidizing on CO_2 adsorption of nitrogen-doped carbons produced from coconut shell, ACS Sustainable Chemistry & Engineering, **2016,** 4, 2806. DOI:10.1021/acssuschemeng.6b00327.

[37] Dawei Wu, Jing Liu, Yingju Yang, Ying Zheng, Nitrogen/oxygen co-doped porous carbon derived from biomass for low-pressure CO_2 capture, Industrial & Engineering Chemistry Research, **2020,** 59, 14055. DOI:10.1021/acs.iecr.0c00006.

[38] C. Pevida, M. G. Plaza, B. Arias, J. Fermoso, F. Rubiera, J. J. Pis, Surface modification of activated carbons for CO_2 capture, Applied Surface Science, **2008,** 254, 7165. DOI:10.1016/j.apsusc.2008.05.239.

[39] Jaroslaw Serafin, Urszula Narkiewicz, Antoni W. Morawski, Rafal J. Wrobel, Beata Michalkiewicz, Highly microporous activated carbons from biomass for CO_2 capture and effective micropores at different conditions, Journal of CO_2 Utilization, **2017,** 18, 73. DOI:10.1016/j.jcou.2017.01.006.

[40] Kalidas Mainali, Sohrab Haghighi Mood, Manuel Raul Pelaez-Samaniego, Valentina Sierra-Jimenez, Manuel Garcia-Perez, Production and applications of N-doped carbons from bioresources: A review, Catalysis Today, **2023,** 423. DOI:10.1016/j.cattod.2023.114248.

[41] Xiangzhou Yuan, Shuangjun Li, Sunbin Jeon, Shuai Deng, Li Zhao, Ki Bong Lee, Valorization of waste polyethylene terephthalate plastic into N-doped microporous carbon for CO_2 capture through a one-pot synthesis, Journal of Hazardous Materials, **2020,** 399. DOI:10.1016/j.jhazmat.2020.123010.

[42] Mohammad Saleh Shafeeyan, Wan Mohd Ashri Wan Daud, Amirhossein Houshmand, Ahmad Shamiri, A review on surface modification of activated carbon for carbon dioxide adsorption, Journal of Analytical and Applied Pyrolysis, **2010,** 89, 143. DOI:10.1016/j.jaap.2010.07.006.

[43] Pooya Lahijani, Maedeh Mohammadi, Abdul Rahman Mohamed, Metal incorporated biochar as a potential adsorbent for high capacity CO_2 capture at ambient condition, Journal of CO_2 Utilization, **2018,** 26, 281. DOI:10.1016/j.jcou.2018.05.018.

[44] Ali Zaker, Samia ben Hammouda, Jie Sun, Xiaolei Wang, Xia Li, Zhi Chen, Carbon-based materials for CO_2 capture: Their production, modification and performance, Journal of Environmental Chemical Engineering, **2023,** 11. DOI:10.1016/j.jece.2023.109741.

[45] Baogen Liu, Xiancheng Ma, Rui Shi, Ke Zhou, Xiang Xu, Jingting Qiu, Huijun Wang, Zheng Zeng, Liqing Li, Synthesis of alkali metals functionalized porous carbon for enhanced selective adsorption of carbon dioxide: A theoretically guided study, Energy & Fuels, **2021,** 35, 15962. DOI:10.1021/acs.energyfuels.1c02313.

[46] Xiuli Han, Haixia Jiang, Yong Zhou, Weifeng Hong, Yangfan Zhou, Ping Gao, Rui Ding, Enhui Liu, A high performance nitrogen-doped porous activated carbon for supercapacitor derived from pueraria, Journal of Alloys and Compounds, **2018,** 744, 544. DOI:10.1016/j.jallcom.2018.02.078.

[47] Yiju Li, Guiling Wang, Tong Wei, Zhuangjun Fan, Peng Yan, Nitrogen and sulfur co-doped porous carbon nanosheets derived from willow catkin for supercapacitors, Nano Energy, **2016,** 19, 165. DOI:10.1016/j.nanoen.2015.10.038.

[48] Jiali Li, Kaizhuo Lim, Haitao Yang, Zekun Ren, Shreyaa Raghavan, Po-Yen Chen, Tonio Buonassisi, Xiaonan Wang, AI applications through the whole life cycle of material discovery, Matter, **2020,** 3, 393. DOI:10.1016/j.matt.2020.06.011.

[49] V. Timoshevskii, Wei Ji, Hakima Abou-Rachid, Louis-Simon Lussier, H. Guo, Polymeric nitrogen in a graphene matrix: An ab initio study, Physical Review B, **2009,** 80. DOI:10.1103/PhysRevB.80.115409.

[50] Ryther Anderson, Jacob Rodgers, Edwin Argueta, Achay Biong, Diego A. Gomez-Gualdron, Role of pore chemistry and topology in the CO_2 capture capabilities of MOFs: From molecular simulation to machine learning, Chemistry of Materials, **2018,** 30, 6325. DOI:10.1021/acs.chemmater.8b02257.

[51] Hao Li, Dan Yang, Zhien Zhang, Eric Lichtfouse, Prediction of CO_2 absorption by physical solvents using a chemoinformatics-based machine learning model, Environmental Chemistry Letters, **2019,** 17, 1397. DOI:10.1007/s10311-019-00874-0.

[52] Xiancheng Ma, Wenjun Xu, Rongkui Su, Lishu Shao, Zheng Zeng, Liqing Li, Hanqing Wang, Insights into CO_2 capture in porous carbons from machine learning, experiments and molecular

simulation, Separation and Purification Technology, **2023,** 306. DOI:10.1016/j.seppur.2022. 122521.

[53] A. K. Priya, Balaji Devarajan, Avinash Alagumalai, Hua Song, Artificial intelligence enabled carbon capture: A review, Science of the Total Environment, **2023,** 886. DOI:10.1016/j. scitotenv.2023.163913.

[54] Mobin Safarzadeh Khosrowshahi, Hossein Mashhadimoslem, Hadi Shayesteh, Gurwinder Singh, Elnaz Khakpour, Xinwei Guan, Mohammad Rahimi, Farid Maleki, Prashant Kumar, Ajayan Vinu, Natural products derived porous carbons for CO_2 capture, Advanced Science, **2023.** DOI:10.1002/advs.202304289.

[55] M. Maheri, C. Bazan, S. Zendehboudi, H. Usefi, Machine learning to assess CO_2 adsorption by biomass waste, Journal of CO_2 Utilization, **2023,** 76. DOI:10.1016/j.jcou.2023.102590.

[56] Sarvesh Namdeo, Vimal Chandra Srivastava, Paritosh Mohanty, Machine learning implemented exploration of the adsorption mechanism of carbon dioxide onto porous carbons, Journal of Colloid and Interface Science, **2023,** 647, 174. DOI:10.1016/j.jcis.2023.05.052.

[57] Xinzhe Zhu, Daniel C. W. Tsang, Lei Wang, Zhishan Su, Deyi Hou, Liangchun Li, Jin Shang, Machine learning exploration of the critical factors for CO_2 adsorption capacity on porous carbon materials at different pressures, Journal of Cleaner Production, **2020,** 273. DOI:10.1016/j. jclepro.2020.122915.

[58] Mengmeng Du, Yu Zhang, Sailei Kang, Chao Xu, Yingxin Ma, Lejuan Cai, Ye Zhu, Yang Chai, Bocheng Qiu, Electrochemical production of glycolate fuelled by polyethylene terephthalate plastics with improved techno-economics, Small, **2023,** 19. DOI:10.1002/smll.202303693.

[59] Xinzhe Zhu, Daniel C. W. Tsang, Lei Wang, Zhishan Su, Deyi Hou, Liangchun Li, Jin Shang, Machine learning exploration of the critical factors for CO_2 adsorption capacity on porous carbon materials at different pressures, Journal of Cleaner Production, **2020,** 273. DOI:10.1016/j. jclepro.2020.122915.

[60] Xiangzhou Yuan, Manu Suvarna, Sean Low, Pavani Dulanja Dissanayake, Ki Bong Lee, Jie Li, Xiaonan Wang, Yong Sik Ok, Applied machine learning for prediction of CO_2 adsorption on biomass waste-derived porous carbons, Environmental Science & Technology, **2021,** 55, 11925. DOI:10.1021/acs.est.1c01849.

[61] Xiangzhou Yuan, Jie Li, Juin Yau Lim, Ashkan Zolfaghari, Daniel S. Alessi, Yin Wang, Xiaonan Wang, Yong Sik Ok, Machine learning for heavy metal removal from water: Recent advances and challenges, American Chemical Society Environmental Science & Technology Water, **2023.** DOI:10.1021/acsestwater.3c00215.

[62] Xuan Liu, Zhongying Fang, Xue Teng, Yanli Niu, Shuaiqi Gong, Wei Chen, Thomas J. Meyer, Zuofeng Chen, Paired formate and H_2 productions via efficient bifunctional Ni-Mo nitride nanowire electrocatalysts, Journal of Energy Chemistry, **2022,** 72, 432. DOI:10.1016/j. jechem.2022.04.040.

[63] Federica Raganati, Riccardo Chirone, Paola Ammendola, CO_2 capture by temperature swing adsorption: Working capacity as affected by temperature and CO_2 partial pressure, Industrial & Engineering Chemistry Research, **2020,** 59, 3593. DOI:10.1021/acs.iecr.9b04901.

[64] Helena Matabosch Coromina, Darren A. Walsh, Robert Mokaya, Biomass-derived activated carbon with simultaneously enhanced CO_2 uptake for both pre and post combustion capture applications, Journal of Materials Chemistry A, **2016,** 4, 280. DOI:10.1039/c5ta09202g.

[65] D. Mallesh, J. Anbarasan, P. Mahesh Kumar, K. Upendar, P. Chandrashekar, B. V. S. K. Rao, N. Lingaiah, Synthesis, characterization of carbon adsorbents derived from waste biomass and its application to CO_2 capture, Applied Surface Science, **2022,** 596. DOI:10.1016/j. apsusc.2022.153669.

[66] Xiangzhou Yuan, Jong Gyu Lee, Heesun Yun, Shuai Deng, Yu Jin Kim, Ji Eun Lee, Sang Kyu Kwak, Ki Bong Lee, Solving two environmental issues simultaneously: Waste polyethylene terephthalate plastic bottle-derived microporous carbons for capturing CO_2, Chemical Engineering Journal, **2020,** 397. DOI:10.1016/j.cej.2020.125350.

[67] M. Oschatz, M. Antonietti, A search for selectivity to enable CO_2 capture with porous adsorbents, Energy & Environmental Science, **2018**, 11, 57. DOI:10.1039/c7ee02110k.

[68] Balpreet Kaur, Raj Kumar Gupta, Haripada Bhunia, Chemically activated nanoporous carbon adsorbents from waste plastic for CO_2 capture: Breakthrough adsorption study, Microporous and Mesoporous Materials, **2019**, 282, 146. DOI:10.1016/j.micromeso.2019.03.025.

[69] Alexander W. Dowling, Sree R. R. Vetukuri, Lorenz T. Biegler, Large-scale optimization strategies for pressure swing adsorption cycle synthesis, AIChE Journal, **2012**, 58, 3777. DOI:10.1002/aic.13928.

[70] Jun Zhang, Paul A. Webley, Cycle development and design for CO_2 capture from flue gas by vacuum swing adsorption, Environmental Science & Technology, **2008**, 42, 563. DOI:10.1021/es0706854.

[71] Zahra Rouzitalab, Davood Mohammady Maklavany, Shahryar Jafarinejad, Alimorad Rashidi, Lignocellulose-based adsorbents: A spotlight review of the effective parameters on carbon dioxide capture process, Chemosphere, **2020**, 246. DOI:10.1016/j.chemosphere.2019.125756.

[72] Carlos A. Grande, Rui P. P. L. Ribeiro, Alírio E. Rodrigues, Challenges of electric swing adsorption for CO2 capture, ChemSusChem, **2010**, 3, 892. DOI:10.1002/cssc.201000059.

[73] Fulai Liu, Xutao Gao, Rui Shi, Zhengxiao Guo, Edmund C. M. Tse, Yong Chen, Concerted and selective electrooxidation of polyethylene-terephthalate-derived alcohol to glycolic acid at an industry-level current density over a Pd-Ni(OH)2 catalyst, Angewandte Chemie International Edition, **2023**, 62. DOI:10.1002/anie.202300094.

[74] Yan Zhang, Ziqi Wei, Xing Liu, Fan Liu, Zhihong Yan, Shangyong Zhou, Jun Wang, Shuguang Deng, Synthesis of palm sheath derived-porous carbon for selective CO2 adsorption, RSC Advances, **2022**, 12, 8592. DOI:10.1039/d2ra00139j.

[75] Federica Raganati, Francesco Miccio, Paola Ammendola, Adsorption of carbon dioxide for post-combustion capture: A review, Energy & Fuels, **2021**, 35, 12845. DOI:10.1021/acs.energyfuels.1c01618.

[76] Yifan Yan, Hua Zhou, Simin Xu, Jiangrong Yang, Pengjie Hao, Xi Cai, Yue Ren, Ming Xu, Xianggui Kong, Mingfei Shao, Zhenhua Li, Haohong Duan, Electrocatalytic upcycling of biomass and plastic wastes to biodegradable polymer monomers and hydrogen fuel at high current densities, Journal of the American Chemical Society, **2023**. DOI:10.1021/jacs.2c11861.

[77] Maureen Puettmann, Kamalakanta Sahoo, Kelpie Wilson, Elaine Oneil, Life cycle assessment of biochar produced from forest residues using portable systems, Journal of Cleaner Production, **2020**, 250. DOI:10.1016/j.jclepro.2019.119564.

[78] ASME International Design Engineering Technical Conferences Computers and Information in Engineering Conference, Life cycle assessment of pyrolysis-derived biochar from organic wastes and advanced feedstocks, **2020**. Available from https://asmedigitalcollection.asme.org/IDETC-CIE.

[79] Snehanjali Behera, Soumitra Dinda, Rajat Saha, Biswajit Mondal, Quantitative electrocatalytic upcycling of polyethylene terephthalate plastic and its oligomer with a cobalt-based one-dimensional coordination polymer having open metal sites along with coproduction of hydrogen, ACS Catalysis, **2022**, 13, 469. DOI:10.1021/acscatal.2c05270.

CHAPTER 6

Hydrothermal Carbonization of Biomasses for Photocatalytic Application

Chengyu Duan, Mengdi Sun, Ruilin Wang,
Quan Zhou, Yinglong Lu, Zheshun Ou,
Huimin Liu, Guanghui Luo, and Zhuofeng Hu*

ABSTRACT

Hydrothermal carbonization stands as a versatile method for treating biomass waste. By subjecting biomass to hydrothermal treatment, the majority of biomass can be transformed into hydrothermal carbonization carbon (HTCC). Unlike carbon-based conductors formed by conventional pyrolysis (> 500 °C), HTCC has been confirmed to be a type of semiconductor with strong light absorption in the wavelength range of 420 to 800 nm. This excellent photoresponse property is much better than that of traditional non-metallic carbon nitride photocatalysts. Considering the demand for high-performance and low-cost photocatalysts, it is necessary to make a comprehensive summary of HTCC. In this chapter, a polyfuran structure is recognized as the main photosensitive unit of HTCC, regardless of the HTCC morphology. Various related photocatalytic applications including hydrogen peroxide synthesis, disinfection, Cr(VI) reduction, organic pollutant degradation, reactive chlorine species generation, water splitting, ammonia recovery from nitrate, and CO_2 reduction are introduced. These findings suggest that doping heteroatoms into HTCC structures, transforming bulk particles into 2D nanosheets, and creating a heterojunction with another semiconductor represent three viable strategies for boosting the performance of HTCC-based photocatalysts. This chapter focuses on hydrothermal carbonization and the corresponding biomass-derived carbon products, hoping to contribute to further research.

DOI: 10.1201/9781003520566-6

6.1 APPEARANCE AND MORPHOLOGY OF HTCC

Hydrothermal carbonization serves as a versatile method for managing biomass waste. During this procedure, biomass is commonly heated alongside water or organic solvents in a sealed container at temperatures ranging from 150 to 300 °C. The carbonization process capitalizes on the vapor pressure of solvents[1]. This process is seen as a simulation of the natural process of biomass conversion into coal in the Earth's crust, but it effectively shortens the duration from millions of years to just a few hours. Hydrothermal carbonization enables the transformation of low-calorific-value biomass into liquid fuels characterized by high calorific value and energy density through hydrothermal liquefaction[2]. Despite the growing interest in hydrothermal biomass treatment since its inception in 1913 by Bergius, who initially experimented with cellulose, most attention has been devoted to the resulting liquid or gaseous products, such as bio-oil or biofuels[3,4]. However, there is insufficient attention and research on solid products formed in hydrothermal carbonization[5,6].

In decades, solid products have received more and more attention. Scanning electron microscopy reveals that the products are microspheres or particles, which contain many pores ranging from nanopores to mesopores. Importantly, these solid products have been discovered to be able to serve as catalysts such as electrochemical catalysts. There is an increasing number of studies on using hydrothermal carbonization technology to fabricate biomass-derived carbonaceous catalysts. Based on these studies, the carbonaceous catalysts prepared by hydrothermal carbonization are generally described as hydrothermal carbonization carbon, which can be named HTCC. And hydrothermal carbonization exhibits three advantages compared to dry pyrolysis.

6.1.1 Higher Utilization Efficiency of Energy

Hydrothermal carbonization, depicted in Figure 6.1(a), operates as an exothermic process and can be likened to a "wet pyrolysis" process. This method contrasts with the dry pyrolysis of biomass, wherein a significant portion of the energy is allocated to moisture elimination rather than carbonization[7]. In this sense, hydrothermal carbonization this method is economic duo for less consumption and more efficient utilization of energy.

6.1.2 Fewer Greenhouse Gas By-Products

On the one hand, hydrothermal carbonization is a non-combustion process that avoids large-scale generation of CO, CO_2, and other carbonaceous gases. On the other hand, the carbonization occurs in a closed reaction vessel, and the gas products produced during the conversion of biomass into carbon will not leave the solvent surroundings. As a result, applying hydrothermal carbonization to produce biomass-derived carbon helps mitigate greenhouse gas release and achieve almost zero net emissions.

6.1.3 Simpler Operating Procedures and Higher Product Yield

Unlike high-temperature thermochemical treatments, the hydrothermal process is characterized by ease of operation and can be conducted at low temperatures and pressures. At a preset temperature, hydrothermal carbonization carbon can be prepared in an appropriate solvent, and certain additives are easily introduced for the modification. In addition, hydrothermal carbonization is a relatively mild conversion technology for biomasses, where biomasses undergo less quality loss during the carbonization process, thus the yield of biomass-derived carbon is often higher than that via pyrolysis or sintering (Figure 6.1(b)).

During hydrothermal carbonization, carbohydrate in biomasses often undergoes hydrolysis, intramolecular condensation, dehydration, and decarboxylation reactions[10,11]. The majority of carbon element and a portion of oxygen element in carbohydrate molecular structure will be retained in HTCC while the other elements are dissolved in the solvent phase. After hydrothermal carbonization, the dried solid HTCC may show various appearances and morphologies due to different raw biomass materials and the degree of hydrolysis.

As shown in Figure 6.2, the HTCC powders range in colour from brown or tan to black[9]. And the morphology is observed as irregular particle, sphere, or nanosheet (Figure 6.3).

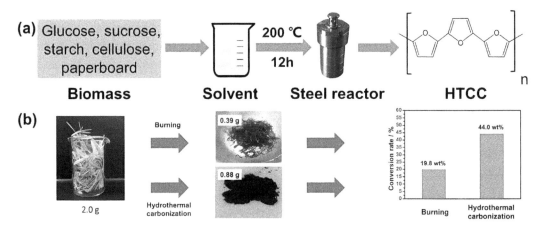

FIGURE 6.1 (a) Hydrothermal carbonization for HTCC products. (b) Higher product yield of biomass-derived carbon[8,9].

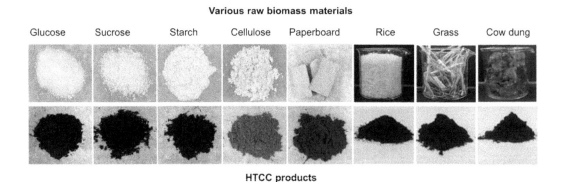

FIGURE 6.2 Different raw materials and their HTCC products[9].

FIGURE 6.3   Cellulose HTCC micron particles formed in (a) distilled water and (b) dilute sulphuric acid. Glucose HTCC (c) microspheres and (d) nanosheets[10,12,13].

6.2 CHEMICAL STRUCTURE OF HTCC

Glucose, one of the simplest carbohydrates, was previously attempted to produce solid carbon-based products through hydrothermal carbonization. In a typical process, glucose is dissolved in water to form a solution, which is sealed in a pressure-resistant and heat-resistant container. By heating the container at a temperature not exceeding 200 °C for 10 to 16 hours, the glucose in the solution will transform into dark brown solid precipitates. In C 1s XPS spectra (Figure 6.4(a)) and O 1s XPS spectra (Figure 6.4(b)) of the glucose HTCC, the C-O-C peaks are obviously stronger than the C=O peaks, suggesting that a class of polymeric chains consisting of unsaturated sp2-hybridized furan units are prevalent in the structure of HTCC (Figure 6.4(c))[9]. The peaks observed at 152 ppm and within the range of 115–127 ppm are attributed to the O-C=C and C=C-C bonds of the furan rings, respectively, and are designated as distinct domains in the spectrum. The broad peak centered around 40 ppm is associated with aliphatic C-H groups. Furthermore, peaks at 175 and 205 ppm are identified with the -COOH and -C=O groups, respectively. These three peaks (40, 175, and 205 ppm) collectively indicate the presence of open-ring domains within the polyfuran structure.

The biomass feedstocks and solvent acidity can affect the structure of the HTCC products. Besides glucose, polysaccharide (cellulose) is also tried to prepare HTCC photocatalysts in pure water. Compared with glucose HTCCs, the cellulose HTCC has more C-O bonds but fewer C-C bonds (Figure 6.4(d)). However, the photoactivity of cellulose HTCC formed in neutral aqueous solvent is not satisfying. After their efforts, Hu et al. found that introducing a small amount of sulphuric acid into the solvent would promote the carbonization of cellulose and significantly enhance the photocatalytic activity of HTCC. This hydrothermal carbonization method in an acidic solvent environment successfully extends carbon-containing raw materials from monosaccharide molecules to polysaccharides and a rich variety of biomass. Compare with cellulose HTCC, the acid-HTCC conversed from cellulose has more C–C bonds in its structure (Figure 6.4(e)), suggesting a higher polymerization degree of furan units. In cellulose HTCC, the primary bonding types of carbon are C–OH or C–O–C, suggesting a polyfuran structure with limited polymerization. Conversely, in the acid-HTCC polyfuran structure, the binding energies of C–O and C=O are 0.8 eV and 0.9 eV lower, respectively, compared to those in cellulose HTCC. The negative

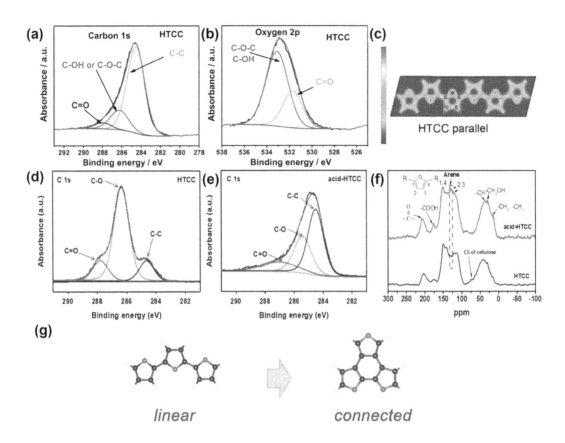

FIGURE 6.4 (a) C 1s XPS spectra and (b) O 1s XPS spectra of glucose HTCC. (c) Charge density map of plane parallel with polyfuran chains of glucose HTCC. C 1s XPS spectra of (d) cellulose HTCC and (e) acid-cellulose HTCC. (f) ^{13}C Solid-state CP-MAS NMR spectra. (g) Linear structure and connected structure polyfuran in HTCC[9,10].

shift indicates a higher negative charge on the carbon atoms in acid-HTCCs. Furthermore, Figure 6.4(f) illustrates that while both cellulose HTCC and acid-HTCC feature O–C=C and C=C–C structures in their furan rings, acid-HTCC exhibits an additional distinct peak at 127 ppm. This peak suggests the formation of aromatic rings from connected furans, a feature not observed in glucose HTCC. It is worth noting that while the polyfuran in cellulosic HTCC adopts a linear structure, in acid HTCC, it undergoes a transformation into a connected structure, ultimately transitioning into aromatic rings (Figure 6.4(g)). The conversion of cellulose to HTCC primarily occurs through direct intramolecular condensation, dehydration, and decarboxylation reactions. As a result, the conversion pathway involving the formation of glucose is considerably restricted[11,14].

6.3 CHEMICAL PROPERTY OF HTCC

One of the most interesting properties of HTCC is its semiconductive nature. Unlike carbon-based conductors formed by conventional pyrolysis (>500 °C), HTCC has been confirmed to be a kind of semiconductor. For example, the cellulose HTCC has significant advantage in photo-responsive activity compared with classical g-N_3N_4. Cellulose HTCC can strongly absorb light in the wavelength range of 420 nm to 800 nm, while g-C_3N_4 shows an extremely low absorption efficiency for light above 475 nm (Figure 6.5(a)). This high utilization efficiency of solar energy is attributed to the photosensitive polyfuran structure in HTCC, which is a conjugated plane containing a large π bond composed of six electrons derived from C and O atoms. The polyfuran structure has been confirmed to possess semiconductive properties, while other functional groups like aliphatic C–H and C–OH are less capable of absorbing visible light.

By comparing the Fermi levels of HTCC and g-C_3N_4, it was found that their reducing abilities are similar (Figure 6.5(b)). However, considering that the VB maximum (E_{VB}) of HTCC and g-C_3N_4 are 1.69 and 1.97 eV lower than their corresponding Fermi levels (Figure 6.5(c)), respectively, the E_{VB} of HTCC (1.18 eV) is 0.22 eV lower than that of g-C_3N_4 (1.4 eV). A lower E_{VB} results in weaker thermodynamic oxidation power of HTCC compared to g-C_3N_4. Notably, the photocurrent density of HTCC is significantly higher than that of g-C_3N_4, indicating an efficient separation of photogenerated carriers of HTCC (Figure 6.5(d)). In addition, the detected fluorescence intensity of HTCC is much lower than that of g-C_3N_4 (Figure 6.5(e)), suggesting the much slower recombination of electrons and holes of HTCC[8]. Consequently, the polyfuran structures give HTCC distinct semiconducting characteristics that enable it to absorb solar energy from the ultraviolet to the near infrared for photocatalytic applications. And Figure 6.5(f) shows a summary of the band structure of HTCC and g-C_3N_4.

6.4 THE APPLICATION OF HTCC IN PHOTOCATALYTIC SYSTEMS

Under visible light irradiation, photo-induced electrons and holes are produced on HTCC and tend to trigger a sequence of redox reactions, which can directly or indirectly oxidize or reduce many substrates in environment. Considering the application of HTCC photocatalyst, the target substrates may be dissolved oxygen, pathogenic micro-organisms, emerging contaminants, or even water molecules.

6.4.1 Photocatalytic Synthesis of Hydrogen Peroxide

Presently, the predominant method for producing commercial hydrogen peroxide (H_2O_2) involves the anthraquinone (AQ) process, characterized by high energy consumption and significant generation of wastewater, exhaust gases, and solid waste. In comparison, photocatalytic oxygen reduction is a good method for H_2O_2 synthesis. For this technology, the substrate dissolved oxygen in solution can be efficiently utilized and reduced in situ.

It is found that under visible light irradiation, HTCC shows a comparable reducing ability to g-C_3N_4, theoretically enabling the photocatalytic synthesis of H_2O_2. Due to the CB edge being higher than the Fermi level, the reaction of O_2 reduction to O_2^- (-0.33V vs NHE) is thermodynamically favorable for HTCC, marking the initial step in H_2O_2 production. At the same time, the E_{VB} of HTCC is more negative by 0.22 volts than g-C_3N_4, so the oxidation power of HTCC is weaker, which is beneficial for reducing the decomposition and consumption of H_2O_2 produced. The theoretical calculation suggests that the polyfuran structure is conducive to the adsorption and activation of dissolved oxygen,

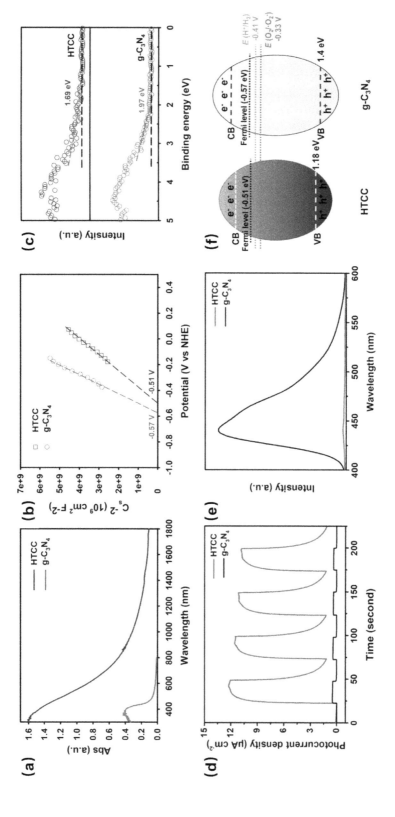

FIGURE 6.5 (a) UV-vis-NIR diffuse-reflectance spectra, (b) Mott-Schottky plots, (c) VB XPS spectra, (d) transient photocurrent responses under visible-light irradiation, (e) photoluminescence spectra, (f) band structure of HTCC and g-C$_3$N$_4$[8].

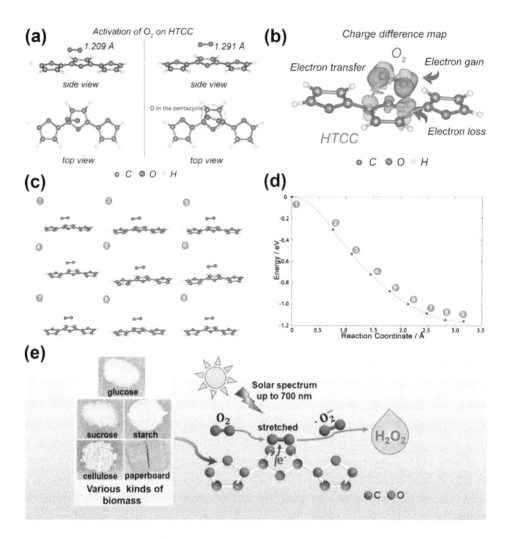

FIGURE 6.6 Density functional calculations for activation of O_2 on HTCC: (a) Charge difference map of the O_2-HTCC system. (b) Process for activation of the O_2 molecule on HTCC. (c) Structure change and (d) energy change for the O_2 activation in the NEB calculations on the HTCC system. (e) The mechanism for photocatalytic synthesis of H_2O_2.[8]

and exhibits excellent electron transfer properties (Figure 6.6(a)). The adsorption energy of O_2 by HTCC is −0.69 eV, resulting in the elongation of the O=O bond length from 1.209 to 1.291 Å. (Figure 6.6(b)). Meanwhile, the O_2 activation process is also found to be energy-favorable (Figure 6.6(c, d)). Therefore, the furan units in HTCC are considered active sites for oxygen reduction. Mechanism studies have shown that the production of H_2O_2 on HTCC is mainly through two-step oxygen reduction reactions by $\cdot O_2^-$ radical as intermediates (Figure 6.6(e)). Finally, using HTCC, the apparent quantum yield and synthesis rate of H_2O_2 in the photocatalytic system can reach 18.2% (420 nm) and 1.16 mmol/(g_{cat} h), respectively. Importantly, the hydrothermal carbon photocatalysts converted from sucrose, starch, cellulose, and cardboard also exhibit the performance of photocatalytic oxygen reduction to synthesize H_2O_2.

6.4.2 Photocatalytic Disinfection

In general, the chemical inactivation of pathogenic micro-organisms is mainly achieved by a series of oxidation processes. During the oxidation processes, the oxidants are generally holes on HTCC catalyst or reactive oxygen species (such as h^+, $\cdot O_2^-$, $\cdot OH$) indirectly generated by photogenerated electrons:

A kind of glucose-derived HTCC nanosheets is effective to photocatalytic disinfection. The experimental setup includes a 300 W xenon lamp with a UV cutoff filter ($\lambda > 420$ nm) to ensure stable light irradiation, while maintaining a constant visible-light intensity of 100 mW/cm². Before the disinfection experiment, the initial concentration of bacterial cells (*E. coil K-12*) is adjusted to 1×10^7 colony forming units per milliliter (cfu/mL), and the dosage of HTCC nanosheet is 200 mg/L. After 3 hours, as shown in Figure 6.7(a), all the bacteria are killed. In comparison, over the course of a 3-hour control experiment conducted in darkness, there is minimal change observed in the bacterial population, and a particular type of bulk HTCCs demonstrates weak activity, with the cell density decreasing by less than 0.5. Even when N_2 was continuously bubbled into the system, there is no significant change in activity, indicating that the performance of the system was due to h^+ oxidation[14].

In addition to h^+ on HTCC, enriched photogenerated electrons can promote the reduction of dissolved oxygen to produce superoxide radicals ($\cdot O_2^-$), hydrogen peroxide (H_2O_2), and hydroxyl radicals ($\cdot OH$) in the photocatalytic system. These active species are also effective oxidants for the inactivation of pathogenic micro-organisms. A MoS_2-encapsulated HTCC microsphere is reported to generate plenty of mainly $\cdot O_2^-$ for rapid inactivation bacterial cells (Figure 6.7(b))[15].

HTCC-based photocatalyst can also be electron donor to combining persulfate (PS) activation for bacterial inactivation. For example, the photoexcited carriers on a composite of magnetic iron oxide and HTCC can efficiently activate PS under visible-light irradiation. Regarding disinfection efficacy, the complete inactivation of 8.0-log E. coli cells occurs within 40 minutes, as illustrated in Figure 6.7(c). The primary reactive oxygen species (ROS) responsible for this process is the sulfate radical, which oxidizes the cytoplasm of the cell, disrupts genomic DNA, and ultimately induces irreversible cell death[16].

6.4.3 Photocatalytic Reduction of Cr(VI)

$\cdot O_2^-$ radicals are an active species with both electron gaining and electron losing abilities, and there are some reports on the reduction and removal of Cr(VI) by $\cdot O_2^-$ radicals. Considering $\cdot O_2^-$ radicals are the necessary intermediate products for photocatalytic reduction of dissolved oxygen, it is possible to use HTCC to removal Cr(VI) in a photocatalytic system.

To verify the removal effect of $\cdot O_2^-$ radicals on Cr(VI), cellulose waste is reported to prepare acid-HTCC for the photocatalytic produce of $\cdot O_2^-$ via molecular O_2 reduction. First, the formation of XTT formazan from XTT is used to quantify $\cdot O_2^-$ radical. After 2 hours of irradiation, the XTT-formazan concentration generated by acidic HTCC is six

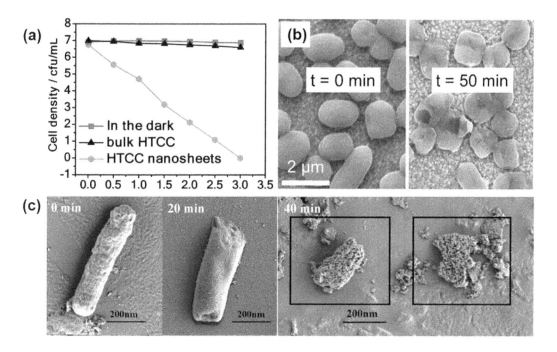

FIGURE 6.7   (a) The disinfection of *E. coli K-12* by HTCC nanosheets. (b) FE-SEM images of bacterial cells after the photocatalytic treatment by HTCC@MoS$_2$ for 0 min and 50 min. (c) And SEM images of *E. coli* cell treated by the VL/PS/MHC system with different irradiation time[14–16].

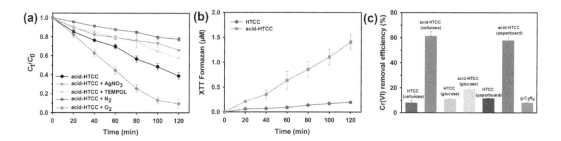

FIGURE 6.8   (a) The formation of XTT formazan, representing the formation of ·O$_2^-$. (b) Photocatalytic Cr(VI) reduction with acid-HTCC in the presence of different kinds of scavengers. (c) Cr(VI) removal efficiency with different catalysts after 2 hours irradiation[10].

times greater than that produced by HTCC (Figure 6.8(a)), suggesting there is a higher efficiency of O$_2$ activation by acid-HTCC to obtain more ·O$_2^-$. Secondly, the promotion of Cr(VI) reduction by acid-HTCC can only be attributed to the reducing agents present in the photocatalytic system. To clear the role of reactive species in Cr(VI) reduction, AgNO$_3$ and TEMPOL are applied to quench e$^-$ and ·O$_2^-$, respectively. The quenching experiments reveal a clear inhibition of the photocatalytic reduction of Cr(VI) by both AgNO$_3$ and TEMPOL, which make k are reduced by 59% and 44%, respectively, suggesting that both e^- and $\cdot O_2^-$ play important roles (Figure 6.8(b)). Following this, it was observed that the reduction of Cr(VI) is notably impeded when dissolved O_2 in the aqueous solution is removed through N_2 purging. Conversely, introducing additional O_2 through bubbling greatly enhances the reduction of Cr(VI). These confirmed the importance of $\cdot O_2^-$ during the photocatalytic reduction of Cr(VI) removal[10]. It is worth noting that acid-HTCC generally performs better than HTCC, especially cellulose acid-HTCC and paperboard acid-HTCC (Figure 6.8(c)).

6.4.4 Photocatalytic Degradation of Organic Contaminants

Visible-light driven advanced oxidation is one important technology for water pollution treatment. In recent years, this technology has received considerable attention to deal with the emerging contaminants in aqueous media. Therefore, HTCC-based photocatalyst is worthy of study for its application in photocatalytic advanced oxidation processes.

If metal ions are present in the hydrothermal environment, they can be incorporated into the matrix, thereby altering the structure, semiconductor properties, and activity of the HTCC.

Metal-hydrothermal carbon composites can also be synthesized through the hydrothermal carbonization process. For instance, Cu^{2+} can be reduced and immobilized within

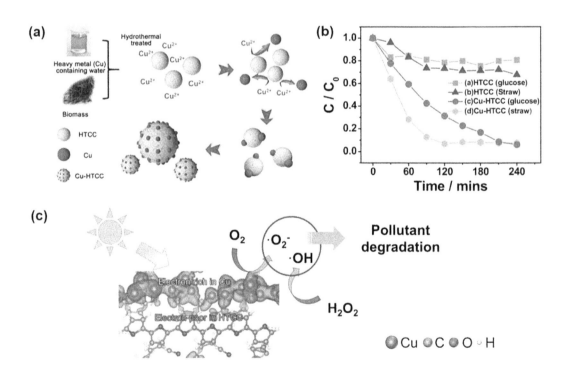

FIGURE 6.9 (a) The synthesis of Cu-HTCC. (b) The degradation constants of Ibuprofen degradation over pure HTCC (glucose), HTCC (straw), Cu-HTCC (glucose) and Cu-HTCC (straw). (c) Catalytic mechanism for pollutants degradation by hydrogen peroxide activation over Cu-HTCC under visible light illumination[17].

the HTCC matrix to produce Cu-HTCC composite materials, which exhibit high photocatalytic activity in the degradation of ibuprofen. In the initial stage of hydrothermal carbonation process, As small HTCC particles nucleate and grow, nearby Cu ions undergo reduction to form Cu metal particles. Over time, these two processes occur concurrently, eventually leading to the formation of Cu-HTCC composite particles (Figure 6.9(a)). This Cu-HTCC photocatalyst demonstrates a light absorption range extending up to 1300 nm, surpassing that of many metal oxides or sulfides. As shown in Figure 6.9(b, c), due to the reduction of oxygen occurs in this photocatalytic system, $\cdot O^-_2$ radicals are detected during the Ibuprofen degradation process. Importantly, Cu in Cu-HTCC can be efficient activator for H_2O_2 converting into $\cdot OH$. Thus, within the system, reactive species such as $\cdot OH$, e^-, h^+, and $\cdot O_2^-$, play pivotal roles, with ECB-mediated reactions being essential pathways for Ibuprofen degradation within a 60-minute timeframe. Moreover, the applicability of the metal-HTCC composite in real water environments suggests its potential as a promising photocatalyst candidate[17].

6.4.5 Photocatalytic Generation of Reactive Chlorine Species (RCSs)

HTCC can also be used to generate RCSs by using oxygen reduction strategy. Unlike reactive oxide species, RCSs are less influenced by solution chemistry and exhibit greater selectivity in pollutant degradation. Based on the performance of Cu-HTCC, Hu et al. believe that HTCC photocatalyst can also be used to produce RCS by the reaction between $\cdot OH$ and common chloride ions. A single copper chloride-loaded hydrothermal carbonaceous carbon is synthesized through hydrothermal carbonization in an aqueous solution containing copper sulfate and sodium. Efficient generation of RCS is facilitated by O_2 activation on CuCl-HTCC. In particular, the cuprous (Cu(I)) species formed within CuCl-HTCC facilitate the adsorption and activation of O_2, which subsequently transforms into superoxide radicals, leading to the generation of H_2O_2 and hydroxyl radicals. Additionally, chloride ions present on the surface of CuCl-HTCC undergo a series of reactions, resulting in the production of RCS (as shown in Figure 6.10(a)). The advantage of CuCl-HTCC photocatalytic system is recognized as the stable existence of low valent copper ions for oxygen reduction. The abundant presence of Cu(I) within CuCl-HTCC facilitates heterogeneous activation of O_2, resulting in significantly higher reduction efficiency compared to homogeneous reactions where stable Cu(I) cannot exist. Consequently, pollutants such as phenol or ibuprofen at a concentration of 10 mg/L were almost completely degraded within 120 min. Specifically, RCS contributes approximately 30% to the degradation of phenol, and the contribution of $\cdot ClO$ radicals should be given due consideration (Figure 6.10(b)).

6.4.6 Photocatalytic Water Splitting

HTCC could couple with other metal oxide semiconductors for photo-electro chemical water splitting by constructing a photoelectrochemical cells (PECs). There are many studies that claim that coupling two-dimensional carbon materials on photoelectrodes results in high-performance PECs, making a simple bottom-up approach to the synthesis of carbon-based two-dimensional materials very attractive. Carbohydrates are ideal precursors for synthesis, but previous reports required pyrolysis at high temperatures (>700 °C).

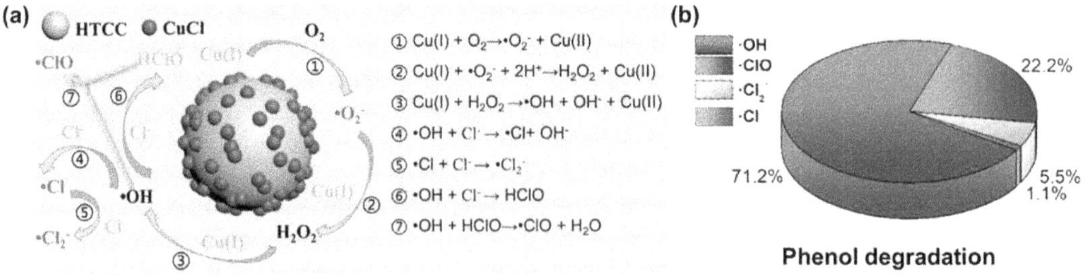

FIGURE 6.10 (a) Mechanism diagram of active substance production on CuCl-HTCC. (b) Contribution of different radicals in phenol degradation[18].

Hu et al. developed a method for synthesizing carbon nanosheets at low temperature (210 °C) in glucose aqueous solution, starting from glucose. With the assistance of ethylenediamine and Fe^{3+} ions, nanosheets can grow in very close contact on a hematite nanorod array (Figure 6.11 (a, b)).

The HTCC structure primarily consists of polyfuran. In this study, a model was constructed with Fe_2O_3 at the base and four polyfuran chains (Figure 6.11(a)). Upon full relaxation, the contact area between the polyfuran chain and Fe_2O_3 became distorted, with the polyfuran chain near Fe_2O_3 losing its planar configuration. This resulted in deviation of Fe and O atoms from the rhombohedral structure, creating a metallic region at the interface (Figure 6.11(b)), which promotes charge transfer. The schematic representation of the photoelectrochemical (PEC) setup is depicted in Figure 6.11(c). The onset potential of HTCC-Fe_2O_3 shifted negatively by 0.12 to 0.82 V vs. RHE, as shown in Figure 6.11(d). This 0.12 V improvement is attributed to a lower kinetic energy barrier for charge transfer from the photoanode to the electrolyte, reducing the external electric power demand required for driving the PEC process. Rapid charge transfer enhances the activity of hematite by approximately 500% when using the nanosheets, surpassing the efficacy of a physical or chemical mixture of graphene and hematite (typically <200%). This enhancement primarily results from the deformation zone between the nanosheets and hematite. Ultrafast transient absorption spectrum analysis confirms an increase in the effective hole diffusion length from 2 to 8 nm, along with an extension of the carrier lifetime. The electronic properties of HTCC-Fe_2O_3 are briefly outlined in Figure 6.11(e).

This method offers further potential for the simple, mild, and cost-effective preparation of carbon-based two-dimensional materials using a bottom-up approach.

6.4.7 Photocatalytic Ammonia Recovery from Nitrate

Presently, ammonia production primarily relies on the conventional Haber-Bosch process, which consumes substantial quantities of fossil fuels and results in significant carbon dioxide emissions. Photocatalytic nitrate reduction is emerging as a promising alternative technology. This technology can be used not only to remove pollutants from water but also to recover ammonia.

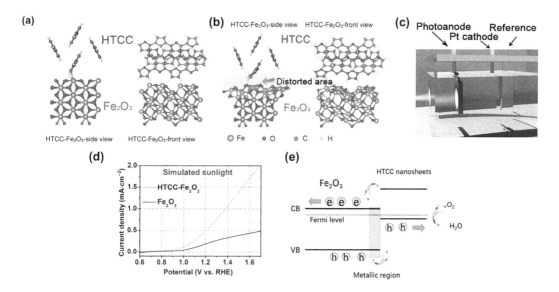

FIGURE 6.11   (a) Crystal structure at the interface between HTCC-Fe$_2$O$_3$ before and (b) after relaxation. (c) Schematic picture of PEC experiments using HTCC nanosheets-Fe$_2$O$_3$ nanorods photoanode. (d) Voltammograms of pristine Fe$_2$O$_3$ photoanode (olive) and HTCC-Fe$_2$O$_3$ photoanode (orange) under simulated sunlight; (e) And electronic property of HTCC-Fe$_2$O$_3$[13].

HTCC, as low-cost and green catalysts, is also investigated for photocatalytic reduction of nitrate to ammonia production in the presence of formic acid. Wei et al. grind dry pomegranate peel into powder and mix it with perovskite (LaFeO$_3$) for hydrothermal treatment at 180 °C under microwave condition. A nanostructured photocatalyst is obtained by assembling LaFeO$_3$ on HTCC (Figure 6.12(a)). The morphology of LaFeO$_3$/HTCC is characterized by two-dimensional nanosheets, featuring a substantial specific surface area and an abundance of surface functional groups. The incorporation of Fe/La salt facilitates surface modification of HTCC, promoting the exposure of oxygen-containing functional groups and aromatic structures, thereby enhancing nitrate adsorption. At optimized mass ratios of pomegranate peel to Fe/La salt, the LaFeO$_3$/HTCC photocatalyst achieves a remarkable nitrate removal efficiency of 94.6% and an impressive ammonia selectivity of 88.7% under visible light irradiation (Figure 6.12(b)). Additionally, the combination of HTCC and LaFeO$_3$ forms a p-n heterojunction, enhancing photocatalytic activity by facilitating the rapid separation of photogenerated electrons and holes. (Figure 6.12(c)).

6.4.8 Photocatalytic CO$_2$ Reduction

The rapid rise of CO$_2$ levels in the atmosphere has led to significant environmental challenges, with combustion of carbonaceous substrates like plant residues and biomass waste being a major source of CO$_2$ emissions. Hydrothermal carbonization presents an ideal approach for converting these biomass wastes into functional materials for CO$_2$ reduction. It has been reported that HTCC can be effectively employed for photocatalytic CO$_2$ reduction.

By incorporating Cu ions, Cu-HTCC can be synthesized through hydrothermal treatment of glucose or straw. In TEM images focusing on the sphere edges, it is observed that 5–15 nm Cu particles are immobilized on the HTCC spheres (Figure 6.13(a)). The as-prepared Cu-HTCC photocatalysts are uniformly dispersed to cover a glass substrate. In the photocatalytic experiment, the glass substrate and $KHCO_3$ are put into the reactor, where the air is removed by vacuum. The 6 M HCl solution is used to react with $KHCO_3$ to provide CO_2 gas for its reduction. Under visible light irradiation, HTCC shows a high activity of CO production (382.1 μmol/(g h)), which is 19 times as high as that of commercial TiO_2 and much higher than that of g-C_3N_4 (15 μmol/(g h)). Furthermore, with the introduction of Cu, the CO yield of Cu-HTCC (643.5 μmol/(g h)) is further improved to 32 and 1.7 times that of commercial TiO_2 and HTCC, respectively. And the Cu-HTCC derived from straw also shows a high activity of 564.8 μmol/(g h) (Figure 6.13(b)). The process of CO formation involves multiple steps: Initially, a hydrogen atom is added to an oxygen atom, resulting in the formation of a

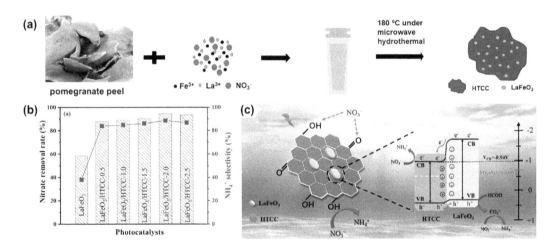

FIGURE 6.12 (a) Schematic for the synthesis of the $LaFeO_3$/HTCC nanocomposite. (b) Photocatalytic conversion test curve of pristine $LaFeO_3$ and $LaFeO_3$/HTCC-0.5-2.5 to nitrate at pH = 2. (c) Schematic of photocatalytic reduction of nitrate by the $LaFeO_3$/HTCC composite[19].

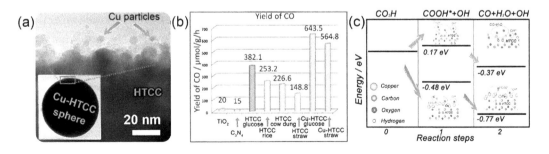

FIGURE 6.13 (a) TEM images of Cu-HTCC prepared from glucose and $CuSO_4$. (b) Yield of CO from photocatalytic CO_2 reduction on different photocatalysts. (c) Energy diagram of the two steps during the formation of CO[20].

COOH* species and an OH* species. Subsequently, the COOH* species reacts with a second hydrogen atom to produce CO, H_2O, and another OH* species. The resulting OH* species can further react with another CO_2 molecule to initiate another round of CO_2 reduction. In comparison to HTCC, the formation of a COOH* intermediate and the subsequent generation of CO are more favorable, with the formation energy being more negative. Therefore, a significant carbon dioxide reduction performance is achieved by the Cu-HTCC photocatalyst (Figure 6.13(c)). This study presents a novel "trash-to-treasure" strategy that transforms biomass waste into photocatalysts for CO_2 reduction[20].

6.5 STRATEGIES FOR ENHANCED PERFORMANCE OF HTCC APPLICATION

The semiconductor properties of HTCC mainly come from the sp^2-hybridized polymer chains. Based on this, pure HTCC will face three limitations when pursuing higher photocatalytic activity:

6.5.1 Inefficient Electron Transfer Between Polymer Chains

While charge transfer within a single polymer chain of HTCC is smooth, inter-chain charge transfer is inefficient. To enhance activity in photocatalytic applications, the primary strategy is to establish a bridge for electronic cross-chain transfer (Figure 6.14(a)).

6.5.2 Long Migration Distance of Electron

In the absence of other supports, HTCC tends to grow isotropically in all directions, resulting in the formation of spherical or irregular large HTCC particles (>1 μm). However, these irregular particles may encounter challenges related to charge transfer due to long migration distance, while a two-dimensional (2D) nanosheet is workable to physically shortens the distance for electron transfer (Figure 6.14(b)).

6.5.3 Recombination of Photo-induced Electrons and Holes

Although the six electrons conjugated to the furan ring are easily excited by visible light, they always recombine with holes in the homogeneous activated carbon. In many current

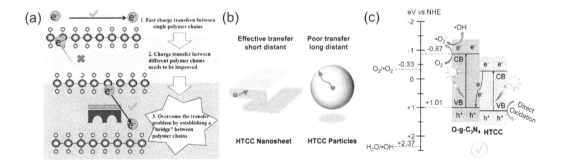

FIGURE 6.14   (a) A bridge for electronic cross chain transfer. (b) A 2D nanosheet for short transfer distant of electron. (c) A heterojunction for prolonged recombination of photo-induced electrons and holes[12,14,21].

photocatalytic systems, the photocatalyst often has a heterojunction for prolonged recombination of electrons and holes. Therefore, it is recommended to combine biochar with other semiconductors to form a heterojunction, which will be beneficial to obtain higher photocatalytic activity (Figure 6.14(c)).

Herein, three strategies are summarized to overcome the mentioned limitations. These strategies include doping heteroatoms into HTCC structures, and forming a heterojunction with another semiconductor

6.5.4 Doping Heteroatoms into HTCC Structures

For HTCC, its sp^2-hybridized units are discrete, leading to poor charge transfer and conductivity in large particles. If a heteroatom is inserted into the HTCC structure to alter the distribution of surrounding electrons, the electron clouds of different polyfuran chains may overlap or be connected, creating a bridge for electron transfer.

It is reported that iodine dopants are efficient to twist and optimize the structures of the sp^2-hybridization in HTCC and leading to an enhanced photon-induced excitation. By

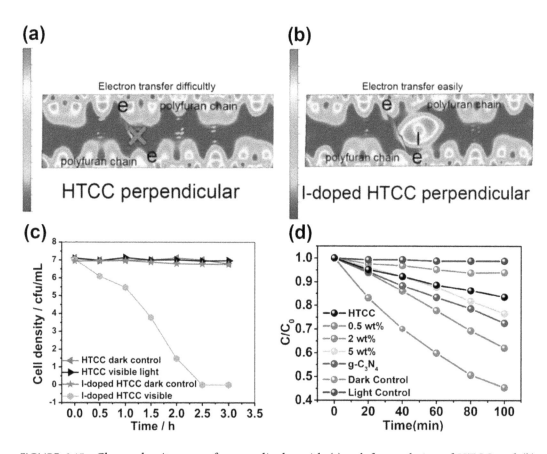

FIGURE 6.15 Charge density map of perpendicular with (a) polyfuran chains of HTCC and (b) I-doped HTCC. (c) Disinfection of *E. coli* K-12 by glucose HTCC without and with iodine doped. (d) Photo-degradation of RhB in the presence of different samples under visible-light irradiation[9,22].

adding elemental iodine during the hydrothermal treatment of glucose, an iodine doped HTCC (I-HTCC) is prepared. For HTCC without iodine, in the charge density map perpendicular to the polyfuran chain, a vacuum field of electrons is found between two chains (Figure 6.15(a)). However, for I-HTCC, an obvious electron cloud surrounds the iodine atom and connects to the electron cloud of adjacent polyfuran chains (Figure 6.15(b)). Based on this "bridge", electrons can transfer between different polyfuran chains and lead to faster charge transfer, thus increasing the conductivity and activity of the HTCC. Comparing with HTCC, I-HTCC gets an exciting preferment on the performance of photocatalytic disinfection (As shown in Figure 6.15(c)).

In addition of iodine, chlorine is also studied to be doped into HTCC structure. Zhang et al. mix glucose and sodium chloride in mass ratios of 100:1, 100:2, and 100:5, respectively, followed by hydrothermal carbonation in absolute ethanol. Chlorine doped HTCC (Cl-HTCC) also obtain an enhanced charge transfer and its photocatalytic activity is 4.53 times higher than undoped glucose HTCC. As shown in Figure 6.15(d), it is apparent that the optimum doping amount of chlorine is 2 wt%, 55% of RhB degradation is achieved within 100 min.

6.5.5 Converting Bulk Particles into 2D Nanosheets

In contrast to bulk irregular HTCC particles, HTCC nanosheets demonstrate superior charge transfer efficiency in the direction perpendicular to their surface.

Using a Fe_2O_3 nanorod substrate or a Si substrate immersed in an ethylenediamine solution, hydrothermal carbonization at 210°C successfully converts glucose, straw, rice, and starch into HTCC nanosheets. These nanosheets, as depicted in Figure 6.16(a, b), exhibit a layered structure with thickness ranging from 2 to 8 nm, akin to graphene. Under light irradiation, HTCC nanosheets display remarkable improvements in catalytic pollutant degradation compared to bulk HTCC. While bulk HTCC degrades less than 10% of MO after 350 minutes, HTCC nanosheets degrade the majority of MO within the same duration (Figure 6.16(c)). Specifically, the degradation kinetics constant of bulk HTCC stands at 0.025 h^{-1}, while that of HTCC nanosheets reaches 0.430 h^{-1} marking a 17.3-fold increase in activity. Likewise, HTCC nanosheets exhibit significantly higher efficiency in the photocatalytic reduction of Cr(VI) (Figure 6.16(d)).

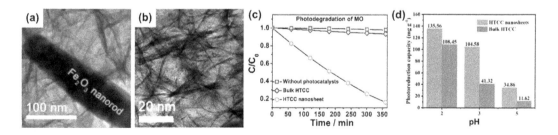

FIGURE 6.16 (a) The typical TEM image of a HTCC-Fe_2O_3 nanorod. (b) The typical TEM image of HTCC nanosheets. (c) Photocatalytic degradation of methyl orange on HTCC nanosheets and bulk HTCC. (d) Cr (VI) photoreduction on bulk HTCC and HTCC nanosheets under illumination[13,14].

6.5.6 Forming a Heterojunction with Another Semiconductor

Generally, the heterojunction of composite photocatalysts is formed by different semiconductors. For HTCC, another semiconductor can be metallic compound or nonmetallic material.

6.5.6.1 Heterojunction with Metallic Compounds

There are many metallic compounds are accessible to combine with HTCC for enhanced photocatalytic activity in various applications. For instance, it has been verified that MoO_2 can establish a chemical bond with the carboxyl functional groups originating from the sp^2-hybridized structures within HTCC. As a result, MoO_2 has been engineered to serve as a charge transport bridge between the polymer chains of HTCC in previous studies[23]. Besides this, there are other metallic compounds that are being studied to combine with HTCC.

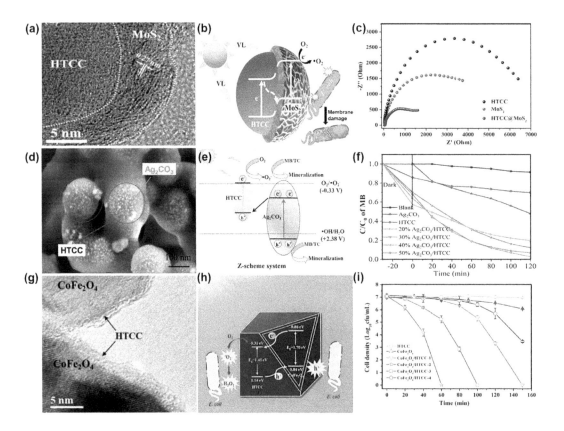

FIGURE 6.17   (a) High-resolution TEM image of the HTCC@MoS_2 nanospheres. (b) The mechanism of photocatalytic bacterial inactivation in HTCC@MoS_2. (c) EIS Nyquist plots of HTCC, MoS_2, and HTCC@MoS_2. (d) SEM of Ag_2CO_3/HTCC. (e) Photoinduced charge separation and transfer for Z-scheme heterojunction. (f) Photocatalytic degradation curves of MB. (g) HRTEM images of the $CoFe_2O_4$/HTCC composite. (h) The type of heterojunction in $CoFe_2O_4$/HTCC. (i) Photocatalytic cell inactivation under visible light irradiation[15,24,25].

A MoS$_2$-encapsulated HTCC heterojunction (HTCC@MoS$_2$) is constructed via a one-pot route (Figure 6.17(a)). The Z-scheme mode significantly enhances electron transfer and charge separation in the HTCC@MoS$_2$ system, where electrons (enrich in the CB of HTCC) are transferred to the VB of MoS$_2$ through the heterojunction, and then they are excited again to reach the CB of MoS$_2$ (Figure 6.17(b)). And the EIS Nyquist plots (Figure 6.17(c)) revealed a lower charge-transfer resistance than HTCC. Based on the enhanced optical response, ·O$_2^-$ radicals are efficiently produced by oxygen reduction under visible light irradiation, allowing the HTCC@MoS$_2$ nanospheres to efficiently eliminate 7-log bacterial cells within a 50-minute timeframe.

To improve the charge transfer capability of HTCC, other Z-scheme heterojunction photocatalyst of silver carbonate (Ag$_2$CO$_3$) and HTCC (Ag$_2$CO$_3$/HTCC) is developed (Figure 6.17(d, e)) for photocatalytic degradation of methylene blue (MB) and tetracycline (TC) from wastewater. The synergistic effect between Ag$_2$CO$_3$ and HTCC facilitates the transfer of photogenerated electrons and holes at the interfaces of these two materials, thereby enhancing the photocatalytic efficiency of their composite. Ag$_2$CO$_3$/HTCC exhibits superior photocatalytic activity and stability compared to both Ag$_2$CO$_3$ and HTCC individually. Radical scavenging experiments reveal that ·O$_2^-$ and h$^+$ are the primary reactive species induced by visible light on the heterojunction. High MB degradation efficiency can be obtained by adjusting the proportion of Ag$_2$CO$_3$ in the composite before the hydrothermal process (Figure 6.17(f)).

One HTCC-coated CoFe$_2$O$_4$ composite is fabricated by mixing and hydrothermal carbonation of CoFe$_2$O$_4$ nanoparticle and glucose. The composite photocatalyst (CoFe$_2$O$_4$/HTCC) adopts a nearly cubic shape, with HTCC coating thickness varying from 0.62 to 4.38 nm. Thin and semi-transparent layers are distinctly visible on the surfaces of the CoFe$_2$O$_4$ nanoparticles, indicating the formation of a heterojunction between CoFe$_2$O$_4$ and HTCC (illustrated in Figure 6.17(g)). Findings reveal that the intimate coating of HTCC on CoFe$_2$O$_4$ nanoparticles enhances electron transfer and charge separation, resulting in a significant enhancement in photocatalytic efficiency (Figure 6.17(h)). As depicted in Figure 6.17(i), CoFe$_2$O$_4$/HTCC demonstrates superior photocatalytic inactivation of *E. coli* K-12 under visible light irradiation, achieving complete inactivation of 7 log 10 cfu/mL of bacterial cells within 60 min. Conversely, HTCC alone exhibits no discernible photocatalytic bacterial inactivation within 150 min.

According to the mentioned and other studies, the Z-scheme heterojunction can be recommended to be constructed by HTCC and another semiconductor for a longer lifetime of photoinduced charges and enhanced separation of carriers.

6.5.6.2 Heterojunction with Nonmetallic Materials

g-C$_3$N$_4$ represents a common nonmetallic semiconductor that responds to visible light irradiation. Like HTCC, g-C$_3$N$_4$ can be easily synthesized from readily available precursors such as urea and exhibits stability with minimal toxicity. Consequently, the combination of HTCC and g-C$_3$N$_4$ presents an appealing approach to create a novel metal-free heterojunction photocatalyst for various photocatalytic applications.

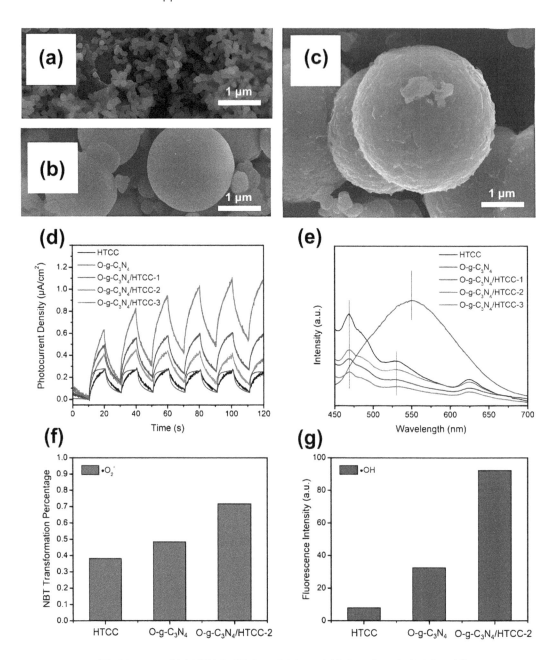

FIGURE 6.18 SEM images of (a) HTCC, (b) O-g-C₃N₄, and (c) O-g-C₃N₄/HTCC-2. (d) PC responses and (e) PL emission spectra of HTCC, O-g-C₃N₄, and different O-g-C₃N₄/HTCC samples. The production of (f) ·O₂⁻ and (g) ·OH[21].

Zhang et al. select a O-doped g-C₃N₄ to construct the heterojunction with HTCC. In a typical procedure, O-g-C₃N₄ microspheres are first prepared by solvothermal treatment of cyanuric chloride and dicyandiamide in acetonitrile. Such microspheres are mixed with glucose, followed by hydrothermal carbonization in deionized water, and then O-g-C₃N₄/HTCC heterojunction can be obtained. As shown in Figure 6.18(a–c), O-g-C₃N₄ are covered by HTCC, and Z- scheme heterojunction is successfully constructed. Under visible light irradiation, O-g-C_3N_4/HTCC, O-g-C_3N_4, and HTCC generate photocurrent, indicating their responsiveness to visible light for the generation of electron-hole pairs. The O-g-C_3N_4/HTCC composites demonstrate significantly higher photocurrent densities compared to HTCC and O-g-C_3N_4 alone (Figure 6.18(d)). In addition, a stronger PL intensity showed by HTCC indicates it suffers a faster radiative charge recombination than other photocatalysts (Figure 6.18(e)). These demonstrate that the heterojunctions dramatically mitigate the photogenerated charge recombination between HTCC and O-g-C_3N_4 and enhance carrier separation. In results as shown in Figure 6.18(f, g), g-C_3N_4/HTCC generates more $\cdot O_2^-$ and $\cdot OH$ for photocatalytic disinfection of human adenovirus.

6.6 CONCLUSION

The hydrothermal treatment of biomass yields photoactive HTCC, capable of absorbing and harnessing solar spectra ranging from ultraviolet to near-infrared light. HTCC is advantageous compared to most metal oxides. HTCC contains a substantial number of sp^2 hybrid units, and these polyfuran structures exhibit excellent light absorption properties. The photoexcited electrons can be generated under visible light irradiation. Photo-induced electrons and holes can achieve various applications by triggering appropriate redox reactions. HTCC photocatalysts have been used for in-situ hydrogen peroxide synthesis, photocatalytic disinfection, hexavalent chromium reduction, organic pollutant degradation, water splitting, ammonia recovery from nitrate, and carbon dioxide reduction. These non-metallic biomass-derived carbons are easy to prepare and have a low environmental risk. Importantly, their surfaces have many functional groups that can be easily coupled with other catalysts or semiconductors to form heterojunctions, thereby achieving higher activity. Overall, conversion of biomass via hydrothermal method has great potential, and HTCC is a good candidate as efficient photocatalyst for environmental treatment.

NOTE

*Corresponding author

REFERENCES

[1] A. Méndez, G. Gascó, B. Ruiz, E. Fuente, Hydrochars from industrial macroalgae "Gelidium Sesquipedale" biomass wastes, Bioresource Technology, **2019**, 275, 386. DOI:10.1016/j.biortech.2018.12.074.

[2] Xiangdong Zhu, Yuchen Liu, Feng Qian, Chao Zhou, Shicheng Zhang, Jianmin Chen, Role of hydrochar properties on the porosity of hydrochar-based porous carbon for their sustainable application, ACS Sustainable Chemistry & Engineering, **2015**, 3, 833. DOI:10.1021/acssuschemeng.5b00153.

[3] Yingdong Zhou, Javier Remón, Xiaoyan Pang, Zhicheng Jiang, Haiteng Liu, Wei Ding, Hydrothermal conversion of biomass to fuels, chemicals and materials: A review holistically connecting product properties and marketable applications, Science of the Total Environment, **2023**, 886. DOI:10.1016/j.scitotenv.2023.163920.

[4] Jude A. Onwudili, Paul T. Williams, Hydrothermal reforming of bio-diesel plant waste: Products distribution and characterization, Fuel, **2010**, 89, 501. DOI:10.1016/j.fuel.2009.06.033.

[5] Jiaqiu Xu, Zepeng Fan, Qilin Yang, Guoyang Lu, Pengfei Liu, Dawei Wang, Hydrothermal carbonization of waste wood: Sustainable recycling of biomass by-products and novel performance enhancer for bitumen, Construction and Building Materials, **2023**, 404, 133307. DOI:10.1016/j.conbuildmat.2023.133307.

[6] Bingbing Qiu, Jicheng Shi, Wei Hu, Jian Gao, Sitong Li, Huaqiang Chu, Construction of hydrothermal liquefaction system for efficient production of biomass-derived furfural: Solvents, catalysts and mechanisms, Fuel, **2023**, 354, 129278. DOI:10.1016/j.fuel.2023.129278.

[7] Farihahusnah Hussin, Nur Nadira Hazani, Munawar Khalil, Mohamed Kheireddine Aroua, Environmental life cycle assessment of biomass conversion using hydrothermal technology: A review, Fuel Processing Technology, **2023**, 246, 107747. DOI:10.1016/j.fuproc.2023.107747.

[8] Liangpang Xu, Yang Liu, Lejing Li, Zhuofeng Hu, Jimmy C. Yu, Fabrication of a photocatalyst with biomass waste for H_2O_2 synthesis, ACS Catalysis, **2021,** 11, 14480. DOI:10.1021/acscatal.1c03690.

[9] Zhuofeng Hu, Zhurui Shen, Jimmy C. Yu, Converting carbohydrates to carbon-based photocatalysts for environmental treatment, Environmental Science & Technology, **2017,** 51, 7076. DOI:10.1021/acs.est.7b00118.

[10] Liangpang Xu, Yang Liu, Zhuofeng Hu, Jimmy C. Yu, Converting cellulose waste into a high-efficiency photocatalyst for Cr(VI) reduction via molecular oxygen activation, Applied Catalysis B: Environmental, **2021,** 295. DOI:10.1016/j.apcatb.2021.120253.

[11] Joana S. Gomes-Dias, José A. Teixeira, Cristina M. R. Rocha, Valorization of residual biomass from the hydrocolloid industry: The role of hydrothermal treatments in the recovery of high-value compounds, Bioresource Technology Reports, **2024,** 25. DOI:10.1016/j.biteb.2023.101720.

[12] Xi He, Ningchao Zheng, Ruiting Hu, Zhuofeng Hu, Jimmy C. Yu, Hydrothermal and pyrolytic conversion of biomasses into catalysts for advanced oxidation treatments, Advanced Functional Materials, **2020,** 31. DOI:10.1002/adfm.202006505.

[13] Guosheng Li, Zhuofeng Hu, Bottom-up synthesis of semiconductive carbonaceous nanosheets on hematite photoanode for photoelectrochemical water splitting, Nano Research, **2021,** 15, 627. DOI:10.1007/s12274-021-3529-2.

[14] Zhuofeng Hu, Yizhe Huang, Xi He, Weiqing Guo, Kai Yan, Solution-phase conversion of glucose into semiconductive carbonaceous nanosheet photocatalysts for enhanced environmental applications, Chemical Engineering Journal, **2022,** 427. DOI:10.1016/j.cej.2021.131464.

[15] Tianqi Wang, Mingzhe Sun, Hongli Sun, Jin Shang, Po Keung Wong, Efficient Z-scheme visible-light-driven photocatalytic bacterial inactivation by hierarchical MoS_2 encapsulated hydrothermal carbonation carbon core-shell nanospheres, Applied Surface Science, **2019,** 464, 43. DOI:10.1016/j.apsusc.2018.09.060.

[16] Wanjun Wang, Hanna Wang, Guiying Li, Po Keung Wong, Taicheng An, Visible light activation of persulfate by magnetic hydrochar for bacterial inactivation: Efficiency, recyclability and mechanisms, Water Research, **2020,** 176. DOI:10.1016/j.watres.2020.115746.

[17] Yu Peng, Xi He, Ningchao Zheng, Ruiting Hu, Weiqing Guo, Zhuofeng Hu, Transferring waste of biomass and heavy metal into photocatalysts for hydrogen peroxide activation, Chemical Engineering Journal, **2021,** 420. DOI:10.1016/j.cej.2021.129867.

[18] Ningchao Zheng, Xi He, Quan Zhou, Ruilin Wang, Xinran Zhang, Ruiting Hu, Zhuofeng Hu, Generation of reactive chlorine species via molecular oxygen activation on a copper chloride loaded hydrothermal carbonaceous carbon for advanced oxidation process, Applied Catalysis B: Environmental, **2022,** 319. DOI:10.1016/j.apcatb.2022.121918.

[19] Leyan Wei, Yuying Zhang, Chunyan Zhang, Chao Yao, Chaoying Ni, Xiazhang Li, In situ growth of perovskite on 2D hydrothermal carbonation carbon for photocatalytic reduction of nitrate to ammonia, ACS Applied Nano Materials, **2023,** 6, 13127. DOI:10.1021/acsanm.3c01826.

[20] Zhuofeng Hu, Weiwei Liu, Conversion of biomasses and copper into catalysts for photocatalytic CO_2 reduction, ACS Applied Materials & Interfaces, **2020,** 12, 51366. DOI:10.1021/acsami.0c13323.

[21] Chi Zhang, Mengyang Zhang, Yi Li, Danmeng Shuai, Visible-light-driven photocatalytic disinfection of human adenovirus by a novel heterostructure of oxygen-doped graphitic carbon nitride and hydrothermal carbonation carbon, Applied Catalysis B: Environmental, **2019,** 248, 11. DOI:10.1016/j.apcatb.2019.02.009.

[22] Yuting Zhang, Zhurui Shen, Zekun Xin, Zhuofeng Hu, Huiming Ji, Interfacial charge dominating major active species and degradation pathways: An example of carbon based photocatalyst, Journal of Colloid and Interface Science, **2019,** 554, 743. DOI:10.1016/j.jcis.2019.07.077.

[23] Zhuofeng Hu, Gang Liu, Xingqiu Chen, Zhurui Shen, Jimmy C. Yu, Metallic photocatalysts: Enhancing charge separation in metallic photocatalysts: A case study of the conducting molybdenum dioxide, Advanced Functional Materials, **2016,** 26, 4444. DOI:10.1002/adfm.201670159.

[24] Hao Xu, Yangyuan Ou, Xinjiang Hu, Daihui Chen, Xingong Li, Chunfang Tang, Xia Zheng, Preparation of reed-based hydrothermal carbonized carbon photocatalyst and effective degradation of methylene blue and tetracycline, Environmental Science and Pollution Research, **2023,** 30, 48048. DOI:10.1007/s11356-023-25739-6.

[25] Tianqi Wang, Zhifeng Jiang, Taicheng An, Guiying Li, Huijun Zhao, Po Keung Wong, Enhanced visible-light-driven photocatalytic bacterial inactivation by ultrathin carbon-coated magnetic cobalt ferrite nanoparticles, Environmental Science & Technology, **2018,** 52, 4774. DOI:10.1021/acs.est.7b06537.

CHAPTER 7

Biomass-Derived Porous Carbon for Farmland Restoration

Nishu, Juntao Yang, and Gaixiu Yang*

ABSTRACT

The utilization of biomass-derived porous carbons (BDPCs) is extremely essential for farmland restoration. This chapter comprehensively discusses the source, production methods, physical/chemical properties, farmland restoration applications, and corresponding environmental/economic impacts of these BDPCs. Firstly, various natural biomass origins and fabrication strategies are summarized. Important parameters of BDPCs related to farmland restoration are subsequently described. Moreover, specific applications of farmland restoration using BDPCs, such as soil conditioner, nutrient retention, crop yield, and growth improvement, are systematically listed. The environmental and economic evaluation is also provided for optimization. Finally, the challenges and prospects of BDPCs in farmland restoration are proposed and outlooked.

7.1 INTRODUCTION

The alterations in the scope and intensity of farmland exert a significant impact on life on Earth. While farmlands rich in biodiversity hold conservation significance, promoting them occurs at the expense of preserving and restoring the natural environment. Revitalizing the potential for the restoration of natural habitat and farmlands rich in biodiversity can be achieved by bringing back the productive potential of the current land. Farmland restoration refers to the process of revitalizing and improving agricultural land that has been degraded or damaged over time. The goal of farmland restoration is to enhance the productivity, sustainability, and ecological health of the land. This process typically involves various agricultural and environmental practices aimed at reversing

DOI: 10.1201/9781003520566-7

the negative impacts of soil erosion, nutrient depletion, contamination, and other factors that can reduce the fertility and functionality of farmland. However, rapid land degradation and soil erosion, on the other hand, are serious environmental issues that lead to the destruction of ecosystem services, including reduced soil productivity and the sustainability of agricultural lands[1–3].

Soils are the largest terrestrial reservoir of organic carbon[4]. It is estimated that the top one meter contains 1,500–1,600 Pg of organic carbon, which is approximately double the atmospheric carbon[5]. When farmlands are converted from natural or unmanaged vegetation to agricultural cultivation, significant carbon is released from both standing biomass and soil. As a result, there are continuous global projects focused on repairing and increasing organic carbon content in soils, such as the "4 per 1,000" initiative (4p1000). Soil productivity and health are critical for agricultural sustainability, in addition to their involvement in global biogeochemical cycles. The expanding global population has put additional strain on the agricultural industry, reducing soil fertility and resulting in lower agricultural output, particularly in tropical countries[6]. Reduced per-capita landholding and poor soil quality have necessitated increased inorganic fertilizer application rates to maintain crop productivity. However, excessive use of chemical fertilizers such as potassium, phosphorus, and nitrogen can upset nutrient balance, deplete soil fertility, and contribute to global warming by causing fast mineralization of soil organic matter, and a subsequent decrease in soil carbon content[7]. Agriculture-induced soil carbon loss has been estimated to be the second-highest anthropogenic source of global carbon emissions, next to the energy sector, accounting for 20% of total greenhouse gas (GHG) emissions[8]. As a result, it is critical to develop crop management strategies that are both ecologically beneficial and economically viable. These activities have the potential to increase soil organic matter while also mitigating climate change by sequestering carbon and lowering GHG emissions[9].

Biochar has significant potential to resolve these issues by improving soil properties, particularly in highly nutrient-poor, acidic, or weathered soils common in semi-arid or arid climates. Furthermore, it benefits underprivileged households or communities in rural areas of middle- and low-income countries[10,11]. Biochar or biomass-derived porous carbon (BDPC) is a carbonaceous material derived from the thermal decomposition of a variety of organic biomass sources such as agricultural residues, forestry residues, or organic waste under controlled, oxygen-limited thermal conditions[12]. The commonly used thermochemical methods in the production of BDPC are pyrolysis, gasification, hydrothermal carbonization, and torrefaction[13,14]. In the last few decades, BDPC has gained considerable attention due to its multiple benefits in various fields such as bioenergy production, agriculture, wastewater treatment, and especially environmental remediation[15]. It has unique physicochemical characteristics that make it an effective candidate in the restoration of farmland. These physicochemical properties include high porosity, large surface area, the presence of various functional groups, stability, and high cation exchange capacity (CEC)[16]. The framework of BDPC exhibits a network of micro, meso, and macropores that give it an extremely porous structure. This porous configuration provides BDPC a substantial surface area, enabling the retention and adsorption of contaminants as well as creating a hospitable environment for beneficial microorganisms, fostering a balanced soil microbiome[17]. The high carbon content of BDPC often more than 70%, makes it stable and resistant to degradation, ensuring efficiency in remedial applications for an extended period of time[18]. Furthermore, BDPC's surface is equipped with functional groups including carboxyl (-COOH), phenolic (-Ph-OH) groups, and hydroxyl (-OH), all of which are essential for chemical reactions and the sorption of contaminants. Interestingly, BDPC has a remarkable potential for exchanging ions with the surrounding environment, which allows it to store and release nutrients.

In the field of farmland restoration, BDPC has become a valuable tool due to its numerous advantages in improving soil quality and advancing sustainable agriculture. When applied to degraded soils, its porosity and high carbon content contribute to retaining more water and making nutrients more readily available. According to the European Union, the application of BDPC as a legal soil amendment is regarded as a way to support worldwide agricultural soils, prevent land degradation, and be a potential geoengineering tool to mitigate climate change through soil sequestration of carbon[19,20]. Studies show that as soils are amended when BDPC is incorporated into the soil, it acts as a "charged material" by enhancing moisture holding capacity, boosting soil fertility, attracting beneficial microbes and fungi, and improving the quality of soil by raising soil pH, and its mobile matter decomposes into plant nutrients. Plants and soil microbes receive nutrients from raw organic materials, and BDPC acts as a catalyst that boosts plants' absorption of water and nutrients. Unlike other soil supplements, BDPC's large surface area and porosity allow it to absorb and hold onto water and nutrients, fostering the growth of beneficial microbes in the soil[8]. According to research, carbon components in BDPC have considerable recalcitrant in soils; for instance, residence times reported for BDPC derived from woody biomass ranged from hundreds to thousands of years, which is nearly 10–1,000 times longer than that of most soil organic matter residence times, owing to its resistance to microbial degradation and mineralization and making it a resilient amendment for addressing soil degradation over the long term[21]. The functional groups on the surface of BDPC, actively participate in nutrient cycling and promote plant growth. Moreover, the significant cation exchange capacity (CEC) of biochar facilitates the retention and gradual release of essential nutrients, contributing to sustained soil fertility. By harnessing these properties, biochar proves to be a versatile and sustainable solution for farmland restoration, aligning with the principles of ecological agriculture and soil health management.

The purpose of this chapter is to investigate the complex role of biochar in farmland restoration, revealing its potential as a long-term solution to soil degradation and improving agricultural productivity. By delving into the production methods and various types of BDPC, the chapter aims to provide a comprehensive understanding of the material's properties and how they contribute to soil health. Furthermore, the chapter seeks to outline specific goals for farmland restoration, emphasizing the environmental and ecological benefits of BDPC in mitigating challenges such as soil erosion, greenhouse gas emissions, soil pollution, and contamination. The chapter intends to provide readers with practical insights into the deployment of BDPC as a transformative tool in sustainable farming.

The chapter is organized systematically, beginning with an introduction to BDPC and its historical context in the restoration of farmland. It then moves on to numerous methods of producing biochar and different modification procedures to optimize its efficacy for specific uses. The physiochemical features of BDPC that affect soil structure, water dynamics, and nutrient availability are investigated. Moving forward, the chapter addresses the specific goals of using BDPC in the context of creating the foundation for understanding its application in farmland restoration and further exploring its environmental and economic benefits that have attracted intensive interest among researchers. Finally, the chapter concludes by highlighting challenges, future directions and the need for ongoing research to harness the full potential of biochar in sustainable agriculture. Thus, this chapter organization aims to provide a comprehensive understanding of BDPC, its production, properties, and the specific benefits it can bring to farmland restoration initiatives. It also illustrates the significance of considering various factors, challenges, and future research needs in the application of BDPC in farmland restoration.

7.2 BIOMASS: SUSTAINABLE SOURCE OF CARBON

Biomass is a combination of "bio + mass", which implies animal and plant origins, including trees, algae, and crops. Biomass, a sustainable source of carbon, plays a crucial role in the evolution of cleaner and eco-friendly energy systems. Biomass energy, derived from organic resources such as forestry, agricultural wastes, and energy crops, is a carbon-neutral energy source. Furthermore, as compared to fossil fuels, biomass has the potential to reduce greenhouse gas emissions while also promoting the land management and utilization of waste. Its potential to sequester carbon via sustainable forestry and afforestation techniques emphasizes its importance in the mitigation of climate change. To ensure the sustainability of biomass, however, careful resource management is required such as preventing deforestation, protecting ecosystems, and maintaining an ecological balance between environmental preservation and energy production.

7.2.1 Biomass Sources

Globally, biomass sources are available from various categories of organic materials due to the significant disparities in quantity and variety of biomass and diverse compositional properties. Biomasses are composed of organic, inorganic and biological materials.

Biomasses can be categorized into two major classes (i) non-woody biomass and (ii) woody biomass. Agricultural solid waste and animal waste are considered under nonwoody biomass whereas tree and forest residues are categorized as woody biomass. The general composition of biomass comprises hemicellulose (10–30% in woody biomass and 20–40% in herbaceous biomass on a dry basis), cellulose (nearly 50% on a dry basis), and lignin (10–40% in herbaceous feedstock and 20–40% in woody biomass on dry basis)[22,23]. The other components are organic extractives (salts, protein, acids, etc.) and inorganic compounds which vary from 1% in woods to 15% in herbaceous biomass and up to 25% in agricultural and forestry residue[23]. Similarly, aquatic biomass is made up of carbohydrates (3–30%), proteins (14–65%), and lipids (1–51%), according to the species[24,25]. Directly or indirectly the yield of biomass depends on the composition of hemicelluloses, cellulose, and lignin. The lignocellulosic biomass mainly consists of hemicellulose, cellulose, and lignin with varying percentages (Table 7.1), while cereals are primarily composed of starch, and cattle manure is high in protein content. However, different chemical structures result in varied chemical characteristics[26].

However, there is no univocal manner of classifying biomass, therefore it can be classified differently depending on purpose and scope or origin and properties. Researchers employ various ways to categorize the various forms of biomass but one simple approach is to categorize four main types, namely; herbaceous biomass, woody biomass, human and animal waste biomass, aquatic biomass, and municipal solid waste biomass. Table 7.2 shows the general classification of different biomass groups.

7.2.1.1 Herbaceous Biomass

According to The European standard EN 14961-1, "Herbaceous biomass is from plants that have a non-woody stem and which die back at the end of the growing season. It includes grains or seeds crops from the food processing industry and their by-products such as cereal straw"[36].

Generally, agricultural residues and energy crops are the two primary categories of herbaceous biomass. Agricultural residues mainly consist of by-products of food or fodder production such as cereal straws, cobs, husks, and leaves. A massive amount of crop residue or agricultural waste is generated during farming activities. These horticultural wastes are either left untreated or burned openly on the farmland resulting in loss of energy that could be converted into an essential source of energy. On the other hand, energy crops such as miscanthus, poplar, switchgrass, and willow are only exploited in the sector of bioenergy. Unlike conventional crops which are cultivated for fiber/food, energy crops are cultivated to enhance the yield of biomass and make them suitable for the production of bioenergy. Herbaceous biomass is considered a sustainable and renewable source that can help mitigate greenhouse gas emissions when utilized for bioenergy production.

7.2.1.2 Woody Biomass

An organic material derived from trees and woody plants is termed woody biomass and generally consists of lignin and carbohydrates. It includes branches, bark, logs, sawdust, other woody residues, and wood chips, which can be obtained from tree trimming, lumber milling, forest management practices, and forestry operations. It is used as a renewable source of energy for various applications and is converted via different conversion techniques[37], when it is harvested sustainably, with an emphasis on afforestation and reforestation to sustain or enhance forest cover, ensuring a constant supply of woody biomass while sequestering carbon dioxide and maintaining healthy ecosystem.

7.2.1.3 Human and Animal Waste Biomass

The common sources of these types of biomasses are animal meat, bones, and various types of animal and human dung. These wastes are directly used as fertilizers on agricultural land, under proper waste management. Anaerobic digestion is the most commonly used method to convert such waste into useful products. For instance, biogas, generated

TABLE 7.1 Chemical Composition of Various Lignocellulosic Biomass

| Feedstock | Composition (%) | | | Ref. |
	Hemicellulose	Cellulose	Lignin	
Ailanthus wood	26.6	46.7	26.2	[27]
Albizia	6.7	6.7	33.8	[27]
Bamboo residue	19.00	19.00	22.61	[28]
Barley straw	29.7	29.7	27.7	[27]
Banana waste	14.8	14.8	14	[27]
Birchwood	40	40	15.7	[27]
Bagasse	22.6	22.6	18.3	[27]
Coir	0.15–0.25	0.15–0.25	41–45	[29]
Coconut shell	25.1	25.1	28.7	[27]
Cashew shell	18.6	18.6	40.1	[27]
Cotton, flax	5–20	5–20	-	[29]
Comcob	35–45	35–45	5–15	[30]
Cotton waste	5–20	5–20	-	[30]
Comstalk	24.3	24.3	12.5	[31]
Costal Bermuda grass	35.7	35.7	6.4	[27]
Eucalyptus	18	18	22	[32]
Elephant grass	24	24	23.9	[27]
Esparto grass	27–32	27–32	17–19	[27]
Groundnut shell	19	19	30	[33]
Hardwood bark	20–38	20–38	30–55	[33]
Millet husk	26.9	26.9	14	[27]
Maple	17.3	17.3	20.7	[34]
Miscanthus	18	18	25	[35]
Nutshell	25–30	25–30	30–40	[30]
Olive stone	20–30	20–30	20–25	[30]
Oak	35.9	35.9	24.1	[31]
Pine	24	24	27	[33]
Poplar wood	17	17	26	[33]
Rice husk	20.5	20.5	17.7	[27]
Rice straw	20.48	20.48	20.03	[35]
Rye straw	27–30	27–30	16–19	[27]
Spruce	23	23	28	[32]
Sugarcane bagasse	25–35	25–35	19–24	[30]
Switchgrass	27.5	27.5	17.5	[31]
Sweet sorghum Bagasse	18–28	18–28	14–22	[29]
Sorghum straw	24	24	13	[29]
Tea waste	19.9	19.9	40	[27]
Walnut	16.5	16.5	21.9	[34]
Wheat straw	27.3–50	27.3–50	15–16.4	[27]
White oak	18	18	23.2	[34]
Oak	21.9	21.9	35.4	[27]
Orchard grass	32	32	4.7	[27]

TABLE 7.2 General Classification of Different Biomass Groups

Biomass Group	Varieties and Species	Ref.
1. Herbaceous biomass	Grasses and flowers (alfalfa, bamboo, brassica, cane, switchgrass, miscanthus, timothy, others); straws (barley, bean, flax, corn. mint, oat, rape, rice, rye, sesame, sunflower, wheat, others); other residues (fruits, shells, husks, hulls, pits, pips, grains, seeds, coir, stalks, cobs, kernels, bagasse, food, fodder, pulps, cakes, etc.)	[37]
2. Wood biomass	Coniferous or deciduous; Angiospermous or gymnospermous, Stems, branches, foliage, bark, chips, lumps, pellets, briquettes, sawdust, sawmills and others from various wood species	[37]
3. Human and animal waste biomass	Bones, meat-bone meal; various manures, etc.	[37]
4. Aquatic biomass	Marine or freshwater algae; macroalgae (green, blue, blue-green, brown, red) or microalgae; seaweed, kelp, lake weed, water hyacinth, etc.	[37]
5. Municipal solid waste biomass	garbage, consist of food waste, cardboard, paper, leaves, wood, agricultural wastes, leather products, plastic, glass, metals and petroleum-based synthetic material, etc.	[37]

during this process, can be burned directly for heating water and rooms or cooking purposes or used to generate electricity[38].

7.2.1.4 Aquatic Biomass

The aquatic biomass includes microalgae, macroalgae, and emerging plants. Currently, it is regarded as an attractive raw material for third-generation biodiesel production since it does not compete with food crops and produces significantly larger biomass per hectare than land crops. It has been estimated that there are nearly 55,000 species and more than 100,000 strains of freshwater, brackish water, and terrestrial algae. The major benefit of using these organisms is their potential to convert carbon dioxide, water, and sunlight into a wide range of chemicals and metabolites, which end up in algal biomass[39]. Besides, there are several parameters such as concentration of oxygen and carbon dioxide, irradiation, pH, salinity, temperature, and nutrients that affect the performance of aquatic biomasses.

7.2.1.5 Municipal Solid Waste Biomass

Municipal solid waste (MSW) generally called garbage, consists of food waste, cardboard, paper, leaves, wood, agricultural wastes, leather products, plastic, glass, metals and petroleum-based synthetic material, etc. which are useless and unwanted for all inhabitants. For example, in our daily life, plastic and cardboard-based packaging materials are usually used for many goods that cause increasing amounts of waste every day. However, direct utilization of this kind of biomass for the production of bioenergy is not possible. Waste segregation is a crucial step in the process of selecting appropriate biomass for application[40].

7.2.2 Biomass Waste: A Potential Feedstock for Porous Carbon

Currently, the global annual production of all biomass waste is around 140 Gt, which poses serious management challenges because discarded biomass could have negative environmental effects. In developing countries, most of the biomass residues are burned openly or left over in the field for natural decomposition, impacting the atmosphere and surface water. For instance, one ton of ash-free wood and landfilled dry produce 0.73 tons of CO_2 whereas approximately 1.4 tons of CO_2 is produced via one ton of fuel wood. It has been estimated burning biomass is about 18% of the total emissions. Crop residues/wastes such as corn stovers, leaves, straws, and seed pods of wheat rice, barley, and oat are some of the wastes generated from agriculture. Currently, the production of crop residue is expected to be 2802 Mt/year[41]. Moreover, due to the multi-cropping system, the norm in modern agriculture, farmers only have a limited time to prepare the field for the next crop. Due to a lack of alternatives for utilizing straws, some farmers choose to burn them in their fields. This open burning of these crop straws emits toxic and harmful gases and aerosols, which greatly enhance the amount of greenhouse gas. MSW due to its global economic impact has become worldwide sustainable development agenda that not only created a financial burden but also the waste's wide-ranging pollutants have a negative impact on both the environment and human health. Conventional methods including incineration and landfill are widely used for the management and treatment of biomass wastes, which are energy intensive and environmentally unfriendly methods, respectively. Waste disposal via landfills causes serious environmental consequences, including the uncontrolled release of methane gas into the atmosphere, a gas with 20–23 times higher greenhouse gas (GHG) potential than carbon-di-oxide (CO_2), unpleasant odors, production of leachate which contaminates the groundwater and the soil, and spread of pathogenic organisms[42]. Therefore, such high carbon-containing biomass wastes could be accumulated in large quantities to not only ease waste management but can also be transformed into value-added products. Utilizing such surplus biomass reduces environmental damage and encourages a circular economy by turning agricultural waste, forestry trash, and even municipal organic waste into valuable energy and products.

A biomass-derived porous carbon (BDPC) or bio-char is a carbonaceous solid by-product formed via thermochemical conversions such as pyrolysis, hydrothermal carbonization, gasification, and torrefaction processes. BDPC or bio-char is a term that has emerged in conjunction with renewable fuel, carbon sequestration, and soil improvement[43]. Figure 7.1 shows the chemical structure of BDPC. It is widely accepted that individual BDPC particles consist of two main structural components: organized crystalline graphene and randomly amorphous aromatic structures[44]. Its intricate carbon lattice structure includes both aliphatic and aromatics structure domains formed by the cleavage and recombination of chemical bonds. These aliphatic and aromatics structures are interlinked in a random manner. The size of the BDPC crystallites grows as the processing temperature rises, and the entire framework gets more organized[44,45]. Additionally, oxygen, hydrogen, phosphorus, nitrogen, and sulphur are incorporated within these

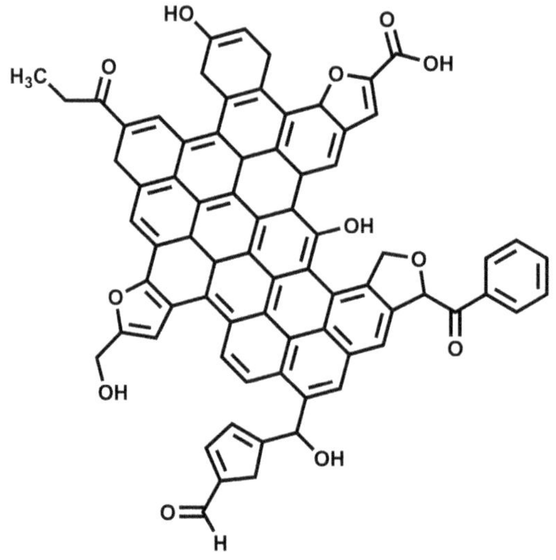

FIGURE 7.1 Chemical structure of BDPC.

aromatic structures as heteroatoms. The presence of heteroatoms is thought to be a major cause of the noticeable variations in the surface chemistry and reactivity of BDPC. According to the International Biochar Initiative (IBI), "the biochar is a solid material obtained from the thermochemical conversion of biomass in an oxygen-limited environment". However, the origin of BDPC is interrelated with the soils in the Amazon region, often referred to as "Terra Preta" soils. These soils have higher levels of stable soil organic matter in addition to higher concentrations of nutrients like calcium, phosphorus, potassium, and nitrogen and have attracted great attention due to their considerably higher crop productivity in contrast to the nearby infertile tropical soils[46]. Following that, biochar has come to be seen as a crucial tool for creating sustainable energy production and environmental management.

7.2.3 Biomass-Derived Porous Carbon: Historical Context in Farmland Restoration

The historical roots of BDPC in farmland restoration trace back to ancient agricultural practices, revealing a deep-seated awareness of its benefits among early civilizations. Archaeological evidence suggests that indigenous communities in the Amazon Basin employed a precursor to modern biochar known as "terra preta" as far back as 2,000 years ago[47]. These dark, fertile soils were enriched with a high amount of carbon and nutrient content together with an increased microbial population, creating an environment conducive to plant growth[48]. BDPC, on the other hand, should not be confused with Terra preta. Terra preta, often known as "Amazonian dark earth", exists naturally in nature. Terra preta differs from BDPC in both its composition and carbon structure. It is crucial to remember that the discovery of Terra preta's nutritional benefits stimulated interest in the use of BDPC[47]. The intentional application of BDPC in ancient farming system demonstrates an early knowledge of its ability to improve productivity and soil fertility. The preservation of terra preta soils demonstrates the durability and adaptability of these ancient agricultural systems, providing a compelling historical background for the modern investigation of BDPC in farmland restoration[49].

In the late 20th and early 21st centuries, BDPC got considerable attention as scientists rediscovered its potential benefits. Its unique characteristics, versatile applications and promising development prospects in various fields, renewed its emphasis on multidisciplinary research. Scientists' pioneering research propelled BDPC to the cutting-edge of agricultural concerns. They concentrated on determining the physical and chemical properties of BDPC, as well as its impact on soil fertility. Their findings raised interest in BDPC as a tool for farmland restoration, motivating additional research and trials. BDPC's historical context as an age-old agricultural practice connected to existing concerns regarding to soil degradation and the need for sustainable farming practices, The historical context of biochar as an age-old agricultural practice resonated with modern concerns about soil degradation and the necessity for environmentally friendly farming methods, resulting in increased interest and investment in BDPC research and implementation.

In the 21st century, BDPC has transformed from a historical curiosity to a practical solution in farmland restoration. The intensified concerns about green farming and climate change and the historical context of BDPC as an established, age-old method of improving soil productivity give credibility to its modern applications. Today, BDPC is being investigated and used in a variety of agricultural contexts by researchers and farmers throughout the world. Its capacity to increase nutrient retention, strengthen soil structure, and sequester carbon is consistent with the objectives of ecological restoration and sustainable agriculture. As we navigate the problems of feeding a growing global population while conserving our planet's health, BDPC poses as a historical ally, providing insights from the past to design a more sustainable future for farmland restoration.

7.3 PRODUCTION OF BIOMASS-DERIVED POROUS CARBON

7.3.1 Biomass-to-Carbon Conversion Methods

The developing interest in porous carbon or biochar derived from biomass for various applications has enhanced its conversion process. Biochar is a carbonaceous solid by-product formed via thermochemical conversion Typically, the porous carbon materials produced via the thermochemical degradation process have a high energy density[50]. The physiochemical property such as yield, pore structure pore size, specific surface area, functional groups, cation exchange capacity, and ash content of synthesized porous carbon or biochar, mainly depends on the preparation procedure followed. Although the principle of operating the reactors in various products is identical yet the other operating variables for instance heating rate, final temperature and amount of oxygen which required for operating the reactor vary. These conditions are vital as affect the quality and quantity of the biochar generated[43]. Table 7.3 shows the thermochemical

TABLE 7.3   Thermochemical Routes and Process Conditions for the Production of BDPC

Routes	Temperature (°C)	Residence Time	Bio-oil (%)	Bio-char (%)	Syngas (%)	Ref.
Pyrolysis	Slow: 300–700	>450 s	30	35	35	[51]
	Fast: 400–800	<2 s	50	20	30	
Hydrothermal carbonization	180–300	1–16 h	5–20	50–80	2–5	[52]
Gasification	750–900	10–20 s	5	10	85	[53]
Torrefaction	290	10–60 min	0	80	20	[54]

routes and process conditions for the production of BDPC. There are four main thermochemical processes: pyrolysis, hydrothermal carbonization, gasification, and torrefaction which are applied for biochar production. The principles and the reaction mechanisms implicated in different thermochemical conversion methods for biochar synthesis are as follows.

7.3.1.1 Pyrolysis

The term "pyrolysis" originated from the Greek words "pyro" (fire) and "lysis" (decomposition or breaking down into basic elements). Biomass pyrolysis is the thermochemical decomposition of biomass into syngas, bio-oil, and bio-char that occurs in the absence of oxygen. Biochar, a carbon-rich solid product, is an excellent absorbent and soil amender that can allow the soil to retain more water, nutrients, agricultural chemicals, store carbon and protect soil from erosion, and increase its fertility either in its original or activated form[55]. Chemically and biologically, the bio-char is more stable compared to the original carbon because of its molecular structure, making it more challenging to decompose[56]. During the pyrolysis process, the macromolecular constituents (hemicellulose, cellulose, and lignin), in aquatic and lignocellulosic biomass show overlapping thermal behavior. For instance, the pyrolysis of proteins and cellulose occurs between 200 and 350 °C, carbohydrates and hemicellulose occur between 150 and 200 °C, whereas the decomposition of lipids and lignin occurs in the range of 200–450 °C[57]. The pyrolysis of lignin produces around 65% of BDPC compared to the pyrolysis of hemicellulose and cellulose. Generally, the pyrolysis of hemicellulose and cellulose produces volatile chemicals, whereas lignin pyrolysis produces more solid BDPC[58]. During the pyrolysis process series of chemical reactions dehydration, decarboxylation, and depolymerization occurs[59]. The final yields and the characteristics of the biomass pyrolysis process depend on the composition of the raw material, pyrolysis condition, and pyrolysis technique. This can alter the direction of reactions and hence have a substantial impact on the quality and the yield of the products. Table 7.4 represents the characteristics of BDPC produced at different pyrolytic conditions According to the heating method, residence time, and heating rate, pyrolysis could be divided into slow and fast pyrolysis.

TABLE 7.4   Characteristics of BDPC Produced at Different Pyrolytic Conditions

Biomass Sample	Pyrolysis Temperature(°C)	Yield of BDPC (%)	Component (wt. %)								Ref.
			C	H	N	O	P	K	Ash	pH	
Arundo donax	300	44.4	65.26	4.51	0.65	21.03	0.12	3.70	7.69	8.42	[60]
Arundo donax	350	41.7	66.97	4.46	0.64	21.67	0.12	3.8	7.73	8.09	[60]
Arundo donax	400	40.4	72.25	4.09	0.69	18.72	0.13	4.18	8.45	8.06	[60]
Arundo donax	500	33.4	73.12	3.01	0.63	11.54	0.16	4.77	10.70	9.73	[60]
Arundo donax	600	30.6	78.61	2.22	0.55	11.24	0.17	5.02	11.27	0.11	[60]
Bamboo culms	350	52.10	68.46	4.50	0.33	26.71	—	—	—	8.00	[61]
Bamboo culms	450	34.30	70.90	3.94	0.25	24.91	—	—	—	9.50	[61]
Bamboo culms	550	31.00	73.75	3.63	0.25	22.38	—	—	—	9.50	[61]
Broiler litter	350	—	45.60	4.00	4.50	18.30	—	—	—	—	[62]
Broiler litter	700	—	46.0	1.42	2.82	7.40	—	—	—	—	[62]
Buffalo-weed	300	50.00	78.09	4.26	10.2	7.44	—	—	20.36	—	[63]
Buffalo-weed	700	29.00	84.96	1.09	17.40	6.56	—	—	32.34	—	[63]
Corn straw	400	35.46	56.46	2.86	1.29	39.21	—	—	22.34	—	[64]
	450	32.23	57.14	2.54	1.20	38.95	—	—	24.39	—	[64]
Corn straw	500	31.00	58.85	2.81	1.17	37.02	—	—	25.51	—	[64]
Corn straw	550	29.68	59.29	2.50	1.13	36.91	—	—	26.62	—	[64]
Corn straw	600	28.65	60.84	2.15	1.00	35.87	—	—	27.56	—	[64]
Cotton stalk	300	46.71	71.32	4.17	1.41	22.80	—	—	4.98	—	[65]
Cotton stalk	350	44.15	72.56	3.07	1.37	22.69	—	—	6.45	—	[64]
Cotton stalk	400	37.38	73.40	2.86	1.47	22.02	—	—	7.03	—	[64]
Cotton stalk	450	36.99	75.61	2.78	1.48	19.78	—	—	7.71	—	[64]
Cotton stalk	500	33.15	76.25	2.01	1.50	19.94	—	—	8.31	—	[64]
Cotton stalk	550	30.26	78.02	1.97	1.38	18.35	—	—	8.55	—	[64]
Poplar bark	400	42.20	60.80	4.30	—	23.90	—	—	11.00	—	[66]
Poplar bark	600	30.90	68.10	2.50	0.30	14.50	—	—	14.70	—	[66]
Poplar leaves	400	33.80	54.30	4.40	—	20.10	—	—	21.20	—	[66]
Poplar leaves	600	25.40	53.20	2.10	—	17.50	—	—	27.20	—	[66]
Poplar wood	400	31.80	66.60	3.70	—	26.90	—	—	2.80	—	[66]
Poplar wood	600	25.40	75.90	2.30	0.10	17.90	—	—	3.80	—	[66]
Swine manure	200	88.5	37.3	4.65	3.17	29.30	—	—	24.8	—	[67]
Swine manure	300	63.00	41.90	3.52	3.70	15.40	—	—	34.8	—	[67]
Swine manure	400	50.80	39.40	2.59	3.68	13.20	—	—	40.5	—	[67]

(Continued)

TABLE 7.4 *(Continued)* Characteristics of BDPC Produced at Different Pyrolytic Conditions

| Biomass Sample | Pyrolysis Temperature(°C) | Yield of BDPC (%) | Component (wt. %) | | | | | | | | |
			C	H	N	O	P	K	Ash	pH	Ref.
Swine manure	500	46.20	38.60	1.85	3.28	11.90	—	—	43.8	—	[67]
Swine manure	600	44.20	35.80	1.22	2.62	7.70	—	—	52.1	—	[67]
Swine manure	700	44.00	33.10	0.88	2.01	8.69	—	—	54.7	—	[67]
Spent mushroom substrate	450	25.70	44.0	0.60	0.55	8.31	—	—	47.9	—	[68]
Miscanthus	500	—	82.10	2.67	0.31	11.00	—	—	3.92	—	[69]
Yak manure	300	—	41.60	1.90	3.20	27.40	4.52	4.52	—	7.6	[70]
Yak manure	500	—	41.30	1.70	3.00	24.40	5.41	5.41	—	10.4	[70]
Yak manure	700	—	41.20	1.40	2.70	20.70	5.65	5.65	—	11.8	[70]

Slow pyrolysis is also known as conventional pyrolysis or carbonization. This process occurs at moderate temperature, typically 300–700 °C, low heating rate >1 °C/s, and a longer residence time of >450 s[23]. The size of the biomass is kept comparatively large to favor the high vapor residence time. It does not allow the condensation of the pyrolysis product, which enhances the secondary cracking of biomass. The produced vapor during this process could be used for providing heat for the process[27]. It is stated that the biochar produced during slow pyrolysis is rich in carbon, highly functional, and generally used for BDPC production with a specific surface area of <400 m²/g and a yield of 30–60%[71]. Due to the presence of phosphate, carbonate, and other inorganics and ash formed during pyrolysis, biochar is alkaline in nature. Biomass with large particle size and high lignin and ash content generally results in a higher yield of biochar[72]. To conduct the slow pyrolysis, on an industrial level generally screw reactors, and continuous and batch, pyrolyzers are used[23].

Fast pyrolysis is typically performed at a temperature range of 400–800 °C, with a high heating rate of 10–200 °C/s and at a shorter residence time usually for several seconds in contrast to slow pyrolysis[23]. In this process the size of biomass particle is usually smaller that favors the quick escape of volatiles preventing further interaction with char. During fast pyrolysis the major produced product is bio-oil together with bio-char and syngas. The operating conditions in fast pyrolysis is favorable to a low yield of biochar (10–20%). The obtained biochar has high oxygen content and low calorific value might be caused due to short residence time[71]. Table 7.4 summarized the production of biochar from various biomass waste under different reactor conditions. Moreover, some pyrolysis characteristics, including moderate pyrolysis temperature, longer residence time, lower heating rate, particle size, presence of the catalyst, and pyrolysis environment, must be taken into consideration to produce biochar with a high carbon content[12]. For fast pyrolysis the reactor configurations comprise fluidized bed reactor, ablative and vacuum pyrolysis systems[23].

Flash carbonization is an effective technology for manufacturing porous carbon materials compared to conventional carbonization due to high yield of porous carbon (28–32%) in a short reaction time (30 min). During, flash carbonization, the biomass is packed into a packed bed reactor and then pressurized to 1–2 bar using air. Then the bottom of the pressurized vessel is heated via a flame. Flash carbonization usually needs a specific level of pressure. During carbonization, the whole packed bed is heated for less than 30 min through the upward movement of the flame as air flows downstream[71].

Microwave-assisted pyrolysis is a remarkable selective heating process that provides microwave dielectric heating i.e., internal and external heating effects to a biomass which increase the frequency of chemical reaction at lower temperature. Compared to conventional pyrolysis, microwave-assisted pyrolysis is fast, easy to control, gives high energy efficiency, improves final product, and less operating cost. Ionic conduction and dipolar polarization are the two main mechanisms that occur during microwave-assisted pyrolysis[73]. Studies revealed that the chemical properties of porous carbon produced via microwave-assisted pyrolysis are more uniform than those produced via conventional pyrolysis. However, the primary constraints of microwave-assisted pyrolysis involve substantial capital costs for biomass pretreatment during scaling up and for microwave equipment. The mesoporous porous carbon material can be fabricated with a high surface area and regulated aperture via simple activation followed by microwave pyrolysis[74,75].

7.3.1.2 Hydrothermal Carbonization

Hydrothermal carbonization is also termed pyrolysis as it is accomplished by submersing the biomass into the water in a sealed reactor followed by heating at a low temperature range of 175–300 °C for 13 h[33]. The produced porous carbon material (hydro-char) via hydrothermal carbonization has increased the number of chemical functional groups due to the presence of water and operational conditions. Furthermore, the different parameters such as pressure, temperature, and residence time also define its unique properties. When the temperature of hydrothermal carbonization increased, the yield of porous carbon material decreased. In addition, the oxygen and carbon content also decreased with an elevated hydrothermal temperature[76]. This carbonization process is exothermic and spontaneous; transfers the existing carbon from the original material into the final product[77]. Notably, the hydro-char produced via hydrothermal carbonization has a smaller surface area, lower carbon stability, and fewer pores compared to porous carbon produced via other processes[33,78]. Hydro-char is characterized by a high cation exchange capacity, abundant oxygen functional groups, and a need for additional energy for its production[79].

7.3.1.3 Gasification

Gasification is the process that occurs at a higher temperature ranging from 600 to 1200 °C, at the fast-heating rate where incomplete combustion of biomass takes place in a different gasifying medium like air, nitrogen, carbon dioxide, steam, and oxygen that forms a gaseous mixture generally known as syngas. The process can be used to produce porous carbon materials and the quality of produced porous carbon materials is examined on the basis of the carbon content present in it[32]. The important factor that affects its carbon content is the properties of the biomass, gasifying agent, and pressure, and the most important affecting factor is the equivalent ratio (ER—the ratio of the actual air/fuel ratio to the stoichiometric air/fuel ratio)[32,80]. The ideal value of ER is mainly based on the chemical and physical properties of the biomass which ranged between 0.25 and 0.28. It has been stated that if the value of ER increases, the oxygen content increases and the carbon content decreases, resulting in a lesser yield of porous carbon material[80]. Modifying the ER values during gasification can have both positive and negative outcomes on the production efficiency of porous carbon. An elevation in ER values corresponds to an increased oxygen level, which can impact the quality of produced porous carbon. As ER values increase, the conversion of carbon from a solid state into gaseous species becomes more pronounced, facilitating increased surface area of the porous carbon due to the formation of micropores. However, it's essential to note that, elevation in oxygen content causes reductions in yield, that in turn reduces the strength of porous carbon, and heightened ash content[80].

7.3.1.4 Torrefaction

Torrefaction is the thermochemical newly emerging process for the production of porous carbon material. The process occurs in the absence of oxygen at a low heating rate of 50 °C/min at 200–300 °C temperature for 20 to 120 min. converts the biomass[43]. The physical and chemical properties of the biomass and the torrefaction temperature greatly influenced the quality of the produced torrefied porous carbon material. However, it is an eco-friendly and economical process. The porous carbon material produced at increased torrefaction temperature had a lower hydrogen content, was richer in oxygen content, and had a higher carbon content. The torrefied porous carbon containing a higher oxygen content, is considered a good-quality promising adsorbent[80,81]. Moreover, it has been stated that the yields of solid porous carbon material obtained via fast pyrolysis and gasification processes are considerably lower compared to the yields of solid porous carbon material produced via slow pyrolysis, flash and hydrothermal carbonization, and torrefaction[39].

7.3.2 Modification Methods

The modification is performed via various physical, chemical, and biological methods to prepare modified or activated porous carbon material. Figure 7.2 shows the different modification methods.

To date, biochar as a porous carbon material has received considerable attention as an additive, adsorbent, and soil amendment because of its excellent chemical and physical characteristics including high surface area, porosity, chemical stability, moderate mineral content, and abundant functional groups. It could be produced from various sources of organic matter including agricultural waste, manure, lignocellulosic energy crops, organic kitchen waste, sewage sludge, and so on. However, the direct application of pristine/conventional porous carbon material, prepared via pyrolysis of biomass directly, was hindered because surficial properties, porosity, and surface area were often limited. Thus, to enhance the properties of porous carbon material several post modifications or activations have been performed that can improve its efficacy for specific applications. Table 7.5 shows the physical, chemical, and biological modification of BDPC and its effect.

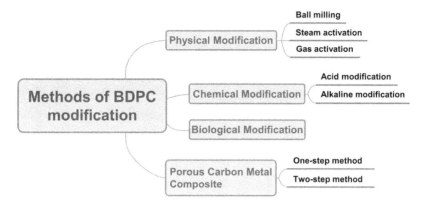

FIGURE 7.2   Different modification methods.

TABLE 7.5   Physical, Chemical, and Biological Modification of BDPC and Its Effect

Activation Method	Biomass Sample	Activation Reagent	Activation Conditions	BET Surface Area(m²/g)	Pore Volume(cm³/g)	Ave. Pore Diameter (nm)	Ref.
Physical activation	Rice straw	Steam	T = 700 °C 60 min	105.21	0.062	1.33	[82]
	Miscanthus	Steam	T = 800°C	332	—	—	[69]
	Spent mushroom	Stem	T = 800°C 2 h	332	0.29	—	[68]
	Rice straw	N₂	T = 600°C, 15 °C/min	10.90	0.04	—	[83]
	Rice straw	CO₂	T = 600°C, 15 °C/min	1.80	0.008	—	[83]
	Paper mill sludge	N₂	T = 600°C, 15 °C/min	27.20	0.07	—	[83]
	Paper mill sludge	CO₂	T = 600°C, 15 °C/min	31.60	0.08	—	[83]
	Wheat straw	Ball milling	T = 600°C, 1 h	130.14	—	—	[84]
Chemical activation	Alfalfa hays	NaOH	T = 800°C, 2 h	769.50	0.087	—	[85]
	Bamboo	H₃PO₄	T = 700–900 °C, 2 h	743.3–815.6	0.34–0.42	—	[86]
	Rice straw	CH₃COOK	T = 700°C 60 min	255.88	0.183	1.04	[82]
	Soybean oil cake	K₂CO₃	T = 600°C 1 h	643.54	0.336	1.04	[87]
		K₂CO₃	T = 800°C 1 h	1352.86	0.680	1.01	[87]
		KOH	T = 600°C 1 h	600.05	0.299	0.99	[87]
		KOH	T = 800°C 1 h	618.54	0.291	0.94	[87]
	Safflower seed press cake	ZnCl₂	T = 600°C 1 h	249.3	0.151	2.42	[50]

(Continued)

TABLE 7.5 *(Continued)*   Physical, Chemical, and Biological Modification of BDPC and Its Effect

Activation Method	Biomass Sample	Activation Reagent	Activation Conditions	BET Surface Area(m²/g)	Pore Volume(cm³/g)	Ave. Pore Diameter (nm)	Ref.
			T = 700°C 1 h	491.9	0.249	2.02	[50]
			T = 800°C 1 h	772.0	0.358	1.85	[50]
			T = 900°C 1 h	801.5	0.393	1.96	[50]
		5 M ZnCl₂	T = 900°C 1 h	555	0.752	2.26	[50]
Biological modification	Softwood bark and aspen wood	biofilm: microbial community	Cultured	4–973	0.01–0.3	-	[88]
	Water hyacinths	Chlorella	Cultured	-	-	-	[89]

7.3.2.1 Physical Modification

Physical modification can improve the pore volume, specific surface area, and pore volume of a porous carbon material. In addition, it can alter the surface chemical properties such as functional groups, polarity, and hydrophobicity, is environmentally friendly as it does not need any chemical reagents, and is also a simple and unexpansive process[33]. Physical modification mainly includes ball milling, steam activation, and gas activation.

7.3.2.1.1 Ball Milling   Ball-milling technique has been extensively used in several industries, including the wide-scale production of nanomaterials[90]. Notably, this technique is eco-friendly, which necessitates no chemical additives and requires relatively low energy consumption during the milling process[91]. Research indicates that ball milling significantly enhanced the specific surface area increasing from 3 m²/g to 25–194 m²/g of the porous carbon, depending upon the ball milling technique[92]. Although the technique is adaptable and simple, still it is a relatively novel concept in the realm of biochar engineering. Compared to conventional activation or modification techniques for porous carbon material, the ball milling technique offers distinct advantages. It has been stated that the grinding action of porous carbon material using the ball milling technique, the particle size, and the specific surface area can be improved, facilitating increased adsorption of both inorganic and organic ions[93]. Additionally, the introduction of different chemicals during the process of ball milling can further improve the functional groups. Nano-sized porous carbon materials can also be achieved via the ball milling technique.

7.3.2.1.2 Steam Activation   Steam modification is one of the most utilized physical modification techniques and a simple and inexpensive modification technique. In this process, the biomass is firstly pyrolyzed for 1–2 h at a temperature range of 300–700 °C under a nitrogen environment or insufficient supply of air. After this, the pyrolyzed biomass is steam activated for 0.5–3 h at the temperature range 800–900 °C with 2.2–5 ml/min steam pass[94]. It is generally performed to introduce the oxygen-containing functional group such as carboxylic (R-COOH), carbonyl (C=O), phenolic hydroxyl, and ether (R-O-R) group on the surface of porous carbon material and increase the porosity and specifics surface area. During steam modification, firstly the oxygen present in the steam molecules reacts with the carbon molecule present on the surface of the porous carbon materials to form hydrogen and surface oxides. Secondly, the surface carbon undergoes a reaction with hydrogen to form an intricate surface hydrogen and lastly, the surface carbon sites are oxidized via steam molecules, which in turn, release the carbon dioxide (CO_2) and hydrogen (H_2) molecule. This facilitates the activation of porous carbon material and reduces the gasification reaction which is also known as water-gas shift reactions on the carbon sites[43]. During steam modification there are several factors such as activation temperature, time and flow of water vapor influence the modification process[33]. However, by increasing the surface oxygen-containing functional groups, the hydrophilic behavior of prepared porous carbon can be increased. Generally, the existence of acid-containing functional groups is responsible for the hydrophilicity[95].

7.3.2.1.3 Gas Activation   Gas activation involves gases like nitrogen, ammonia, oxygen, and carbon dioxide or their mixtures to activate the porous carbon materials. Volatile chemicals are removed and pores are expanded during gas activation[96]. Thus, this method offers a significant advantage in terms of enhancing both the specific surface area and pore volume. In contrast to nitrogen-activated porous carbon, oxygen-activated porous carbon materials had a considerably higher H/C ratio which signifies that the oxygen-activated porous carbon is more carbonized compared to the nitrogen-activated porous carbon. In the case of activation via carbon dioxide, elevating the activation time and temperature can lead to an increase in the pore size and the specific surface area. Furthermore, the introduction of nitrogen molecules in the carbon matrix of porous carbon is accomplished via the process of ammonia purging[33].

7.3.2.2 Chemical Modification

Chemical modification is usually performed by adding chemical activators to the biomass precursors followed either simultaneously with carbonization (i.e. in situ activation) or by incorporating chemical activators into pre-synthesized porous carbon followed by a consequent thermal treatment. The post-chemical modification method is particularly useful for porous carbon formed via HTC[46]. This modification can create porous structures generally at a lower temperature for a shorter duration compared to physical modification and is cost-effective[33]. Chemical activation enhances the microporosity, lowers the mineral matter, activates the carbon content, and improves the surface functional groups (cation and anion exchange characteristics) of the porous carbon material[12]. However, several issues, such as chemical recycling, product purification equipment corrosion, still need to be considered while using chemical modification. Common activators used for chemical modification include $ZnCl_2$, K_2CO_3, KOH, and H_3PO_4. Moreover, different chemical activators may have different activation mechanisms.

7.3.2.2.1 Acid Modification   Acid modification enhances the acidic properties by eliminating mineral elements, thus it also enhances the hydrophilicity of the porous carbon.
In addition, acid treatment can improve the pore structure probably due to the removal
of some surface impurities via acid treatment[95,97]. In this modification method, the commonly utilized acids are HNO_3, HCl, and H_3PO_4 and it increases the oxygen-containing
functional groups (–OH, –COOH) on the surface of the porous carbon. For instance,
H_3PO_4 can decompose the aliphatic and aromatic components of biomass and produce
polyphosphate/phosphate cross bridges that prevent any shrinkage or contraction that
would occur as the pore system develops[98]. During the modification process, the porous
carbon material is soaked in an acid solution with a ratio of 1:10 for 1 h a day or 1 h at
room temperature to 120 °C[94]. The generally used acids are HNO_3, HCl, and H_3PO_4. The
porous carbon modified via H_3PO_4 improved the specific surface area and pore volume[99].
Whereas, the porous carbon treated with HNO_3, and HCl shows little difference in the
specific surface area but has a more acidic oxygenated functional group[94]. Increasing the
concentration of H_2SO_4, it had been stated that the specific surface area of the porous carbon material increased, but when an excess amount of H_2SO_4 was added, it was noticed
that the specific surface area of the porous carbon material decreased[43].

7.3.2.2.2 Alkaline Modification   Similar to acid modification, alkali modification can
improve the surface alkalinity of the porous carbon and change its porous structure.
The most commonly used alkali reagents are KOH, NaOH, and NH_4OH which show different modification effects[71]. For instance, alkali modification using KOH and NaOH
can enhance the oxygen content and surface basicity. Additionally, the KOH and NaOH
can also dissolve the condense organic components like cellulose lignin and ash thereby
facilitating the subsequent modification processes[100]. During modification, K^+ ions
might be integrated into the crystalline layers that create the condensed carbon structure, resulting in the formation of K_2O and K_2CO_3. These potassium species have the
potential to diffuse throughout the inner structure of the porous carbon matrix, intensifying the existing pores and creating new ones[101]. It's worth noting that compared to
KOH, NaOH is considered less corrosive and more cost-effective for modification of
porous carbon. Nevertheless, the porous carbon material modified via NaOH at low
temperatures is micropores and exhibits a smaller surface area than the porous carbon
material modified via KOH[101]. However, the synthesis of porous carbon material using
alkaline modification also enhanced its hydrophobicity. In this modification method,
the porous carbon material is generally soaked for one day or 6 h in a basic solution at
room temperature or 100 °C depending on the raw material used. Then the alkali-acti-
vated porous carbon is washed and dried. Further, the dried activated porous carbon was
pyrolyzed at 300–700 °C under a limited oxygen environment or in a nitrogen environment for 1 to 2 h to acquire the desired modified porous carbon[94]. The porous carbon
materials produced via alkaline modification have a higher ratio of H/C, and N/C, while
the ratio of O/C was lower and possesses advanced specific surface area[33].

7.3.2.3 Biological Modification

The term "biological modification" refers to the biological pre-treatment of porous carbon using microorganisms like bacteria or anaerobic digestion. The obtained porous carbon material via this type of biomass efficiently removes pollutants like nickel, lead, cadmium, and copper from an aqueous solution[43]. During anaerobic digestion, with the help of anaerobic bacteria, the biomass is transformed into methane and also produces a residue known as solid digestate. Pyrolysis of these digestates enhances the physio-chemical properties of produced carbon such as the pH, exchange capacity of anion and cation, hydrophobicity, and specific surface area (more negative charge). For instance, the value of zeta potential generated on the surface of porous carbon material after anaerobic digestion is highly negative, and thus the cations of the adsorption via functional groups are enhanced[71]. The alteration in pH of the biomass occurs during the anaerobic digestion due to the presence of bacteria acting on it. The main benefit of having porous carbon with a high anion and cation exchange capacity can be utilized as an ion exchanger during wastewater treatment[12]. The biological modification includes the formation of biofilm and colonization by microorganisms on the surface of porous carbon material which improves the physico-chemical and functional characteristics. This process allows the simultaneous adsorption of contaminants and their degradation via introduced microorganisms. For example, microorganisms, such as Paenibacillus, Clostridium, Aeromonas, Shewanella, Chloroflexi, and Cellulosimicrobium, show bio-adsorbent properties for heavy metals. Combining fungi or bacteria with porous carbon material is an effective and innovative approach for remediating contaminated soils. When the porous carbon mixed with microorganisms such as Actinomycetes *Bacillus*, or *Lactobacillus*, applied to soil, it considerably enhanced the soil nutrient and the growth or diversity of soil microbial communities[102].

7.3.2.4 Porous Carbon Metal Composites

Porous carbon metal composites are prepared by mixing the metal salts or oxides with porous carbon to facilitate metal ions to adhere to the surface of the porous carbon through chemical and physical attraction. The most commonly used impregnation salts are carbonate, nitrates, ammonium chloride, hydrogel, and CNT and commonly used metal ions are zinc, silver, magnesium, iron, etc.[94]. These metal composites are mainly synthesized using the following two methods

7.3.2.4.1 One-Step Method   In the step method, the porous carbon is impregnated with different concentrations of the metal precursor at various temperate ranging between 25 and 120 °C without or with stirring for 1 to 6 h. Further, the metal-impregnated porous carbon was subjected to pyrolysis at 300 to 900 °C for the synthesis of modified porous carbon material under an inert environment[94]. The one-step approach has several advantages. Firstly, it is simple to operate. Secondly, the produced solid or gaseous products can act as a reducing agent, that avoids the use of additional reagents. Lastly, the mixed metal nanoparticles with porous carbon itself can catalyze the thermal decomposition and improve the quality of the porous carbon[45,103]. Various porous carbon metal composites were synthesized using this method. Richardson et al.[103] synthesized Ni nanoparticles (Ni NPs) supported porous carbon metal composites by pyrolyzing Ni^{2+} dopped wood between 400 and 500 °C and the formation mechanism of Ni NPs was deeply examined. Results showed that compared to pre-formed Ni^0NPs (inserted into the biomass before) pyrolysis, in situ-produced Ni^0NPs revealed a higher catalytic performance for tar conversion.

7.3.2.4.2 Two-Step Method   In two-step methods, first, the pyrolysis of biomass occurs to synthesize the porous carbon. Further, the synthesize of the porous carbon is impregnated with the metal salt precursors of different concentrations. followed by the conversion of metal ions into metals via surface redox reactions[104] or carbothermal reduction[105]. Yao et al.[106] prepared a Ni-supported porous carbon by impregnation method by dipping the porous carbon in an aqueous solution of $Ni(NO_3)_2.6H_2O$, followed by drying at 105 °C overnight and further calcinated at 800 °C under nitrogen atmosphere. Their results revealed that the impregnated Ni was effective for hydrogen production via biomass steam gasification. Shen et al.[105] prepared Ni/Fe bimetallic porous carbon catalysts by two-step approach with $Ni(NO_3)_2.6H2O$ and $Fe(NO_3)_3.9H_2O$ as metal precursors.

7.4 PROPERTIES OF POROUS CARBON RELEVANT TO FARMLAND RESTORATION

Restoration of farmland using porous carbon refers to a sustainable agricultural practice that involves the use of BDPC material to improve the fertility and health of or degraded depleted farmland. BDPC materials have several properties that are relevant to soil improvement and farmland restoration. These properties make them valuable materials as soil amendments in agricultural practices. The physical, chemical, and biological properties of BDPC are vital in optimizing its performance in various applications[107]. However, the physicochemical properties of BDPC are significantly influenced by both the source of biomass and the conditions under which it has been prepared. Table 7.6 shows the physicochemical properties of BDPC prepared via different biomass. Various types of biomasses exhibit distinct chemical, physical, and structural properties that can be harnessed to create porous carbon materials, and various techniques are involved to characterize these properties, which play a vital role in optimizing BDPC performance in various applications.

7.4.1 Physical Properties

Physical properties comprise the particle size distribution, bulk density, specific surface area, pore volume, and pore size distribution[113]. These parameters are directly associated with the production conditions of BDPC such as the temperature of the reactor and residence time, the addition of different oxygen-containing media (air, pure oxygen, CO_2, and steam) in the process, and potential post-production treatments to modify the final product[111]. However, the temperature is considered as a specific factor in determining its specific surface area, surface functional group, pore structure, and elemental compositions, attributed to the release of volatile compounds at high temperatures[114]. Surface area is one of the essential factors in determining the adsorption effectiveness of BDPC material.

TABLE 7.6 Physicochemical Properties of BDPC Prepared Via Different Biomass

Biomass sample	Temperature(°C)	pH	CEC(mmol/kg)	Ash (%)	Surface Area (m²/g)	Element (%)						Ref.
						C	H	O	N	P	K	
Cow dung	450	6.15	—	—	—	67.04	—	—	0.12	1.18	2.29	[108]
	500	6.33	—	—	—	67.78	—	—	0.32	2.12	3.44	[108]
	600	6.45	—	—	—	—	—	—	0.45	2.45	3.65	[108]
Maize cob	450	-	—	30	—	66.0	—	—	1.31	1.76	—	[109]
	550	6.70	123	23	—	43.0	—	—	1.32	1.53	—	[109]
	600	6.40	115	24	25.4	23.4	3.5	0.5	2.3	1.45	2.4	[109]
Peanut shell	300	7.76	—	1.24	3.14	68.27	3.85	25.8	1.91	—	—	[110]
	700	10.57	—	8.91	448.2	83.76	1.75	13.3	1.14	—	—	[110]
Soybean stover	300	7.27	—	10.4	5.61	68.81	4.29	24.9	1.88	—	—	[110]
	700	11.32	—	17.1	420.3	81.98	1.27	15.4	1.30	—	—	[110]
Rice husk	400	6.84	—	—	193.70	44.59	2.50	16.32	2.50	—	—	[111]
	500	8.99	—	—	262.00	45.15	1.27	7.12	1.27	—	—	[111]
	600	9.41	—	—	243.00	40.35	0.85	9.23	0.85	—	—	[111]
	700	9.52	—	—	256.00	38.81	0.46	12.69	0.46	—	—	[111]
	800	9.62	—	—	295.57	40.41	0.28	2.69	0.28	—	—	[111]
Rice straw	400	8.62	—	—	46.60	49.92	2.80	12.02	1.22	—	—	[111]
	500	9.82	—	—	59.91	37.48	0.93	8.64	0.61	—	—	[111]
	600	10.19	—	—	129.00	33.78	0.60	13.68	0.41	—	—	[111]
	700	10.39	—	—	149.00	36.26	0.51	17.38	0.34	—	—	[111]
	800	10.47	—	—	256.96	29.17	0.25	3.71	0.25	—	—	[111]
Wood chips of apple tree	400	7.02	—	—	11.90	70.18	4.13	20.56	0.76	—	—	[111]
	500	9.64	—	—	58.60	79.12	2.65	11.98	0.34	—	—	[111]
	600	10.04	—	—	208.69	81.46	1.96	13.63	0.46	—	—	[111]
	700	10.03	—	—	418.66	82.26	1.21	16.34	0.41	—	—	[111]
	800	10.02	—	—	545.43	84.84	0.68	5.81	0.34	—	—	[111]
Oak tree	400	6.43	—	—	5.60	70.52	3.70	21.47	0.69	—	—	[111]
	500	8.10	—	—	103.17	77.57	2.51	17.73	0.51	—	—	[111]
	600	8.85	—	—	288.58	81.22	1.92	15.96	0.48	—	—	[111]
	700	9.54	—	—	335.61	83.22	1.16	14.97	0.31	—	—	[111]
	800	9.68	—	—	398.15	82.85	0.69	17.29	0.32	—	—	[111]
Urban waste	300	7.21	—	6	—	21.7	2.8	0.36	3.4	29.5	1.6	[112]

The specific surface area and porosity increase as the carbonization process accelerates. Micropores are randomly distributed throughout the BDPC, and most of the porous structures are preserved, leading to a higher porosity[72]. The BDPC materials, produced at varying pyrolytic temperatures, exhibited a unique honeycomb-like porous structure, attributed to the presence of tubular structure initially originating from plant cells[115,116]. Several studies have shown that elevating the pyrolytic temperature leads to an augmentation in the specific surface area of porous carbon and contributes to the development of a more intricate pore structure within it[117–119]. Due to these extensively developed pores, the porous carbon materials possess a notably elevated specific surface area.

The pore volume of BDPC materials is considered to be directly related to its surface area. The pore volume is significantly influenced at high pyrolytic temperatures due to the volatilization of tar components thereby influencing the accessibility of pores on the surface of porous carbon[120]. This, in turn, enhances the physical adsorption capacity of BDPC, which is closely associated with its pore volume[121]. This enhanced adsorption ability is attributed to the presence of abundant pores and elevated surface energy, which allows for efficient attraction and retention of metal ions[122]. Rouquerol et al.[123] based on the size of pores categorized the pores of porous carbon material into three groups: micropores (< 2 nm), mesopores (2–50 nm), and macropores (>50 nm). The elevated temperatures trigger the liberation of inner pores within the raw materials, resulting in the formation of an extensive quantity of unstructured pores and new ones. This, in turn, enhances the porosity and surface roughness of the porous carbon[124]. Moreover, the porous carbon produced at lower temperatures might be well-suited for regulating the release of fertilizer nutrients, similar to activated carbon. Additionally, it's worth mentioning that lower temperature-created porous carbon can have a hydrophobic surface, which could potentially limit its water retention capacity in the soil. Furthermore, despite its greater strength, in contrast to high-temperature porous carbon, low-temperature porous carbon is brittle and likely to abrade into finer particles once mixed into the mineral soil[125]. It has been stated that when hemicellulose and cellulose decomposed at 500°C, and the resulting porous carbon forms a honeycomb-like larger pore structure typically ranging from 5 to 40 mm in diameter, with thin walls[126]. Due to this transformation, the porous carbons produced at high pyrolytic temperatures exhibited less hydrophobicity and their water-holding capacity is higher[127]. In another study, it has been reported that at lower temperatures, less than 500 °C, lignin could not convert into a hydrophobic polycyclic aromatic hydrocarbon and the porous carbon becomes more hydrophilic whereas temperatures higher than 650 °C, the porous carbon becomes thermally stable and more hydrophobic[128]. This highlights the significance of both, the type of biomass and pyrolytic temperature are very vital for making good porous carbon. However, it should be noted that the application of porous carbon produced at high pyrolytic temperatures might be most appropriate for improving the water holding of coarse-textured, arid soils. Furthermore, when exposed to extremely high temperatures and prolonged residence times, the surface area tends to diminish due to the collapse of the porous structure. This reduction could be explained by factors such as structural ordering, pore enlargement, and the merging of adjacent pores. Additionally, the fusing of ash softening, and melting may also block these pores. Literature show that the yield of porous carbon material reduced as pyrolytic temperature elevated, owing to the loss of volatile organic components and water[129–131]. Additionally, the biomass type also determines the yield of porous carbon material. The porous carbon produced via woody biomasses exhibit lower ash content compared to porous carbon produced via non-woody biomasses[132]. The presence of organic matter and alkaline minerals in porous carbon material contributes to its ash content, which tends to increase as pyrolytic temperatures increase[133]. Consequently, porous carbon containing a considerable amount of ash can serve as a promising source for soil amendment for addressing acidic soil conditions[134]. Additionally, it has been noted that increased ash content is associated with higher yield of porous carbon materials[135].

7.4.2 Chemical Properties

Chemical properties such as elemental concentration, cation exchange capacity (CEC), pH, and surface functional group, of porous carbon material are also important aspects in determining its characteristics. As stated above, the physicochemical properties of porous carbon material are closely related to its biomass and production conditions. The production of porous carbon is generally evaluated through changes that might happen in the elemental concentration (C, O, H, N, and S) and their ratios[136]. The physicochemical properties of porous carbon derived from various biomass, differ significantly. For instance, plant-produced porous carbon including wheat stalk, rice husk, rice straw, maize cob, corn stalk, groundnut husk shells, and sugar cane bagasse, contain elements such as carbon (C), nitrogen (N), oxygen (O), hydrogen (H), sulphur (S), potassium (K), magnesium (Mg), calcium (Ca), and phosphorus (P). The production of BDPC from the pyrolysis technique also comprises nutrient elements, and using K and N-rich BDPC into the soil improves the growth of plants. Likewise, the carbon content in urban sewage sludge is present in significant amounts, making it a valuable resource for producing porous carbon materials[112,137]. When the biomass is carbonized and volatile matters are expelled, the remaining solid carbonaceous residue denotes the fixed carbon, which can help to estimate the yield of solid carbonaceous residue[116]. Most of the carbon (C) within porous carbon shows recalcitrant characteristics, thus suggesting its capacity for carbon sequestration. Furthermore, the ratio of O/C and H/C is used to determine the degree of polarity and the degree of aromaticity[127]. The elemental ratios of C/H, O/C, and O/H have proven to be dependable indicators of both the degree of pyrolysis and the extent of oxidative modification of porous carbon in soil[138,139]. The O/C and H/C ratios are influenced by pyrolytic temperature. The elemental composition of porous carbon material is intrinsically linked to the type of biomass used and its production temperature. Biomass has a higher content of lignin has been stated to promote carbonization and also increase the carbon content of porous carbon materials[95,140]. Porous carbon material derived from solid waste feedstocks and animal litter demonstrates reduced surface areas in contrast to porous carbon originating from woody biomass and crop residue, even when subjected to elevated pyrolysis temperatures[141]. This phenomenon can likely be attributed to the comparatively lower carbon content and volatile matter content, along with elevated molar H/C and O/C ratios in woody biomass and crop residue samples, which result in the development of substantial cross-linkages[95]. It has been reported that the porous carbon generated via urban waste at 300 °C exhibited a higher carbon content of 21.7% compared to the plant-based porous carbon at the identical temperature[112]. Another study shows that elevating the pyrolysis temperature the carbon content in porous carbon material of municipal solid waste increased to 48.6%, 59.5%, and 70.1% at 400, 500, and 600 °C, respectively[142]. Moreover, higher temperatures result in more carbonization, which reflects the rise in carbon content. Whereas, lower temperatures show decrease in the O and H content, which is most likely a result of the breakdown of oxygenated bonds, dehydration processes and the release of low molecular weight by-products comprising H and O[127]. The porous carbon materials, especially when derived from poultry manure and livestock, are rich in mineral nutrients like P, Mg, K, N, S, and Ca, which can significantly enhance productivity and soil nutrient levels. Additionally, it can enhance the accessibility of vital cations such as K, Mg, Ca, Mn, Zn, and Cu that are crucial for plant growth after being added to the soil[120]. The application of porous carbon in challenging farmlands with different soil needs could effectively enhance the quality of soil and provide significant potential for mitigating the deterioration of soil fertility.

Compared to other carbon materials (such as: carbon black and activated carbon), BDPC generally has abundant surface functional groups. These surface functional groups are highly beneficial for BDPC functionalization[143]. A variety of phosphorous, sulphur, and nitrogen-based functionalities might also be present that depend upon the composition of the raw material, however, these are present in considerably less amounts than oxygen-containing groups. Numerous investigations have highlighted the pivotal role of surface functional groups like C–O, C=O, and –OH in pristine porous carbon material, attributing their influence to the effective removal of contaminants such as heavy metals and organic ionic compounds[139]. Thus, extensive research efforts have been dedicated to modifying both the type and quantity of these surface functional groups in order to further improve pollutant removal performance. This can be achieved through pre- or post-pyrolysis treatments aimed at introducing specific functional groups, including surface sulfonation, oxidation, amination, and P–O–P insertion[144]. Surface sulfonation is generally introduced via concentrated sulphuric acid (H_2SO_4), gaseous SO_3 or fuming H_2SO_4[145], whereas oxygen and nitrate functional groups accomplished through nitric acid (HNO_3) onto the surface of porous carbon materials which is termed as amination[146]. Chemical oxidizing agents such as $KMnO_4$, HNO_3, H_2O_2, or a mixture of H_2SO_4 and HNO_3 are used for surface oxidation of porous carbon, which increases the carboxylic groups onto the surface of porous carbon. The increase of carboxylic groups on the porous carbon surface has the potential to enhance cation exchange capacity while concurrently reducing the pH[147,148]. The functional groups such as carbonyl groups (–C=O), carboxyl groups (–COOH), phenolic ester groups (–COOR), and hydroxyl groups (–OH) present on the surface of porous carbon helps in diverse chemical attributes such as cation exchange capacity, alkaline biological carbon and surface electrical characteristics[149]. Surface functional groups of porous carbon materials play a vital role in altering the hydrophobic or hydrophilic nature of porous carbon surfaces, as well as in buffering alkaline and acidic conditions, ultimately shaping porous carbon materials' performance in soil. Notably, the pyrolysis temperature exerts a profound impact on the composition of porous carbon materials' surface functional groups. While dehydration processes at lower temperatures (≤150 °C) have minimal influence on these groups, temperatures in the range of 250–350 °C result in a significant reduction in biochar surface functional groups, primarily attributed to esterification and decarboxylation. However, as the temperature was raised to 700 °C, the functional group concentration decreased significantly to 0.3 mmol/g, resulting in an alkaline biochar. Concurrently, it has been observed that within the temperature range of 500–700 °C, there is a consistent rise in the development of aromatic structures, accompanied by the transformation of alcohol groups into phenolic hydroxyl groups. This transformation contributes to an overall extension in the concentration of surface functional groups. Nevertheless, considerable portion of porous carbon is aromatic in structures, the extent of aromatization and the diversity and quantity of surface functional groups can vary considerably based on the source material and the specific preparation conditions.

The capacity to absorb cations by porous carbon materials is evaluated through its CEC. Typically, the CEC is affected by both temperature and the choice of biomass. As mentioned, elevated temperatures boost the formation of aromatic carbon oxidation and carboxyl groups. The CEC follows a similar pattern to temperature-driven functional groups[49,150]. Studies revealed that the CEC of porous carbon decreases with elevating the pyrolysis temperature[151,152], depending on the distribution and nature of the oxygen-containing functional groups[153] whereas the porous carbon with higher surface area has higher microporosity and increased CEC[154]. The dissolution of surface functional groups from porous carbon (such as -OH and -COOH) enhances the negative charge on its surface, consequently enhancing the CEC[155,156]. Thus, the introduction of porous carbon materials into the soil can considerably boost in soil CEC. It has been reported that the porous carbon material produced from animal manure has a lower carbon-to-nitrogen ratio, higher ash content, and higher CEC compared to the porous carbon material generated via plant-based feedstock[157]. However, the porous carbon material considerably advances the CEC of the soil. The effect of porous carbon material on soil CEC mainly depends on the type of soil, type of carbon material, and its duration in soil.

The pH value of porous carbon materials varies with the raw materials used and their production methods. Usually, porous carbon materials are suggested to be alkaline due to the presence of basic functional groups, which can retain alkaline components like hydroxides, carbonates, nitrates, and inorganic minerals. Increasing the pyrolysis temperature, the content of ash increases while the oxygen and nitrogen content decreases This leads to a heightened aromatic structure in the porous carbon and an increase in pH values[132,158]. At high temperatures, the pH value of porous carbon materials generated via forestry and agricultural wastes ranged between 7.0 and 10.4 due to the volatilization of organic acids and the breakdown of acidic functional groups (carboxylic and phenolic hydroxyl)[109,110,159]. The pH range of porous carbon material falls between 4 and 12, and its alkaline properties considerably influence the pH of soil after its application. Incorporation of BDPC into soil effectively adjusts the pH level of soil and increases the soil's saturation base. Through interaction with water, the porous carbon can replace the H^+ and Al^{3+} ions, and thus reduce their concentration in soil[49,155]. The addition of porous carbon to acidic soils enhances the soil's pH levels. Moreover, increasing the porous carbon dosage in the soil gradually increases the pH of the soil but it does not impact alkaline soils[160].

Thus, the physicochemical properties of porous carbon materials significantly influence soil structure and microbial activity, collectively contributing to the improvement of soil quality and fertility. These properties are instrumental in farmland restoration efforts, helping to make the soil more conducive to crop growth and sustainable agriculture.

7.5 APPLICATIONS OF BIOMASS-DERIVED POROUS CARBON IN FARMLAND RESTORATION

Biomass-derived porous carbon (BDPC) represents an alternative source to the other carbon materials and holds great potential in the realm of farmland restoration. This versatile and sustainable material, produced from renewable biomasses via pyrolysis, can significantly contribute to soil improvement and agricultural sustainability. BDPC, with its high porosity and surface area, enhances soil nutrient retention, cation exchange capacity, and water retention, ultimately improving the fertility of the soil. Enhancing soil structure and mitigating soil erosion, helps in preventing land degradation challenges. The application of BDPC in farmland restoration can improve crop yields, reduce the need for chemical inputs, and endorse long-term agricultural suppleness, making it a vital asset in the pursuit of productive and sustainable farmland ecosystems. Figure 7.3 shows BDPC application in farmland restoration.

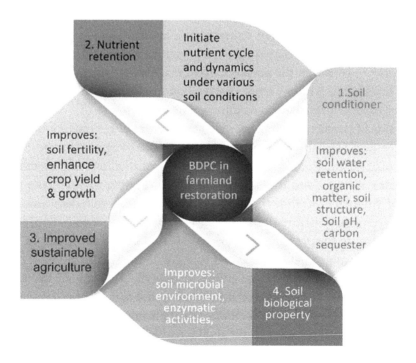

FIGURE 7.3   BDPC application in farmland restoration.

7.5.1 Soil Conditioner

BDPC, a carbon-rich by-product of pyrolysis with good catalytic and sorption potential may boost the amount of organic carbon in the soil. Typically, BDPC was employed as a soil conditioner to improve the overall health of the soil and the quality of soil organic matter[161,162]. Since the late 1990s, the idea of "soil health" has gained popularity, emphasizing priority on the significance of treating soil as a delicate, limited resource that demands particular care from its users. The "capacity of soil to function as a living system" is the definition of soil health.[163]. In addition to recycling vital nutrients for plant growth, forming advantageous symbiotic relationships with plant roots, improving soil structure for nutrient and water retention, and eventually boosting crop production, a healthy soil maintains soil biodiversity. The preservation of soil biota, which is essential for enhancing soil health, is at the center of this definition. Figure 7.4 shows application of BDPC and its effect on soil properties[164].

In modern society, the vitality of soil is constantly in danger due to a variety of anthropogenic activities, such as pollution, erosion acidification, and compaction[163,165–168]. To meet the anticipated needs of a global population of 9 billion by 2050, it is important to increase sustainable food production through responsibly managing and restoring healthy soils. With a long history in agronomy—including the application of Terra Preta de Índio[169] biochar is a useful soil amendment that promotes sustainable crop development via several unique processes. First and foremost, biochar improves soil fertility and crop output by acting as a source of carbon and nutrients[170]. Furthermore, its capacity to hold onto water and nutrients stops them from being lost, and biochar is an efficient way to encapsulate

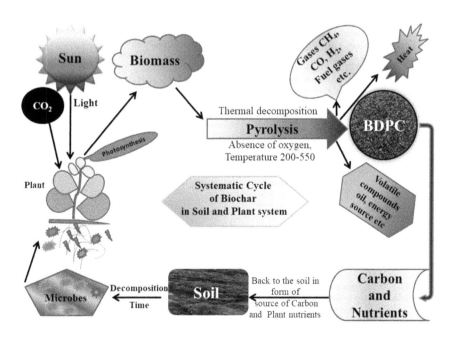

FIGURE 7.4   Application of BDPC and its effect on soil properties[164].

soil pollutants that might otherwise endanger biodiversity[171]. Keeping in mind the previous concept of healthy soil, maintaining and modifying the soil microbiome is essential to improving soil health.

The physical characteristics of soil are significantly altered by BDPC, which increases soil biodiversity. The addition of BDPC alters soil density, available water content, and aggregate characteristics, all of which have a substantial impact on the soil microbiome[139]. Table 7.7 shows the effect of BDPC application on soil properties. As BDPC is a porous material with a lower density than soil, applying it to the soil can reduce the bulk density of the soil[172]. This increase in soil porosity and reduction in density contribute to decreased tensile strength and raised soil oxygen levels, creating a favorable environment for root elongation and microbial growth[139,173]. An essential component of microbial growth is the soil's ability to retain water. Studies show that adding biochar generally improves the soil's ability to retain water. This is because biochar has a large specific surface area and has hydrophilic domains on its surface[174]. Furthermore, through binding to organo-mineral complexes, biochar enhances the stability and encourages the production of soil aggregates. These stable aggregates serve as microhabitats for a variety of soil microorganisms, offering them sources of nutrients and carbon that are customized to meet their requirements[175,176]. Therefore, the creation of more stable aggregates increases the number of habitats available for microbial growth.

Numerous studies have revealed the efforts to increase the amount of organic carbon in soil by adding animal dung and biomass to crops. The application of BDPC in soil not

TABLE 7.7   Effect of BDPC Application on Soil Properties

				Soil Properties				
BDPC	Temperature(°C)	Rate(t/ha)	pH	CEC(mmol/ kg)	SBD (g/ cm³)	Porosity(cm³/ cm³)	SOC(g/kg)	Ref.
Corn cob	Control	0	5.90		1.52	0.43	1.03	[177]
	500–550	10	6.20		1.49	0.44	1.39	[177]
Corn	Control	0			1.66	0.37	3.00	[178]
	350	4			1.30	0.51	30.5	[178]
	650	4			1.32	0.50	7.50	[178]
Orange peel	Control	0	8.30				16.7	[179]
	350	20	8.45				24.2	[179]
Rice husk	Control	0	6.5	112				[109]
	500–550	4	6.2	94				[109]
Straw	Control		5.4					[172]
	525		6.7	1485	1.17	84.3		[172]
Woody shrub	Control	0	4.57	111.9			1.53	[180]
	400–500	4	4.82	163.5			2.10	[180]
Hardwood	Control	0	5.6		1.57	40.75		[181]
	500	30	5.9		0.91	66.0		[181]

SBD: soil bulk density; *SOC*: soil organic carbon

only assists the isolation of carbon in soil but also improves soil quality by balancing its pH, enhancing cation exchange capacity, and encouraging microbial growth[182]. The functional groups including hydroxyl, carboxylic, and phenolic groups in BDPC, interact by hydrogen ions (H^+) in the soil, causing a reduction in concentration of H^+ and thus elevating the pH of soil. Furthermore, bicarbonates, carbonates, and silicates in BDPC react with H^+ ions, neutralizing the soil pH[110]. As a result, the application of BDPC in soil remediation in farmland sectors has gathered great attention due to its elemental composition and surface properties. Moreover, as a soil amendment in farmland restoration, BDPC can: improve the structure and enhance the fertility of soil[183], increase the CEC of soil[184], aid in maintaining water retention, thus improving productivity, support carbon sequestration and mitigate the impact of greenhouse gases[185], and also enhances microbial activity in soil by easing nutrient leaching[186]. Besides, the utilization of BDPC is considered as a promising technique for remediating contaminated soil afflicted with toxic pollutants, such as pesticides, hydrocarbons, heavy metals, and other contaminants.

7.5.2 Nutrient Retention

BDPC amendment can initiate nutrient cycling and dynamics under a variety of soil conditions despite of soil quality or fertility status. It stimulates nutrient interactions through several processes in the soil. Such as: serving as a nutrient source, thus supplying nutrients; serving as a nutrient sink, thus reducing the mobilization and accessibility of nutrients; and by altering the properties of soil, thus changing the nutrient cycles and reactions in soil. In tropical regions, especially in the monsoon season, nutrient leaching (washing of the externally supplied nutrition) and accelerated acidification of agricultural soil, resulted in decreased yield of crops and higher requirements of fertilizer. Application of BDPC to the soil is a feasible way to reduce the leaching of nutrients[187]. BDPC offers lasting benefits, including increased stability of organic matter, decelerated nutrient release from additional organic matter, higher nutrient utilization efficiency, and improved cationic retention due to increased CEC. BDPC production generates a diverse source of nutrients, including nitrogen (N), magnesium (Mg), calcium (Ca), phosphorus (P), potassium (K), and trace elements which are naturally present in the raw biomass[188,189]. A meta-analysis study revealed that BDPC generated through crop residual sources exhibited higher nutrient concentrations, such as Mg, Ca, P, and Si, compared to their feedstocks[190]. Accordingly, most BDPCs possess extensive fertilizing value and can contribute to the recycling and retrieval of essential plant nutrients[191,192]. Though the incorporation of BDPC into farmland can enhance the substantial amount of both available and total nutrients, except for available inorganic nitrogen[190], may control the accessibility of soil nutrients, possibly due to the restrictions in plant nutrient uptake[193]. Although some of the nutrients, for instance, sulphur (S) and nitrogen (N), present in the raw biomass are lost during their production for BDPC, particularly at elevated pyrolytic temperatures by hydrosulfide and ammonia volatilization[176,194] yet most of the nutrients are released gradually as the biochar weathers in the soil, becoming accessible for plant uptake[193,195]. The nutrient content of the BDPC varies with the raw biomass materials used for their production and also on pyrolytic conditions. For instance, biosolid-based and manure-based biomass derived BDPC exhibits higher nutrient levels of P and N, compared to those produced via grass, straw, and wood-based biomass[193,196,197].

Mainly the porosity, large surface area, organic coatings, cation and anion exchange capacity, and the pH of BDPC are responsible for nutrient retention capacity. BDPC, as a nutrient sink, can hold nutrients in soils, minimizing losses due to gaseous emission or leaching. On biochar, certain nutrients can adhere quite easily. More specifically, the surface functional groups of biochar produced at lower pyrolysis temperatures are where NH_4^+–N attachment takes place via cation exchange. On the other hand, increasing the pyrolytic temperatures promotes NO_3^-–N adsorption on the BDPC's fundamental functional groups[198]. In the presence of calcium, phosphorus ions show distinct adsorption tendencies on specific types of biochar; nevertheless, the effect on nutrient retention can be restricted and varies depending on the type of BDPC and soil[199]. Recent mechanistic research indicates that co-composting BDPC, frequently in conjunction with nutrient-rich organic additions like liquid manure, can improve plant development[200]. Furthermore, because the hydrophilic organic coatings lessen the hydrophobic aspect of the BDPC, the synergy between water and co-composted BDPC is enhanced. These interactions with the BDPC can improve agronomic performance and nutrient retention[201]. However, the application of BDPC influences a number of soil parameters, including pH levels, bulk density, cation exchange capacity, water retention, and biological activity, significantly. The interactions of nutrients with soil particles and the microbial conversion of nutrients are anticipated to be significantly impacted by these changes in soil properties. It has been shown that increasing cation exchange capacity reduces cationic nutrient leaching, including potassium (K^+) and ammonium nitrogen (NH_4^+), as well as nitrogen loss through emissions. Additionally, the pH of the BDPC has also a significant impact on modifying plant productivity, with alkaline BDPC proving more effective in enhancing plant growth by modifying soil pH in acidic soils[202]. Furthermore, BDPC-induced alkalinity (liming) can potentially boost soil phosphorus (P) availability while concurrently reducing the mobility of harmful elements like cadmium (Cd) and aluminum (Al)[203,204].

7.5.3 Improved Crop Yields and Growth

Nutrients are typically conserved in soil and remain accessible to crops primarily via adsorption to organic matter and minerals. Generally, organic materials such as manure and compost are added to the soil to improve nutrient retention. In particular, BDPC is known to be a very effective organic amendment for retaining and making nutrients more accessible to plants because of its large surface area and micropore structure, which foster an environment that is ideal for the good bacteria and fungi that plants need to absorb nutrients. The most deliberate application of BDPC to soil has shown enhanced crop yields with increasing BDPC addition quantities, even at very high loadings of 140 MgCha^{-1}. Studies also revealed that in degraded and nutrient-deficient soil its application considerably improves the productivity of crops in contrast to fertile and healthy soil[205–207]. In order to determine the influence of applying BDPC on soil characteristics and plant productivity, a meta-analysis was conducted on a range of diverse soil types and different climatic conditions (primarily in tropical areas). Their results revealed that the application of BDPC increases the crop yield from the next year relatively after its immediate application which proves the long-term effect of BDPC[208,209]. Compared to fresh BDPC, the aged BDPC exerts a more positive impact on the yield of crops over time and nutrient capture. This could be attributed to the slow formation of an organic layer on the surface of BDPC, which occurs when it ages in compost media and subsequently boosts nutrient retention[210]. However, long-term field studies on the application of BDPC are required in order to precisely determine how it affects soil quality and crop yield. Several other meta-analyses have been conducted to survey the effect of the application of BDPC and results revealed that the addition of BDPC to soil had a positive effect on crop yield due to increased pH, CEC, enhanced soil quality enriched with P, N, and K and organic carbon[200,210].

Despite the positive results associated with BDPC application, sometimes its application also shows negative effects on plant growth, crop yield, and nutrient availability. Notably, it has been discovered that applying freshly made BDPC occasionally reduces crop yield by immobilizing nutrients[207,211]. In addition, a negative impact on crop yields was also noticed where the application of BDPC considerably increased the pH generating an overliming effect that immobilized important micronutrients like magnesium (Mg), iron (Fe), zinc (Zn), and phosphorus (P)[212–214]. Moreover, improved crop yield due to the application of BDPC is prominently witnessed in less acidic, fertile, weathered soils due to its ability to neutralize the pH of the soil (liming effect) and enhancement in the physico-chemical properties and biological properties of soil[207,215]. However, the positive and negative impact on the crop yield and plant growth due to the application of BDPC depends on several factors: the distinct properties of soil such as soil pH and texture, raw material for BDPC and its production technique, its application rates, the crop species being targeted, and the local climate conditions[216,217]. Thus, the inconsistent effects of BDPC application on crop yield and productivity necessitate a more thorough comprehension of the fundamental principles underpinning BDPC's ability to increase crop productivity.

7.5.4 Soil Biological Property

Soil hosts intricate micro-organisms communities that are constantly adapting to the climate, land management practices, and soil properties. Crop production, soil respiration, soil nutrient cycling, and organic carbon levels are all closely correlated with the number and activity of soil microbial communities[218]. The microbial activity of soil is influenced by the application of BDPC and the effects vary depending on a number of variables, including soil type, quality of BDPC, and rate of application. Table 7.8 shows some significant aspects influencing the microbiology of soil while application of BDPC to the soil. Research indicates that both rhizobial and mycorrhizal populations are enhanced by BDPC application[219]. With the help of these present microbial populations, a variety of reactions occurs in the soil matrix at the interface between microbes in soil and root hairs[220]. It promotes the soil microbial activity by supplying growth nutrients and carbon substrate. Furthermore, it provides as a suitable habitat for their growth and guards them from predators. It also enhanced the buffering capacity of the soil which in turn minimizes pH fluctuations in microhabitats that present within the BDPC particles. Microbial populations play a vital

TABLE 7.8   Some Significant Aspects Influencing the Microbiology of Soil While Application of BDPC to Soil

Factors Affecting Soil Microbiology	Influencing Aspects	Ref.
Effect on soil pH	a. Application of BDPC causes a change in soil pH (increase) and has a good role for the nitrifying microbes (Nitrosomonas and Nitrospirm) and favors the bacterial community.	[222–224]
	b. Improves the process of soil liming, benefitting soil amendment	[222,225]
	c. In addition to acidic soils, BDPC shows a positive synergistic effect leading to modification of the taxonomic composition of the microbial communities	[222,226]
	d. Negative effects due to the mineralization of organic substances leading to the emission of N_2O and CO_2 into the atmosphere may pose a threat.	[227,228]
Effect of volatilesubstances releasedby BDPC	a. The volatile organic components cause both negative and positive effects on the various microbes present in the soil.b. The toxic effects are generally attributed to the type of biomass being utilized for the preparation of BDPC as well as the procedures adopted. These consequences could be avoided by using adequate hazardous assessment when selecting raw materials and carbonization processes.	[21,225]
Effect of polyaromatichydrocarbons (PAH) released by BDPC	a. Because of the high quantity of aromatic C in BDPC, investigating the influence of PAH on microorganisms is a big challenge. Several studies show a negative association between bacteria and aromatic C, whereas the inverse is shown for aliphatic carbon concentration.	[229,230]
	b. Some PAHs have also been discovered to minimize some of the adverse effects of bacteria, hence improving the soil bacteria's anti-biotic effect attributed to the transfer of PAH degradation genes on similar plasmids The pore structure of BDPC is also essential in this case.	[231,232]
Microorganism basedcolonization of BDPC	a. BDPC colonization by microorganisms is mainly dependent on BDPC composition, indirectly related to the pyrolysis reactor, biomass used for BDPC production, temperature, and residence time.	[229,233]
	b. Addition of aged BDPC compared to newly generated BDPC boosted microbial activity and thereby colonization. The occurrence of toxic substances in freshly formed BDPC is the most obvious cause.	[229,234]
	c. The presence of a pre-existing community of microorganisms and some specific soil features as well as the BDPC-soil contact time, may improve the colonization process.	[229,235]
Effect on soilenzymatic activities	a. The contribution of BDPC in electron transport and the accessibility of surface free radicals has been demonstrated to benefit oxidoreductase enzymes. The sorption of a few trace elements on the surface of BDPC may also serve as an enzyme cofactor.	[236,237]
	b. BDPC can activate particular types of microorganisms (nitroreductase, dehydrogenases, catalase, urease, hydrolase, and enzymes of the oxidoreductase class, etc.), which can interact with plant roots and influence some enzyme secretion and expression levels. While certain research reports some beneficial impacts, other results are controversial.	[229,238]

TABLE 7.8 *(Continued)*   Some Significant Aspects Influencing the Microbiology of Soil While Application of BDPC to Soil

Factors Affecting Soil Microbiology	Influencing Aspects	Ref.
Effect on soil propertiesand the microbialenvironment	a. Application of BDPC increases the porosity of soil as well as enhances its water-holding capacity, which provides a favorable environment to the microorganism communities even if plants can only access a small amount of water. It has been found that BDPC-mixed soils recover from droughts more quickly than soils without it.	[239,240]
	b. Certain chemical and physical characteristics of the BDPC-soil mixture, including as bulk density, pH, water-to-soil ratio, and microaggregate stability, have a direct impact on the plant's root system and a beneficial effect on the rhizosphere's microbial populations. Additionally, the denitrifying and nitrifying bacteria are also affected.	[229,241]

role in the degradation of fertilizers, consequently mitigating the concern of nutrient leaching[221]. Their existence in soil is essential for the strength and health of food crops[65].

BDPC promotes the "plant growth promoting bacteria", present within the rhizosphere and improves soil biodiversity due to its high organic carbon content and established porous structure. An application of relatively small dosages of BDPC can induce substantial changes in soil microbiome. Figure 7.5 shows the complex interaction of BDPC and rhizobial population[242]. One can relate changes in the rhizosphere's microbial structure, diversity, and activity due to the application of BDPC that are directly or indirectly related to BDPC effects on different soil environment components including physical, chemical, and biological aspects. BDPC induced "plant growth promoting bacteria" employ various mechanisms to enhance the growth of plants. Firstly, on the plant roots, the colonization of bacteria generates a protective microbial layer, that performs as a barrier that protects them from the intake of toxic substances. Secondly, particular growth-promoting bacteria such as Bacillus, Arthrobacter, and Stenotrophomonas are capable of generating the phytohormone (indole-3-acetic acid (IAA)), which stimulates plant development. Finally, these bacteria help to mobilize phosphorus and sulphur and break down insoluble, refractory carbon compounds, boosting the availability of these nutrients for plants. BDPC also suppresses the activities of plant pathogens directly via adsorption or indirectly via gene expression. Its application did not harm the pathogen directly while increasing soil biodiversity, inducing plant resistance via BDPC-induced microorganisms, and increasing beneficial bacteria account for the pathogen suppression.

7.6 ENVIRONMENTAL AND ECONOMIC BENEFITS

Globally the demand for energy is increasing as the world population increases since energy is required for all sectors in any nation. Presently, fossil fuels are the dominant source of energy. Biomass is considered a promising source of renewable energy that could be converted through physical, biochemical, mechanical, and thermochemical processes

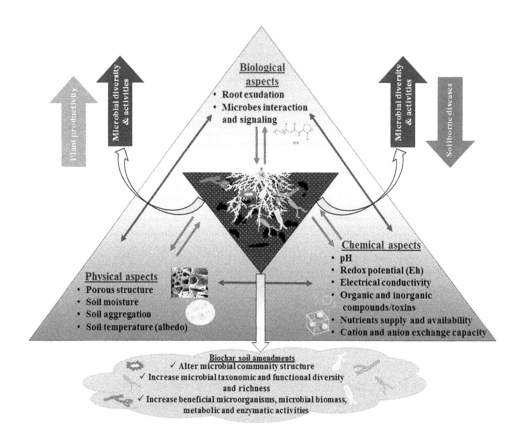

FIGURE 7.5 The complex interaction of BDPC and rhizobial population[242].

into valuable products. Thermochemical conversion of biomass to achieve high-quality and efficient product yields via the breaking of organic matter chemical bonds results in the generation of bio-oil, biochar, and syngas. In recent years, there has been growing concern about converting biomass into BDPC, due to its sustainability advantages, economic benefits, and ever-increasing demand in energy fields and environmental. The cost-effective, eco-friendly, and simple preparation of BDPC from various biomass resources using thermochemical techniques for environmental and economic benefits has attracted intensive interest among researchers.

7.6.1 Reduced Soil Degradation and Erosion

BDPC forms physical connections with soil constituents due to hydrophobic interactions and van der Waals forces. It interacts submolecularly with silt and clay particles as well as soil organic matter (SOM)[243]. These complex interactions control how biochar affects the physicochemical characteristics of soil, and also how it interacts with cations, anions, and other organic soil constituents[244]. It is worth noting that these interactions are highly particular to each type of BDPC, with their precise physical characteristics being altered by factors such as biomass type and pyrolysis conditions. Applying biochar can result in a variety of effects on soil parameters, including specific surface area, bulk density, cation exchange capacity (CEC), and water-holding capacity, all of which can be advantageous or disadvantageous.

The application of BDPC during restoring farmland to prevent and control soil erosion is very important. When restoring farmland, soil erosion could be a pivotal challenge due to the degraded and often exposed nature of the soil. During this phase, biochar can be very helpful in reducing erosion and improving the quality of the soil. According to a study, biochar may have an indirect impact on soil erosion through the following mechanisms: (i) altering plant growth, which may change the impact of raindrops on the soil surface and, in turn, affect the mobilization of soil particles through rain splash erosion; (ii) altering architecture and root growth, which may affect cohesion and accumulation between aggregates; and (iii) enhancing soil roughness by creating physical obstacles to overland flow and altering its velocity. Thus, the application of BDPC can promote plant growth, which promotes a denser network of roots which can stabilize the soil and reduce erosion risk. Enhancing soil structure, or the size, stability, and spatial arrangement of soil aggregates, is one of the primary ways that biochar may influence erosion rates[173]. This can alter the time-to-runoff (or infiltration capacity), the amount and duration of runoff, and the soil's deterioration. The concentration of topsoil biochar and soil physical-hydrological characteristics, such as bulk density, porosity, accessible water capacity, and saturated hydraulic conductivity, have been found to positively correlate in multiple meta-analyses[174,245,246]. However, it is possible to anticipate that biochar will help to mitigate soil erosion as it has been demonstrated to generally promote plant development in soils with a pH within the ideal range.

7.6.2 Carbon Sequesters

Carbon sequestration is a technique to reduce carbon dioxide emissions. The application of BDPC to soils that trap carbon is currently seen to be one of the most promising ways to mitigate climate change, besides its ability to increase crop yield[247]. Carbon sequestration is the long-lasting storing and capturing of atmospheric CO_2, to reduce global warming. The four main carbon pools are geological, atmospheric, terrestrial, and oceanic[248]. The carbon residing in these pools exhibits a flexible lifetime, with interlinked or interconnected carbon flows. Carbon present in the active carbon pool swiftly transitions between diverse pools. However, in order to mitigate atmospheric carbon, it should be moved to a passive pool where it can remain stable or inert[249]. With time, BDPC is being viewed as a viable solution for meeting the global climate change goals because it provides an easy way for carbon to move from the active pool to the passive pool.

According to an estimation of global soil organic carbon stocks, the uppermost one-meter soil layer contains 1505 Pg of carbon[250]. Numerous studies have shown that adding BDPC to soil has priming effects. Priming is the process that causes an alteration in inherent soil organic carbon decomposition due to the application of a new organic substrate like BDPC. Depending on the decrease or increase rate of soil organic carbon mineralization owing to the BDPC addition, priming effects could be negative or positive, respectively[251,252]. Various strategies have been employed to increase soil organic carbon content, including the adoption of crop rotations, incorporation of organic amendments, conservation tillage, and utilization of cover cropping[253,254]. BDPC, an organic amendment, is often resistant to biodegradation due to its highly condensed aromatic structure[247]. Thus, it is anticipated that BDPC will have a beneficial effect on the sequestration of carbon in the soil. Several researches have been conducted to examine the effect of BDPC on sequestrating carbon. However, no consistent results have been observed because both increased and decreased carbon emissions have been recorded[58,255,256]. Notably, carbon sequestration is partially hindered during positive priming, whereas carbon sequestration potential is increased during negative priming[257]. Furthermore, the priming effect of biochar was significantly influenced by the incubation duration[95].

BDPC carbon can be categorized into insoluble and liable carbon[12]. When BDPC is applied to the soil, the microorganisms present in the soil may rapidly consume the available carbon, resulting in an initial increase in carbon mineralization. This explains why adding BDPC improves the mineralization of carbon. Notably, refractory carbon is abundant in BDPC as opposed to labile carbon[258]. The carbon resistant to decomposition might remain in the soil for a long time, meaning that the carbon intake from adding biochar exceeds the carbon output from the mineralization of liable carbon. In soil, BDPC exists in a highly stabilized state which can store carbon for many years, thus converting the soil into a significant carbon reservoir[259]. In contrast to other alternative terrestrial carbon sequestration techniques such as replanting or afforestation, carbon sequestration in BDPC greatly enhances carbon's storage period[260]. Consequently, the application of BDPC can play an essential role in both removing carbon from the environment and simultaneously reducing greenhouse gas emissions. Many authors recommend BDPC as a long-term method of storing carbon in soil[125]. The International Biochar Initiative estimates that BDPC has the potential to trap up to 2.2 billion tons of carbon dioxide by 2050. However, the consequence of adding BDPC to carbon sequestration remained still uncleared. Since the priming effect varies depending on the feedstock and pyrolysis conditions, it is imperative to investigate the relationship between the creation of BDPC and the kind of feedstock. It is also crucial to look into the connection between pyrolysis conditions and BDPC's ability to sequester carbon as these factors have a big impact on the physiochemical characteristics of the material. Examining soil constituents is also necessary when analyzing carbon sequestration caused by BDPC.

7.6.3 Reduced Greenhouse Gas Emissions

An increasing level of greenhouse gases (GHGs) has advanced climate change. Thus, it is important to take appropriate action in order to balance the anthropogenic GHG emissions (CH_4, CO_2, N_2O) with storage in the soil. It has been stated that the application of BDPC into the soil could reduce the emissions of N_2O and CH_4 which have a global warming potential (GWP100) more than 28 and 265 times more than CO_2 over a 100-year time horizon, respectively[8]. BDPC can play a pivotal role in the short-term reduction in CH_4 emission to assist in reaching the 2050 GHG targets because methane's GWP20 (for a 20-year duration) value of 84 is substantially greater than its GWP100 due to its relatively short residence duration in the atmosphere[261]. In contrast, N_2O, an important contributor to GHG, has longer residence in the atmosphere. Soils are responsible for approximately

62% of atmospheric N_2O emissions[262]. The application of nitrogen-based fertilizer to fields at high rates also emits N_2O into the environment. BDPC application to soil effectively reduces soil N_2O emissions, and the reduction can be attributable to the suppression of either the nitrification or denitrification stages, as observed in both field and laboratory investigations. BDPC application improves soil aeration, which reduces denitrification by inhibiting the activity of anaerobic microbes engaged in denitrification. BDPC addition causes microbial immobilization of accessible N in the soil, lowering the soil's N_2O source capacity. The generation of N_2 from N_2O is stimulated by the improved pH which is also caused by the application of BDPC. Furthermore, increased soil fertility via applying BDPC will help farmers adapt to changing climates, thus lowering the severity of climate change[263].

BDPC application suppressed soil N_2O emissions by lowering soil labile nitrogen (N) forms and interfered with the activities of N-cycling enzymes[264]. Improved aeration, particularly in fine-grained soils, increases the sinking ability for CH_4 by increasing the population of methanotrophic proteobacteria, hence increasing CH_4 oxidation and lowering CH_4 emissions[265]. Compared to other soil organic materials, the stability of BDPC is 10 to 100 times greater, attributed to its condensed aromatic structure which makes it resistant to mineralization and microbial decomposition[217]. A meta-analysis performed on the stability of BDPC in soils assessed that the major portion of BDPC (97%) significantly contributes to carbon sequestration in the long term in the soil[260]. According to model assumptions, the incorporation and production of BDPC into soil has the potential to offset up to 1.8 Pg CO_2eq emission per year, accounting for 12% of anthropogenic CO_2eq emission, and over a century, the cumulative net emission offset by BDPC would be 130 Pg CO_2eq[248]. It has been stated that the application of BDPC produced via crop residues and its incorporation into the soil is projected to prevent emissions corresponding to 4,348 to 4,878 kg of CO_2 per hectare per year[266]. Since BDPC contains 60–80% of carbon, each ton of BDPC added to soil can sequester 0.6–0.8 tons of carbon which is equivalent to removing 2.2–2.93 MT of CO_2[267]. For agricultural applications, limestone is usually utilized to reduce the pH of the soil, however, utilizing limestone/ton leads to generate 0.059 MTC or 0.22 MT CO_2 in the atmosphere. These CO_2 emissions could be avoided using BDPC in place of limestone. It is anticipated that if 76.53MT BDPC is used in place of 6.48MT lime per hectare, it will mitigate 225.6 MT CO_2 per hectare through emissions and BDPC carbon sequestration[268].

In addition to being used in agriculture, BDPC is gaining more and more popularity within the waste management sector as a considerable approach to improve landfill gas emission control. The microbial oxidation of methane (CH_4) in bio-windows, biofilters, or bio-covers is one method for a long-term reduction in CH_4 fluxes[269,270]. The efficacy of these constructed methane oxidation systems could be improved by amending the soil in use with BDPC.

7.6.4 Mitigation of Soil Pollution and Contamination

The expansion of anthropogenic activities, including industrialization and urbanization has caused an increased concentration of organic and heavy metals in the water and soil. This directly impairs the health of the soil, increasing stress and the possibility of pollutants entering plants and eventually entering the human body[271]. Thus, BDPC can be employed as a remediation method to lessen soil contamination from heavy metals and organic pollutants. Adsorption is considered the main method by which biochar remediates soil[272]. Surface complexation, electrostatic attractions, hydrogen bonding, π-π interactions, and acid-base interactions are all involved in this adsorption mechanism.

Using BDPC for soil remediation is the cheapest process in contrast to other remediation techniques, such as physical treatment, electro-kinetic remediation, and biological remediation. Heavy metals including Cd, Cu, Zn, and Pb are frequently found in soil; typically present in the form of compounds or individual ions in the soil. It is necessary to remove heavy metal contamination from agricultural land in order to protect human health and the food chain as it poses serious risks to animals and human beings[273]. Chemical reduction, cation exchange, surface complexation, precipitation, and electrostatic attraction are the main processes behind heavy metal adsorption onto BDPC[33]. The efficiency of contamination removal depends on various factors such as cation exchange capacity and specific surface area of BDPC, the presence of interfering ions, the dosage BDPC applied, solution pH, soil pH, soil type, metal concentration, contact time, temperature and as well as the type of BDPC used[95,274]. The pH is the key aspect because in acidic environments, most heavy metals, including Pb, Ni, Cr, Cd, Zn, and Cu, become more bioavailable Since biochar is primarily basic; thereby the BDPC soil mixture may cause metals to become immobilized and generate a liming effect in the soil[275]. To enhance the removal efficiency of heavy metals, several studies have been performed by researchers to increase the functional groups or surface area of the pristine BDPC. Consequently, many different alterations have been examined in depth, such as activation by physical and chemical treatment, doping of elements, coating or loading of functional structure, and modification via chemical grafting[72,116,276,277]. The experimental conditions, chiefly the initial concentration of heavy metals and the types of BDPC, were vastly varied, which makes the comparison challenging.

BDPC possesses the potential to absorb organic pollutants such as pesticides, antibiotics, and polycyclic aromatic hydrocarbons (PAHs), and causes adverse effects on farmland soil. The usage of sewage irrigation and organic pesticides are the main causes of organic contamination. Organochlorine insecticides are widely used in developing nations to boost agricultural productivity and to prevent various vector-borne illnesses including kala-azar and malaria[278]. These persistent organic pollutants enter our bodies via the food chain and cause serious risks to human health[279]. Even after being banned for use in agriculture in 1997, dichlorodiphenyltrichloroethane (DDT), is still one of the most often used persistent organic pollutants[280,281]. The removal of organic pollutants from the soil by biochar was affected by several variables, including the types of feedstocks, the dose applied, the specific contaminants, and their concentration. Therefore, eliminating this pollution from our food chain and environment is very crucial. BDPC increases the adsorption capacity of soil organic contaminants and decreases their desorption activity while also preventing flow in the soil[282].

However, the ability of various BDPCs to adsorb organic and inorganic pollutants varies due to their unique physiochemical characteristics. Accordingly, when it comes to pollutant removal, the choice of feedstock is more important than adjusting the pyrolysis temperature or the properties of the BDPC surface[283]. In addition to increasing soil organic and inorganic pollutants' adsorption capacity, BDPC also reduces their ability for desorption, which stops the pollutants from migrating within the soil[282]. Moreover, various parameters, such as surface functional groups, porosity, mineral composition, pH, and surface charge can affect the adsorption ability of BDPC, as reported in earlier studies[284,285].

7.6.5 Sustainable Agricultural Practices and Cost-Effectiveness

Improving crop productivity in an eco-friendly and sustainable manner offers considerable challenges in current agro-farming systems[286]. Following the green revolution, agricultural techniques intensified their reliance more heavily on organic fertilizers to secure higher yields. Although chemical fertilizers do boost crop yields, they also endanger environmental sustainability by producing ecological imbalances such as global warming, biodiversity loss, and heavy metal accumulation in living beings[287,288]. Therefore, adopting more natural farming methods is important for reducing reliance on organic fertilizers and ensuring the long-term sustainability of productivity and agricultural production.

More recently, BDPC has emerged as a viable soil conditioner for retaining nutrients and carbon in the soil, solving environmental concerns associated with sustainable agricultural fertilizer management. Current research on BDPC is predominantly focused on tailoring its characteristics to improve its ability to eliminate inorganic and organic contaminants[289,290]. BDPC offers a wide range of environmental applications due to its unique features, such as microporosity, large surface area, ion exchange capacity, and greater adsorption capacity which contribute significantly to environmental sustainability. The conversion of biomass into BDPC is carbon-negative, sequestering around 87% of carbon[291]. This not only alleviates the challenges associated with agricultural residual disposal but also provides a feasible and cost-effective technique for converting waste into value-added products. BDPC's excellent surface characteristics, make it a promising candidate in mitigating contaminants such as heavy metals, herbicides, pesticides, antibiotics, and dyes and thus reducing global climate change[73].

Studies reveal that the application of BDPC is a promising way to enhance carbon into the soil, enhance sustainable plant growth, and improve the value of agricultural products. It has the potential to boost crop yield and growth by improving nutrient availability. Soil amendment with BDPC leads to improve the fertility of soil under various agricultural and natural environments[292]. Moreover, higher crop yield and improved soil fertility require a synergistic effect on the accessibility of nutrients and plant uptake, which can be achieved by combining inorganic and organic fertilizers with BDPC[125]. It can also immobilize farm chemicals and rhizospheric heavy metals and hinder their movement within the plants, thus enhancing crop productivity[73]. However, the beneficial effects and usage of BDPC amendment are widely discussed, and more research is needed to deeply understand its magnitudes and perks, as well as its constraints in agroecosystems.

The exploitation and exploration of bioresources to produce new bioproducts that are economically valuable using biotechnology is termed as bio-economy. In such cases, the bioresources are feedstock while the BDPC is the bio-product which is the important aspect of bio-economy. Various bio-products can rapidly increase the value of versatile bioresources, facilitating their use[293]. Production, commercialization, marketing, and awareness campaigns, are critical components for maintaining the sustainability of bio-economy long-term viability. These bio-economic activities generate both direct and indirect job prospects. Furthermore, the bio-economy overall impact is influenced by the quantity, quality, and safety of bio-products[294]. As for BDPC, sustainable large-scale production will affect economic and agronomic benefits. For instance, crop yields influenced by BDPC application and the profit from surplus harvest have a direct impact on the economic balance. This highlights the necessity of carefully examining issues such as soil type, crop selection, and the quantity and quality of BDPC application, as they all have an impact on the economic equilibrium. The economics of BDPC production are greatly influenced by transportation expenses. The cost of producing BDPC in one site and transporting it to another influences the price. The economic sustainability of mobile pyrolysis plants by exploration of two types of BDPC in three states that travel at different times had been investigated which showed that the net present value of BDPC enhanced as the number of times the facility of mobile pyrolysis dropped down. The production of BDPC is gaining significant popularity owing to its estimated energy and ecological benefits[175,289,295–300].

7.7 CHALLENGES AND FUTURE DIRECTIONS

7.7.1 Addressing Scalability and Large-Scale Implementation

One important area of research and development for sustainable agriculture is the scalability and widespread application of BDPC in restoring agricultural land. Lab and small-scale studies have only shown BDPC efficacy in remediation; however, further research is necessary to scale up production and its application methods for real-world circumstances. This entails developing production processes that are both economical and energy-efficient, investigating the best application rates and procedures, and determining if large-scale implementation is feasible. For BDPC to be successfully adopted and used widely, it is imperative that the practical and logistical aspects of incorporating it into current remediation strategies be understood.

One key aspect of scaling up BDPC implementation is the development of cost-effective production techniques. Slow pyrolysis is a common traditional method, which might not be commercially feasible on a big scale. The focus of research endeavors is on refining production procedures, investigating substitute feedstocks, and utilizing cutting-edge technologies to simplify the BDPC manufacturing process. To ensure that BDPC is a feasible and affordable choice for broad use in a variety of agricultural contexts, cost-effectiveness is critical. Another critical aspect of large-scale deployment is determining the BDPC application rates and techniques. Although research conducted on a small scale has shown that BDPC improves soil fertility, it is crucial to customize application methods to fit various soil types, temperatures, and crop varieties. The objective of research endeavors is to develop protocols that harmonize the advantages of BDPC with ecological factors and assure that its utilization conforms to sustainable farming methods. This means being aware of how BDPC affects soil composition, nutrient retention, and water-use efficiency over time.

A comprehensive feasibility study must be carried out before biochar is used extensively in farmland restoration. This means assessing the potential possible social, economic, and environmental effects associated with the widespread use of BDPC. Researchers are looking into possible emissions, the carbon footprint of producing BDPC, and any unanticipated impacts on ecosystem dynamics. Furthermore, determining the acceptability and practicality of incorporating BDPC into conventional agricultural operations requires an understanding of the economic and social consequences for farmers and stakeholders. The practical and logistical challenges of incorporating BDPC into current farmland restoration strategies necessitate careful consideration. This includes tackling significant logistics issues concerning storage, transit, and application. The goal of the research efforts is to create application techniques that are easy to use and blend well with current farming tools and methods. It is imperative to comprehend the practical obstacles and offer solutions to ensure the effortless incorporation of BDPC into standard agricultural practices in order to ensure its widespread adoption.

In conclusion, an extensive approach is required for the scalability and large-scale application of BDPC in farmland restoration. Through tackling concerns associated with optimal utilization, cost-effective production, practical considerations, and feasibility evaluation, researchers aim to facilitate the extensive integration of BDPC as an eco-friendly and efficient instrument for augmenting agricultural yield while stimulating soil health and ecological conservation.

7.7.2 Environmental Considerations and Life Cycle Assessments

BDPC is a highly concentrated form of carbon with nitrogen and oxygen-containing surface functional groups that can be used for various applications as mentioned earlier. However, the most promising application of BDPC is soil amendment[16]. When added to soil, it is intended to improve soil qualities while acting as a durable carbon storage which is influenced by the type of BDPC as well as the composition of soil and condition of the climate. A considerable number of research indicated the beneficial impacts of using BDPC, while some also found negative impacts[301]. The intricacies of the soil systems and the variability in BDPC characteristics required further research to establish a simple mechanistic principle defining the BDPC behavior in soils. To date, an idyllic condition of BDPC production must be determined through experimentation for each feedstock and application. However, the feasibility of utilizing BDPC from spanning its acquisition and production to the final applications ought to result in a positive overall effect in several areas, including financial, environmental aspects, and energy-related[125]. Before launching any significant project, a thorough and careful examination is necessary to determine whether the benefits surpass the disadvantages. The entire system analysis is required to determine its efficacy in achieving its intended goal. Additionally, the economic feasibility of a BDPC project is also an important factor to take into account.

A life-cycle analysis (LCA), is a well-established and standardized method that helps to analyze and understand various aspects of biochar system on the environment over the whole life cycle such as costs allied with the entire process for BDPC production, utilization of energy, logistics, emission of greenhouse gases during the pyrolysis process, and the application points of view[302]. LCA comprises four stages; (a) goal and scope definition, (b) inventory analysis, (c) impact assessment, and (d) interpretation[303]. Sometimes, a process can have more than one output at times. Multifunctionality of the process can be addressed in LCA in a variety of ways. to use system expansion is one way where exclusion of the most likely by-products is excluded from the system. Allocation is another approach, that involves dividing inputs and outputs among different products, ideally using physical parameters[303].

A significant body of literature on the LCA method was proposed for assessing the environmental effect of BDPC-to-soil projects. A comprehensive review of literature on LCA focused on BDPC produced through pyrolysis and its application as a soil amendment has been conducted. The reviewed literature used different kinds of biomass, functional units, locations, techniques of dealing with by-products, and assumptions. While LCA studies share many fundamental aspects, there are some specific differences, such as functional unit selection or system boundaries. Nevertheless, all study concludes that BDPC systems are advantageous regarding climate change. During LCA of BDPC generated via slow pyrolysis and their application for soil amendment from different biomasses such as forest residue, different types of straws, willow, and pig manure. For a large-scale BDPC system, forestry leftover chips had the maximum feasible carbon reduction of 3.9 t CO_2eq/ ton BDPC, followed by sawmilled wastes with 3.7 t CO_2eq/ton BDPC. On a broad scale, the carbon sequestration capacity of three distinct types of straws studied is 2.7 t CO_2eq/ton BDPC. However, LCA with BDPC produced via pig manure and willow, system expansion had been used for the bio-oil and syngas produced during the production of BDPC. Syngas were assumed to be combustible in a combined heat and power plant in order to meet the heat and electricity requirement for the pyrolysis process. They also used system expansion to take into consideration the fertilizer consumption that was reduced. The results of the two scenarios under investigation showed a favorable potential for reducing carbon emissions in the form of a negative global warming potential (-0.472 t CO_2eq/ton of BDPC for manure and -2.06 t CO_2eq/ton of BDPC for willow). The application of BDPC to soil was the main factor contributing to the potential for negative climate change. Another study of LCA was conducted to compare BDPC to soil systems where BDPC was obtained from different wastes and residues via fast and slow pyrolysis and gasification. It was discovered that slow pyrolysis performed the best in terms of lowering or eliminating emissions. Since the yield of BDPC is much higher in the case of slow pyrolysis, carbon storage plays a crucial role in affecting environmental advantages. In contrast, for gasification and fast pyrolysis, the environmental benefits lie in the mitigation of emissions due to the production of electricity, which replaces fossil fuels.

Moreover, in order to enhance the benefits that can be obtained from BDPC, logistical issues regarding biomass collection and biochar distribution must be addressed. Additionally, co-products produced during pyrolysis must be investigated, and BDPC should be strategically applied in potential fields. This comprehensive approach might lead to both economic gains and favorable environmental effects, which would help to fully close the life cycle assessment of BDPC. The initial stage in the LCA of biochar involves the procurement of biomass, the operational costs of which will vary with the source, location, and complexity involved in the collection, transport, and supply at the destined source of use. Figure 7.6 shows the various stages involved during the LCA of BDPC[304]. In the study of LCA of BDPC from corn stover and switchgrass, considered expenses involved the feedstock collection, the farm size, bailing, and harvest machinery. They estimated the transportation cost of a dry metric ton of corn stover and switchgrass was $63–75 and $80–96 respectively for up to 60 km. The difference in cost is mainly affected because corn stover is a by-product of corn crops whereas switchgrass is a dedicated crop used for energy harvesting purposes. Furthermore, procuring forest residue has a higher capital cost for processing at stockpile stations when compared to biomass obtained from agriculture. This is owing to the greater effort required for the management, collection, and transportation of low-bulk and energy-density leftovers. Nevertheless, the net global warming potential created by converting forest wastes into biochar utilizing portable technologies is lower than that of pile burning. As a result, adopting a systematic strategy for generating BDPC from forest leftovers may prove

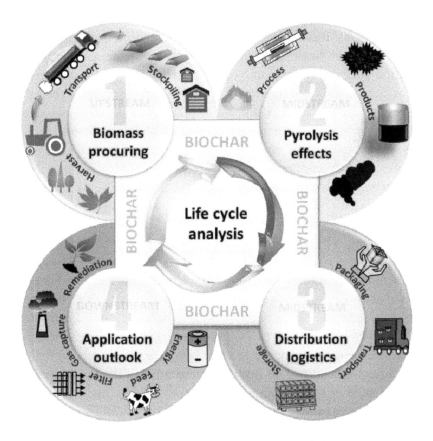

FIGURE 7.6   The various stages involved during the LCA of BDPC[304].

more environmentally advantageous and practicable, even if it is more expensive than using agricultural residues.

7.7.3 Research Gaps and Areas for Further Investigation

BDPC is a carbon-based product that is environmentally friendly and renewable. Despite being a promising and rising research area, there are several uncertainties and gaps in BDPC production and its application for soil improvement, necessitating further study and exploration. Feedstock type, temperature, and application rate all have a substantial impact on the physical and chemical properties of BDPC. A thorough investigation of the biomass-to-BDPC conversion technique is required. To maximize the structure and properties of BDPC, a good grasp of the fundamental features of this conversion process is required. Despite increasing interest in this topic, the precise response routes remain unknown. Tracking the evolution of functional groups during biomass conversion and explaining the production process of BDPC's porous structures should be prioritized. Some BDPC manufacturing costs are currently expensive, thus uniform BDPC production standards are required. Furthermore, kinetic investigations at both the microscopic and macroscopic levels are required to improve our understanding of the biomass-to-BDPC conversion.

The physicochemical structures of BDPC synthesized in situ can be adjusted by carefully considering preparation circumstances such as feedstock compositions, temperatures, and atmospheres. It is critical to first identify the influence mechanisms of these conditions and then investigate their interconnections. Furthermore, approaches such as in situ activation and pretreatment of biomass have shown potential effects for improving BDPC characteristics. Coordination of the BDPC manufacturing process with these technologies allows for further development of BDPC with different characteristics. The post-optimization to enhance the intrinsic value of BDPC via selective integration of heteroatoms, functional groups, and metal atoms into its structure can also refine its specific features. By further altering the surface and regulating pores, a special BDPPC could be designed. Furthermore, the development of novel BDPC matrix composites with increased surface area and specific functionalities using eco-friendly ingredients is a promising approach.

As a result of aging, BDPC's ability to adsorb pollutants decreases over time. Thus, the long-term consequences of adding BDPC to soil necessitate careful assessment and monitoring via extensive field studies done over time. These studies seek to better understand the processes that occur during long-term BDPC application, with a particular emphasis on finding and comprehending any negative environmental effects. Efforts should be made to investigate the adverse aspects of BDPC's impact on the environment, which will lead to the development of innovative approaches that optimize the benefits of BDPC additions. Major research has concentrated on carbon sequestration and soil fertility enhancement. More emphasis should be paid to techniques to optimize BDPC use in order to reduce CO_2 and other greenhouse gas emissions in our environment. To mitigate climate change, it is critical to reduce carbon dioxide pollutants in agricultural soil. These activities are critical for permitting future broad BDPC use with little environmental impact.

Further research should be conducted to improve the prediction of nutrient dynamics in BDPC-amended soils by strengthening kinetic models suitable to both field and laboratory situations. It is critical to understand the mechanisms that influence soil nutrient availability and fertility. The activation processes, as well as desorption and adsorption mechanisms for diverse pollutants, must be identified. The current state of knowledge about microbial communities and their distribution in BDPC-amended soil is limited, particularly concerning BDPC features like particle size, microporosity, pH, ion exchange capacity, and nutrient content. Despite an increasing focus on soil infertility and degradation, the application of BDPC could be beneficial in presenting new opportunities both in remediation as a source of macro and micronutrients in nutrient-lacking soils.

Further deep study on the influence of BDPC on soil ecosystems, which includes soil fauna, plants, and microbes, in their interrelated relationships. While current research on the impacts of BDPC on soil plants has advanced, there is a significant gap in the study of microbes and soil fauna. It is critical to investigate the micro-scale complexities of these components. A thorough analysis of the entire soil ecosystem has the potential to improve the practicality of large-scale BDPC application to soil, ensuring an in-depth awareness of its benefits across the ecosystem's different aspects. For large-scale applications, synergistic optimization of the BDPC synthesis method is required. Coordination of in-situ production and post-optimization is essential for producing BDPC with exceptional features. Reduce the environmental impact by minimizing the use of chemicals in the preparation process and investigating ecologically friendly activators and dopants. Furthermore, for feedback-based improvement of the preparation process, a techno-economic study of the full technical route is recommended.

7.8 CONCLUSION

Land deterioration is a significant concern for farmland soils. BDPC technology is consistent with modern green development ideas such as "low carbon, environmental protection, and sustainable development". BDPC technology is vital for regulating soil contamination, protecting the environment, and maintaining ecosystem and agricultural system balance. Because of its distinct physical and chemical properties, it has enormous potential to enhance soil properties. BDPC with a high carbon content is formed at high temperatures and anoxic conditions and is extremely aromatic. It has emerged as an effective soil amendment material due to its ability to reduce greenhouse gas emissions through carbon immobilization in soil structures. BDPC can be made from a variety of feedstocks with several functional groups, a large specific surface area, and porosity. It is also extensively modified via physical, chemical, biological, and other techniques for numerous purposes that help in improving its properties like enhancing the functional groups and pores. The application of BDPC has several beneficial effects, such as enhancing soil porosity, decreasing soil bulk density, raising soil temperature, optimizing soil hydraulic properties, and improving adsorption capacity. In addition to influencing chemical processes like material mobility and the cation exchange of carbon and nitrogen, the activation of soil enzymes encourages the growth and reproduction of soil organisms and microorganisms. Thereby improving the nutrients of the soil, enhancing agricultural yields, encouraging plant growth and development, and mitigating greenhouse gas emissions. However, it is important to be aware of any possible negative consequences and environmental risks related to BDPC. Heavy metal contamination of soil can happen, particularly if feedstocks contaminated with heavy metals are used in the production of BDPC. The preparation process, which is accompanied by high temperatures and oxygen-deficient conditions, may have an adverse effect on the environment and greenhouse gas emissions. Moreover, an excessive application of BDPC may cause adverse effects on soil microbial activity and crop yield. Thus, examining the detailed properties of BDPC before its application in order to determine how its intrinsic qualities interact with operating conditions, and its physiochemical properties as well as to determine whether it is appropriate for a specific purpose and to detect any contaminants or ecotoxicological characteristics. This chapter provides a comprehensive overview of BDPC, primarily emphasizing its uses in farmland restoration. The conversion of different biomass wastes into BDPC not only offers treatment options for biomasses but also promotes ecological sustainability. Moreover, the economic feasibility of biochar as a feedstock and its uncomplicated manufacturing procedure strengthen its practical application. To attain economic and ecological efficiency, it is essential to optimize the preparation, modification, and reprocessing of BDPC synergistically. In the realm of farmland restoration, BDPC stands out as a viable alternative to commercially activated carbon.

NOTE

* Corresponding author

REFERENCES

[1] Rattan Lal, Managing soils and ecosystems for mitigating anthropogenic carbon emissions and advancing global food security, BioScience, **2010,** 60, 708. DOI:10.1525/bio.2010.60.9.8.
[2] Pasquale Borrelli, David A. Robinson, Larissa R. Fleischer, Emanuele Lugato, Cristiano Ballabio, Christine Alewell, Katrin Meusburger, Sirio Modugno, Brigitta Schütt, Vito Ferro, Vincenzo Bagarello, Kristof Van Oost, Luca Montanarella, Panos Panagos, An assessment of the global impact of 21st century land use change on soil erosion, Nature Communications, **2017,** 8, 2013. DOI:10.1038/s41467-017-02142-7.
[3] H. Eswaran, R. Lal, P. F. Reich, Response to land degradation, **2019,** CRC Press.
[4] Tessa Sophia van der Voort, Frank Hagedorn, Cameron McIntyre, Claudia Zell, Lorenz Walthert, Patrick Schleppi, Xiaojuan Feng, Timothy Ian Eglinton, Variability in 14C contents of soil organic matter at the plot and regional scale across climatic and geologic gradients, Biogeosciences, **2016,** 13, 3427. DOI:10.5194/bg-13-3427-2016.
[5] Esteban G. Jobbágy, Robert B. Jackson, The vertical distribution of soil organic carbon and its relation to climate and vegetation, Ecological Applications, **2000,** 10, 423. DOI:10.1890/1051-0761(2000)010[0423:tvdoso]2.0.co;2.
[6] Rattan Lal, Sequestering carbon and increasing productivity by conservation agriculture, Journal of Soil and Water Conservation, **2015,** 70, 55A. DOI:10.2489/jswc.70.3.55A.
[7] Jonathan A. Foley, Ruth DeFries, Gregory P. Asner, Carol Barford, Gordon Bonan, Stephen R. Carpenter, F. Stuart Chapin, Michael T. Coe, Gretchen C. Daily, Holly K. Gibbs, Joseph H. Helkowski, Tracey Holloway, Erica A. Howard, Christopher J. Kucharik, Chad Monfreda, Jonathan A. Patz, I. Colin Prentice, Navin Ramankutty, Peter K. Snyder, Global consequences of land use, Science, **2005,** 309, 570. DOI:10.1126/science.1111772.

[8] Vandit Vijay, Paruchuri M. V. Subbarao, Ram Chandra, An evaluation on energy self–sufficiency model of a rural cluster through utilization of biomass residue resources: A case study in India, Energy and Climate Change, **2021,** 2, 100036. DOI:10.1016/j.egycc.2021.100036.

[9] Johannes Lehmann, Stephen Joseph, Biochar for environmental management: Science, technology and implementation, **2015,** Routledge.

[10] Simon Shackley, Saran Sohi, Rodrigo Ibarrola, Jim Hammond, Ondřej Mašek, Peter Brownsort, Andrew Cross, Miranda Prendergast-Miller, Stuart Haszeldine, Biochar, tool for climate change mitigation and soil management, Geoengineering Responses to Climate Change, **2013,** 73. DOI:10.1007/978-1-4419-0851-3.

[11] Virendra Kumar Vijay, Rimika Kapoor, Abhinav Trivedi, Vandit Vijay, Biogas as clean fuel for cooking and transportation needs in India, Advances in Bioprocess Technology, **2015,** 257. DOI:10.1007/978-3-319-17915-5_14.

[12] P. R. Yaashikaa, P. Senthil Kumar, Sunita Varjani, A. Saravanan, A critical review on the biochar production techniques, characterization, stability and applications for circular bioeconomy, Biotechnology Reports, **2020,** 28. DOI:10.1016/j.btre.2020.e00570.

[13] A. V. Bridgwater, Review of fast pyrolysis of biomass and product upgrading, Biomass and Bioenergy, **2012,** 38, 68. DOI:10.1016/j.biombioe.2011.01.048.

[14] Wei Cheng Ng, Siming You, Ran Ling, Karina Yew-Hoong Gin, Yanjun Dai, Chi-Hwa Wang, Co-gasification of woody biomass and chicken manure: Syngas production, biochar reutilization, and cost-benefit analysis, Energy, **2017,** 139, 732. DOI:10.1016/j.energy.2017.07.165.

[15] Haowei Yu, Weixin Zou, Jianjun Chen, Hao Chen, Zebin Yu, Jun Huang, Haoru Tang, Xiangying Wei, Bin Gao, Biochar amendment improves crop production in problem soils: A review, Journal of Environmental Management, **2019,** 232, 8. DOI:10.1016/j.jenvman.2018.10.117.

[16] Jianlong Wang, Shizong Wang, Preparation, modification and environmental application of biochar: A review, Journal of Cleaner Production, **2019,** 227, 1002. DOI:10.1016/j.jclepro.2019.04.282.

[17] Quan Chen, Xiangmei Cheng, ShengSheng Liu, Dichen Xia, Yan Liu, Zhen Zhang, Pengcheng Gu, Multi-heteroatom self-doped microporous biochar derived from fish scale for boosting uranium immobilization performance, Diamond and Related Materials, **2023,** 110052. DOI:10.1016/j.diamond.2023.110052.

[18] Md Abdullah Al Masud, Do Gun Kim, Won Sik Shin, Highly efficient degradation of phenolic compounds by Fe (II)-activated dual oxidant (persulfate/calcium peroxide) system, Chemosphere, **2022,** 299, 134392. DOI:10.1016/j.chemosphere.2022.134392.

[19] C. J. Barrow, Biochar: Potential for countering land degradation and for improving agriculture, Applied Geography, **2012,** 34, 21. DOI:10.1016/j.apgeog.2011.09.008.

[20] Behrouz Gholamahmadi, Simon Jeffery, Oscar Gonzalez-Pelayo, Sergio Alegre Prats, Ana Catarina Bastos, Jan Jacob Keizer, Frank G. A. Verheijen, Biochar impacts on runoff and soil erosion by water: A systematic global scale meta-analysis, Science of The Total Environment, **2023,** 871, 161860. DOI:10.1016/j.scitotenv.2023.161860.

[21] Tara Allohverdi, Amar Kumar Mohanty, Poritosh Roy, Manjusri Misra, A review on current status of biochar uses in agriculture, Molecules, **2021,** 26. DOI:10.3390/molecules26185584.

[22] François-Xavier Collard, Joël Blin, A review on pyrolysis of biomass constituents: Mechanisms and composition of the products obtained from the conversion of cellulose, hemicelluloses and lignin, Renewable and Sustainable Energy Reviews, **2014,** 38, 594. DOI:10.1016/j.rser.2014.06.013.

[23] Shruti Vikram, Pali Rosha, Sandeep Kumar, Recent modeling approaches to biomass pyrolysis: A review, Energy & Fuels, **2021,** 35, 7406. DOI:10.1021/acs.energyfuels.1c00251.

[24] Arun K. Vuppaladadiyam, Noemi Merayo, Angeles Blanco, Jingwei Hou, Dionysios D. Dionysiou, Ming Zhao, Simulation study on comparison of algal treatment to conventional

biological processes for greywater treatment, Algal Research, **2018**, 35, 106. DOI:10.1016/j. algal.2018.08.021.

[25] Guangcan Su, Hwai Chyuan Ong, Yong Yang Gan, Wei-Hsin Chen, Cheng Tung Chong, Yong Sik Ok, Co-pyrolysis of microalgae and other biomass wastes for the production of high-quality bio-oil: Progress and prospective, Bioresource Technology, **2022**, 344, 126096. DOI:10.1016/j.biortech.2021.126096.

[26] Antonio Tursi, A review on biomass: Importance, chemistry, classification, and conversion, Biofuel Research Journal, **2019**, 6, 962. DOI:10.18331/brj2019.6.2.3.

[27] Vaibhav Dhyani, Thallada Bhaskar, A comprehensive review on the pyrolysis of lignocellulosic biomass, Renewable Energy, **2018**, 129, 695. DOI:10.1016/j.renene.2017.04.035.

[28] Yingkai Li, Yichen Wang, Meiyun Chai, Chong Li, Nishu, Dominic Yellezuome, Ronghou Liu, Pyrolysis kinetics and thermodynamic parameters of bamboo residues and its three main components using thermogravimetric analysis, Biomass and Bioenergy, **2023**, 170. DOI:10.1016/j.biombioe.2023.106705.

[29] Jitendra Kumar Saini, Reetu Saini, Lakshmi Tewari, Lignocellulosic agriculture wastes as biomass feedstocks for second-generation bioethanol production: Concepts and recent developments, 3 Biotech, **2015**, 5, 337. DOI:10.1007/s13205-014-0246-5.

[30] A. Abdolali, W. S. Guo, H. H. Ngo, S. S. Chen, N. C. Nguyen, K. L. Tung, Typical lignocellulosic wastes and by-products for biosorption process in water and wastewater treatment: A critical review, Bioresource Technology, **2014**, 160, 57. DOI:10.1016/j.biortech.2013.12.037.

[31] Valeriia Chemerys, Edita Baltrėnaitė, a review of lignocellulosic biochar modification towards enhanced biochar selectivity and adsorption capacity of potentially toxic elements, Ukrainian Journal of Ecology, **2018**, 8, 21. DOI:10.15421/2017_183.

[32] Jinju Hou, Xiaotong Zhang, Shujia Liu, Shudong Zhang, Qiuzhuo Zhang, A critical review on bioethanol and biochar production from lignocellulosic biomass and their combined application in generation of high-value byproducts, Energy Technology, **2020**, 8, 2000025. DOI:10.1002/ente.202000025.

[33] S. Rangabhashiyam, P. Balasubramanian, The potential of lignocellulosic biomass precursors for biochar production: Performance, mechanism and wastewater application-a review, Industrial Crops and Products, **2019**, 128, 405. DOI:10.1016/j.indcrop.2018.11.041.

[34] Xuebing Zhao, Lihua Zhang, Dehua Liu, Biomass recalcitrance: Part I: The chemical compositions and physical structures affecting the enzymatic hydrolysis of lignocellulose, Biofuels, Bioproducts and Biorefining, **2012**, 6, 465. DOI:10.1002/bbb.1331.

[35] Nishu, Chong Li, Dominic Yellezuome, Yingkai Li, Ronghou Liu, Catalytic pyrolysis of rice straw for high yield of aromatics over modified ZSM-5 catalysts and its kinetics, Renewable Energy, **2023**, 209, 569. DOI:10.1016/j.renene.2023.04.025.

[36] Ottmar Edenhofer, Ramón Pichs-Madruga, Youba Sokona, Kristin Seyboth, Susanne Kadner, Timm Zwickel, Patrick Eickemeier, Gerrit Hansen, Steffen Schlömer, Christoph von Stechow, Patrick Matschoss, Renewable energy sources and climate change mitigation, **2011**, Cambridge University Press.

[37] Stanislav V. Vassilev, David Baxter, Lars K. Andersen, Christina G. Vassileva, Trevor J. Morgan, An overview of the organic and inorganic phase composition of biomass, Fuel, **2012**, 94, 1. DOI:10.1016/j.fuel.2011.09.030.

[38] N. Wid, N. J. Horan, Anaerobic digestion of screenings for biogas recovery, Anaerobic Digestion Processes: Applications and Effluent Treatment, **2018**, 85. DOI:10.1007/978-981-10-8129-3_6.

[39] Stanislav V. Vassilev, Christina G. Vassileva, Composition, properties and challenges of algae biomass for biofuel application: An overview, Fuel, **2016**, 181, 1. DOI:10.1016/j. fuel.2016.04.106.

[40] Javad Gharechahi, Mohammad Farhad Vahidi, Mohammad Bahram, Jianlin Han, Xuezhi Ding, Ghasem Hosseini Salekdeh, Metagenomic analysis reveals a dynamic microbiome with

diversified adaptive functions to utilize high lignocellulosic forages in the cattle rumen, The ISME Journal, **2021,** 15, 1108. DOI:10.1038/s41396-020-00837-2.

[41] Bhupendra Koul, Mohammad Yakoob, Maulin P. Shah, Agricultural waste management strategies for environmental sustainability, Environmental Research, **2022,** 206, 112285. DOI:10.1016/j.envres.2021.112285.

[42] Leonidas Matsakas, Qiuju Gao, Stina Jansson, Ulrika Rova, Paul Christakopoulos, Green conversion of municipal solid wastes into fuels and chemicals, Electronic Journal of Biotechnology, **2017,** 26, 69. DOI:10.1016/j.ejbt.2017.01.004.

[43] Meenal Gupta, Nishit Savla, Chetan Pandit, Soumya Pandit, Piyush Kumar Gupta, Manu Pant, Santimoy Khilari, Yogesh Kumar, Daksh Agarwal, Remya R. Nair, Dessy Thomas, Vijay Kumar Thakur, Use of biomass-derived biochar in wastewater treatment and power production: A promising solution for a sustainable environment, Science of the Total Environment, **2022,** 825, 153892. DOI:10.1016/j.scitotenv.2022.153892.

[44] Aik Chong Lua, Ting Yang, Jia Guo, Effects of pyrolysis conditions on the properties of activated carbons prepared from pistachio-nut shells, Journal of Analytical and Applied Pyrolysis, **2004,** 72, 279. DOI:10.1016/j.jaap.2004.08.001.

[45] Feng Cheng, Xiuwei Li, Preparation and application of biochar-based catalysts for biofuel production, Catalysts, **2018,** 8. DOI:10.3390/catal8090346.

[46] Harpreet Singh Kambo, Animesh Dutta, A comparative review of biochar and hydrochar in terms of production, physico-chemical properties and applications, Renewable and Sustainable Energy Reviews, **2015,** 45, 359. DOI:10.1016/j.rser.2015.01.050.

[47] Nur Salsabila Kamarudin, Farrah Aini Dahalan, Masitah Hasan, Ong Soon An, Nor Azizah Parmin, Naimah Ibrahim, Myzairah Hamdzah, Nor Azimah Mohd Zain, Khalida Muda, Edza Aria Wikurendra, Biochar: A review of its history, characteristics, factors that influence its yield, methods of production, application in wastewater treatment and recent development, Biointerface Research in Applied Chemistry **2022,** 12, 7914. DOI:10.33263/briac126.79147926.

[48] Bruno Glaser, Jago Jonathan Birk, State of the scientific knowledge on properties and genesis of anthropogenic dark earths in Central Amazonia (terra preta de Índio), Geochimica et Cosmochimica Acta, **2012,** 82, 39. DOI:10.1016/j.gca.2010.11.029.

[49] Bruno Glaser, Ludwig Haumaier, Georg Guggenberger, Wolfgang Zech, The "Terra Preta" phenomenon: A model for sustainable agriculture in the humid tropics, The Science of Nature, **2001,** 88, 37. DOI:10.1007/s001140000193.

[50] Lijian Leng, Qin Xiong, Lihong Yang, Hui Li, Yaoyu Zhou, Weijin Zhang, Shaojian Jiang, Hailong Li, Huajun Huang, An overview on engineering the surface area and porosity of biochar, Science of the Total Environment, **2021,** 763, 144204. DOI:10.1016/j.scitotenv.2020.144204.

[51] Yaning Zhang, Yunlei Cui, Paul Chen, Shiyu Liu, Nan Zhou, Kuan Ding, Liangliang Fan, Peng Peng, Min Min, Yanling Cheng, Sustainable resource recovery and zero waste approaches, **2019,** Elsevier.

[52] Axel Funke, Felix Ziegler, Hydrothermal carbonization of biomass: A summary and discussion of chemical mechanisms for process engineering, Biofuels, Bioproducts and Biorefining, **2010,** 4, 160. DOI:10.1002/bbb.198.

[53] Naomi B. Klinghoffer, Marco J. Castaldi, Ange Nzihou, Influence of char composition and inorganics on catalytic activity of char from biomass gasification, Fuel, **2015,** 157, 37. DOI:10.1016/j.fuel.2015.04.036.

[54] P. C. A. Bergman, A. R. Boersma, R. W. R. Zwart, J. H. A. Kiel, Torrefaction for biomass co-firing in existing coal-fired power stations "Biocoal", **2005,** ECN Publications.

[55] Genyr Kappler, Débora Machado de Souza, Carlos Alberto Mendes Moraes, Regina Célia Espinosa Modolo, Feliciane Andrade Brehm, Paulo Roberto Wander, Luís António da Cruz Tarelho, Conversion of lignocellulosic biomass through pyrolysis to promote a sustainable

value chain for Brazilian agribusiness, Lignocellulosic Biorefining Technologies, **2020,** 265. DOI:10.1002/9781119568858.ch12.

[56] Caecilia R. Vitasari, G. W. Meindersma, André B. de Haan, Water extraction of pyrolysis oil: The first step for the recovery of renewable chemicals, Bioresource Technology, **2011,** 102, 7204. DOI:10.1016/j.biortech.2011.04.079.

[57] Arun Krishna Vuppaladadiyam, Sai Sree Varsha Vuppaladadiyam, Vineet Singh Sikarwar, Ejaz Ahmad, Kamal K. Pant, Murugavelh S, Ashish Pandey, Sankar Bhattacharya, Ajit Sarmah, Shao-Yuan Leu, A critical review on biomass pyrolysis: Reaction mechanisms, process modeling and potential challenges, Journal of the Energy Institute, **2023,** 108. DOI:10.1016/j.joei.2023.101236.

[58] Govindarajan Venkatesh, Kodigal A. Gopinath, Kotha Sammi Reddy, Baddigam Sanjeeva Reddy, Mathyam Prabhakar, Cherukumalli Srinivasarao, Venugopalan Visha Kumari, Vinod Kumar Singh, Characterization of biochar derived from crop residues for soil amendment, carbon sequestration and energy use, Sustainability, **2022,** 14, 2295. DOI:10.3390/su14042295.

[59] Duu-Jong Lee, Jia-Shun Lu, Jo-Shu Chang, Pyrolysis synergy of Municipal Solid Waste (MSW): A review, Bioresource Technology, **2020,** 318, 123912. DOI:10.1016/j.biortech.2020.123912.

[60] Hao Zheng, Zhenyu Wang, Xia Deng, Jian Zhao, Ye Luo, Jeff Novak, Stephen Herbert, Baoshan Xing, Characteristics and nutrient values of biochars produced from giant reed at different temperatures, Bioresource Technology, **2013,** 130, 463. DOI:10.1016/j.biortech.2012.12.044.

[61] Mariana Sbizzaro, Silvio Cesar Sampaio, Ralpho Rinaldo dos Reis, Francielle de Assis Beraldi, Danielle Medina Rosa, Claudia Maria Branco de Freitas Maia, Cleuciane Tillvitz do Nascimento, Edson Antonio da Silva, Carlos Eduardo Borba, Effect of production temperature in biochar properties from bamboo culm and its influences on atrazine adsorption from aqueous systems, Journal of Molecular Liquids, **2021,** 343, 117667. DOI:10.1016/j.molliq.2021.117667.

[62] Minori Uchimiya, Lynda H. Wartelle, Isabel M. Lima, K. Thomas Klasson, Sorption of deisopropylatrazine on broiler litter biochars, Journal of Agricultural and Food Chemistry, **2010,** 58, 12350. DOI:10.1021/jf102152q.

[63] Mahtab Ahmad, Deok Hyun Moon, Meththika Vithanage, Agamemnon Koutsospyros, Sang Soo Lee, Jae E. Yang, Sung Eun Lee, Choong Jeon, Yong Sik Ok, Production and use of biochar from buffalo-weed (Ambrosia trifida L.) for trichloroethylene removal from water, Journal of Chemical Technology & Biotechnology, **2014,** 89, 150. DOI:10.1002/jctb.4157.

[64] Xianjun Xing, Fangyu Fan, Wen Jiang, Characteristics of biochar pellets from corn straw under different pyrolysis temperatures, Royal Society Open Science, **2018,** 5, 172346. DOI:10.1098/rsos.172346.

[65] Jie Cheng, Sheng-Chun Hu, Guo-Tao Sun, Zeng-Chao Geng, Ming-Qiang Zhu, The effect of pyrolysis temperature on the characteristics of biochar, pyroligneous acids, and gas prepared from cotton stalk through a polygeneration process, Industrial Crops and Products, **2021,** 170, 113690. DOI:10.1016/j.indcrop.2021.113690.

[66] Li Qiu, Chao Li, Shu Zhang, Shuang Wang, Bin Li, Zhenhua Cui, Yonggui Tang, Xun Hu, Distinct property of biochar from pyrolysis of poplar wood, bark, and leaves of the same origin, Industrial Crops and Products, **2023,** 202, 117001. DOI:10.1016/j.indcrop.2023.117001.

[67] Feiyue Li, Xuan Wu, Wenchao Ji, Xiangyang Gui, Yihan Chen, Jianrong Zhao, Chunhuo Zhou, Tianbao Ren, Effects of pyrolysis temperature on properties of swine manure biochar and its environmental risks of heavy metals, Journal of Analytical and Applied Pyrolysis, **2020,** 152, 104945. DOI:10.1016/j.jaap.2020.104945.

[68] Divine Damertey Sewu, Hwansoo Jung, Seung Soo Kim, Dae Sung Lee, Seung Han Woo, decolorization of cationic and anionic dye-laden wastewater by steam-activated biochar produced at an industrial-scale from spent mushroom substrate, Bioresource Technology, **2019,** 277, 77. DOI:10.1016/j.biortech.2019.01.034.

[69] Taeyong Shim, Jisu Yoo, Changkook Ryu, Yong-Kwon Park, Jinho Jung, Effect of steam activation of biochar produced from a giant Miscanthus on copper sorption and toxicity, Bioresource Technology, **2015,** 197, 85. DOI:10.1016/j.biortech.2015.08.055.

[70] Jianghong Zhang, Bing Huang, Liang Chen, Yang Li, Wei Li, Zhuanxi Luo, Characteristics of biochar produced from yak manure at different pyrolysis temperatures and its effects on the yield and growth of highland barley, Environmental Pollutants and Bioavailability, **2018,** 30, 57. DOI:10.1080/09542299.2018.1487774.

[71] Yunchao Li, Bo Xing, Yan Ding, Xinhong Han, Shurong Wang, A critical review of the production and advanced utilization of biochar via selective pyrolysis of lignocellulosic biomass, Bioresource Technology, **2020,** 312, 123614. DOI:10.1016/j.biortech.2020.123614.

[72] Sohrab Haghighi Mood, Manuel Raul Pelaez-Samaniego, Manuel Garcia-Perez, Perspectives of engineered biochar for environmental applications: A review, Energy & Fuels, **2022,** 36, 7940. DOI:10.1021/acs.energyfuels.2c01201.

[73] Farah Amalina, Abdul Syukor Abd Razak, Santhana Krishnan, Haspina Sulaiman, A. W. Zularisam, Mohd Nasrullah, Advanced techniques in the production of biochar from lignocellulosic biomass and environmental applications, Cleaner Materials, **2022,** 6. DOI:10.1016/j. clema.2022.100137.

[74] Amirreza Talaiekhozani, Shahabaldin Rezania, Ki-Hyun Kim, Reza Sanaye, Ali Mohammad Amani, Recent advances in photocatalytic removal of organic and inorganic pollutants in air, Journal of Cleaner Production, **2021,** 278, 123895. DOI:10.1016/j.jclepro.2020.123895.

[75] Vaishakh Nair, R. Vinu, Peroxide-assisted microwave activation of pyrolysis char for adsorption of dyes from wastewater, Bioresource Technology, **2016,** 216, 511. DOI:10.1016/j. biortech.2016.05.070.

[76] Klaudia Czerwińska, Maciej Śliz, Małgorzata Wilk, Hydrothermal carbonization process: Fundamentals, main parameter characteristics and possible applications including an effective method of SARS-CoV-2 mitigation in sewage sludge: A review, Renewable and Sustainable Energy Reviews, **2022,** 154, 111873. DOI:10.1016/j.rser.2021.111873.

[77] Fatma Mbarki, Taher Selmi, Aida Kesraoui, Mongi Seffen, Philippe Gadonneix, Alain Celzard, Vanessa Fierro, Hydrothermal pre-treatment, an efficient tool to improve activated carbon performances, Industrial Crops and Products, **2019,** 140, 111717. DOI:10.1016/j. indcrop.2019.111717.

[78] Qian Lei, Shrikalaa Kannan, Vijaya Raghavan, Microwave hydrothermal carbonization of orange peel waste, Waste Management, **2018,** 126, 106. DOI:10.1016/j.wasman.2021.02.058.

[79] Yi Wei, Wei Chen, Chuanfu Liu, Huihui Wang, Facial synthesis of adsorbent from hemicelluloses for Cr(VI) adsorption, Molecules, **2021,** 26, 1443. DOI:10.3390/molecules26051443.

[80] Luguang Wang, Ye Chen, Fei Long, Lakhveer Singh, Stephanie Trujillo, Xiang Xiao, Hong Liu, Breaking the loop: Tackling homoacetogenesis by chloroform to halt hydrogen production-consumption loop in single chamber microbial electrolysis cells, Chemical Engineering Journal, **2020,** 389, 124436. DOI:10.1016/j.cej.2020.124436.

[81] Ghizlane Enaime, Abdelaziz Baçaoui, Abdelrani Yaacoubi, Manfred Lübken, Biochar for wastewater treatment—Conversion technologies and applications, Applied Sciences, **2020,** 10, 3492. DOI:10.3390/app10103492.

[82] Anil Kumar Sakhiya, Paramjeet Baghel, Abhijeet Anand, Virendra Kumar Vijay, Priyanka Kaushal, A comparative study of physical and chemical activation of rice straw derived biochar to enhance Zn^{+2} adsorption, Bioresource Technology Reports, **2021,** 15, 100774. DOI:10.1016/j.biteb.2021.100774.

[83] Md Shahinoor Islam, Jin-Hyeob Kwak, Christopher Nzediegwu, Siyuan Wang, Kumuduni Palansuriya, Eilhann E. Kwon, M. Anne Naeth, Mohamed Gamal El-Din, Yong Sik Ok, Scott X. Chang, Biochar heavy metal removal in aqueous solution depends on feedstock type and pyrolysis purging gas, Environmental Pollution, **2021,** 281, 117094. DOI:10.1016/j. envpol.2021.117094.

[84] Yaoyao Cao, Weihua Xiao, Guanghui Shen, Guanya Ji, Yang Zhang, Chongfeng Gao, Lujia Han, Carbonization and ball milling on the enhancement of Pb(II) adsorption by wheat straw: Competitive effects of ion exchange and precipitation, Bioresource Technology, **2019**, 273, 70. DOI:10.1016/j.biortech.2018.10.065.

[85] Hyun Min Jang, Eunsung Kan, Engineered biochar from agricultural waste for removal of tetracycline in water, Bioresource Technology, **2019**, 284, 437. DOI:10.1016/j.biortech.2019.03.131.

[86] Xiaoming Peng, Fengping Hu, Tao Zhang, Fengxian Qiu, Hongling Dai, Amine-functionalized magnetic bamboo-based activated carbon adsorptive removal of ciprofloxacin and norfloxacin: A batch and fixed-bed column study, Bioresource Technology, **2018**, 249, 924. DOI:10.1016/j.biortech.2017.10.095.

[87] Wenya Ao, Jie Fu, Xiao Mao, Qinhao Kang, Chunmei Ran, Yang Liu, Hedong Zhang, Zuopeng Gao, Jing Li, Guangqing Liu, Jianjun Dai, Microwave assisted preparation of activated carbon from biomass: A review, Renewable and Sustainable Energy Reviews, **2018**, 92, 958. DOI:10.1016/j.rser.2018.04.051.

[88] Mathew L. Frankel, Tazul I. Bhuiyan, Andrei Veksha, Marc A. Demeter, David B. Layzell, Robert J. Helleur, Josephine M. Hill, Raymond J. Turner, Removal and biodegradation of naphthenic acids by biochar and attached environmental biofilms in the presence of co-contaminating metals, Bioresource Technology, **2016**, 216, 352. DOI:10.1016/j.biortech.2016.05.084.

[89] Ying Shen, Huan Li, Wenzhe Zhu, Shih-Hsin Ho, Wenqiao Yuan, Jianfeng Chen, Youping Xie, Microalgal-biochar immobilized complex: A novel efficient biosorbent for cadmium removal from aqueous solution, Bioresource Technology, **2017**, 244, 1031. DOI:10.1016/j.biortech.2017.08.085.

[90] Honghong Lyu, Bin Gao, Feng He, Andrew R. Zimmerman, Cheng Ding, Jingchun Tang, John C. Crittenden, Experimental and modeling investigations of ball-milled biochar for the removal of aqueous methylene blue, Chemical Engineering Journal, **2018**, 335, 110. DOI:10.1016/j.cej.2017.10.130.

[91] Tan Xing, Jaka Sunarso, Wenrong Yang, Yongbai Yin, Alexey M. Glushenkov, Lu Hua Li, Patrick C. Howlett, Ying Chen, Ball milling: A green mechanochemical approach for synthesis of nitrogen doped carbon nanoparticles, Nanoscale, **2013**, 5, 7970. DOI:10.1039/C3NR02328A.

[92] Steven C. Peterson, Michael A. Jackson, Sanghoon Kim, Debra E. Palmquist, Increasing biochar surface area: Optimization of ball milling parameters, Powder Technology, **2012**, 228, 115. DOI:10.1016/j.powtec.2012.05.005.

[93] Huimei Cai, Lingyun Xu, Guijie Chen, Chuanyi Peng, Fei Ke, Zhengquan Liu, Daxiang Li, Zhengzhu Zhang, Xiaochun Wan, Removal of fluoride from drinking water using modified ultrafine tea powder processed using a ball-mill, Applied Surface Science, **2016**, 375, 74. DOI:10.1016/j.apsusc.2016.03.005.

[94] Harjeet Nath, Biswajit Sarkar, Sudip Mitra, Sachin Bhaladhare, Biochar from biomass: A review on biochar preparation its modification and impact on soil including soil microbiology, Geomicrobiology Journal, **2022**, 39, 373. DOI:10.1080/01490451.2022.2028942.

[95] Agnieszka Tomczyk, Zofia Sokołowska, Patrycja Boguta, Biochar physicochemical properties: Pyrolysis temperature and feedstock kind effects, Reviews in Environmental Science and Bio/Technology, **2020**, 19, 191. DOI:10.1007/s11157-020-09523-3.

[96] P. R. Yaashikaa, P. Senthil Kumar, Sunita J. Varjani, A. Saravanan, Advances in production and application of biochar from lignocellulosic feedstocks for remediation of environmental pollutants, Bioresource Technology, **2019**, 292, 122030. DOI:10.1016/j.biortech.2019.122030.

[97] Jiawen Wu, Tao Wang, Yongsheng Zhang, Wei-Ping Pan, The distribution of Pb(II)/Cd(II) adsorption mechanisms on biochars from aqueous solution: Considering the increased oxygen functional groups by HCl treatment, Bioresource Technology, **2019**, 291, 121859. DOI:10.1016/j.biortech.2019.121859.

[98] Huayi Chen, Wenyan Li, Jinjin Wang, Huijuan Xu, Yonglin Liu, Zhen Zhang, Yongtao Li, Yulong Zhang, Adsorption of cadmium and lead ions by phosphoric acid-modified biochar generated from chicken feather: Selective adsorption and influence of dissolved organic matter, Bioresource Technology, **2019**, 292, 121948. DOI:10.1016/j.biortech.2019.121948.

[99] Ling Zhao, Wei Zheng, Ondřej Mašek, Xiang Chen, Bowen Gu, Brajendra K. Sharma, Xinde Cao, Roles of phosphoric acid in biochar formation: Synchronously improving carbon retention and sorption capacity, Journal of Environmental Quality, **2017**, 46, 393. DOI:10.2134/jeq2016.09.0344.

[100] Anushka Upamali Rajapaksha, Season S. Chen, Daniel C. W. Tsang, Ming Zhang, Meththika Vithanage, Sanchita Mandal, Bin Gao, Nanthi S. Bolan, Yong Sik Ok, Engineered/designer biochar for contaminant removal/immobilization from soil and water: Potential and implication of biochar modification, Chemosphere, **2016**, 148, 276. DOI:10.1016/j.chemosphere.2016.01.043.

[101] Xiaofei Tan, Shaobo Liu, Yunguo Liu, Yanling Gu, Guangming Zeng, Xinjiang Hu, Xin Wang, Shaoheng Liu, Luhua Jiang, Biochar as potential sustainable precursors for activated carbon production: Multiple applications in environmental protection and energy storage, Bioresource Technology, **2017**, 227, 359. DOI:10.1016/j.biortech.2016.12.083.

[102] Zhijie Bao, Chunzhen Shi, Wenying Tu, Lijiao Li, Qiang Li, Recent developments in modification of biochar and its application in soil pollution control and ecoregulation, Environmental Pollution, **2022**, 313, 120184. DOI:10.1016/j.envpol.2022.120184.

[103] Yohan Richardson, Joël Blin, Ghislaine Volle, Julius Motuzas, Anne Julbe, In situ generation of Ni metal nanoparticles as catalyst for H_2 rich syngas production from biomass gasification, Applied Catalysis A: General, **2010**, 382, 220. DOI:10.1016/j.apcata.2010.04.047.

[104] Xiaoming Sun, Yadong Li, Colloidal carbon spheres and their core/shell structures with noble-metal nanoparticles, Angewandte Chemie International Edition, **2004**, 43, 597. DOI:10.1002/anie.200352386.

[105] Yafei Shen, Peitao Zhao, Qinfu Shao, Dachao Ma, Fumitake Takahashi, Kunio Yoshikawa, In-situ catalytic conversion of tar using rice husk char-supported nickel-iron catalysts for biomass pyrolysis/gasification, Applied Catalysis B: Environmental, **2014**, 152, 140. DOI:10.1016/j.apcatb.2014.01.032.

[106] Dingding Yao, Qiang Hu, Daqian Wang, Haiping Yang, Chunfei Wu, Xianhua Wang, Hanping Chen, Hydrogen production from biomass gasification using biochar as a catalyst/support, Bioresource Technology, **2016**, 216, 159. DOI:10.1016/j.biortech.2016.05.011.

[107] Lijian Leng, Siyu Xu, Renfeng Liu, Ting Yu, Ximeng Zhuo, Songqi Leng, Qin Xiong, Huajun Huang, Nitrogen containing functional groups of biochar: An overview, Bioresource Technology, **2020**, 298, 122286. DOI:10.1016/j.biortech.2019.122286.

[108] Edris Gavili, Ali Akbar Moosavi, Farzad Moradi Choghamarani, Cattle manure biochar potential for ameliorating soil physical characteristics and spinach response under drought, Archives of Agronomy and Soil Science, **2018**, 64, 1714. DOI:10.1080/03650340.2018.1453925.

[109] A. Abukari, A. Abunyewa, H. Issifu, Effect of rice husk biochar on nitrogen uptake and grain yield of maize in the guinea savanna zone of Ghana, UDS International Journal of Development, **2018**, 5, 1. DOI:10.47740/290.UDSIJD6i.

[110] Mahtab Ahmad, Sang Soo Lee, Xiaomin Dou, Dinesh Mohan, Jwa-Kyung Sung, Jae E. Yang, Yong Sik Ok, Effects of pyrolysis temperature on soybean stover- and peanut shell-derived biochar properties and TCE adsorption in water, Bioresource Technology, **2012**, 118, 536. DOI:10.1016/j.biortech.2012.05.042.

[111] K. Jindo, H. Mizumoto, Y. Sawada, M. A. Sanchez-Monedero, T. Sonoki, Physical and chemical characterization of biochars derived from different agricultural residues, Biogeosciences, **2014**, 11, 6613. DOI:10.5194/bg-11-6613-2014.

[112] Huanliang Lu, Weihua Zhang, Shizhong Wang, Luwen Zhuang, Yuxi Yang, Rongliang Qiu, Characterization of sewage sludge-derived biochars from different feedstocks and pyrolysis

temperatures, Journal of Analytical and Applied Pyrolysis, **2013**, 102, 137. DOI:10.1016/j. jaap.2013.03.004.

[113] Hicham Zeghioud, Lydia Fryda, Hayet Djelal, Aymen Assadi, Abdoulaye Kane, A comprehensive review of biochar in removal of organic pollutants from wastewater: Characterization, toxicity, activation/functionalization and influencing treatment factors, Journal of Water Process Engineering, **2022,** 47. DOI:10.1016/j.jwpe.2022.102801.

[114] Muhammad Khalid Rafiq, Robert Thomas Bachmann, Muhammad Tariq Rafiq, Zhanhuan Shang, Stephen Joseph, Ruijun Long, Influence of pyrolysis temperature on physico-chemical properties of corn stover (Zea mays L.) biochar and feasibility for carbon capture and energy balance, PLoS One, **2016**, 11, e0156894. DOI:10.1371/journal.pone.0156894.

[115] The World Congress on Advances in Civil, Environmental, and Materials Research, Characterization and evaluation of biochars derived from agricultural waste biomasses from Gansu, China, **2014**. Available from www.i-asem.org/publication_conf/acem14/4.EST/ T4D.3.ES301_285F.pdf.

[116] Obemah D. Nartey, Baowei Zhao, Biochar preparation, characterization, and adsorptive capacity and its effect on bioavailability of contaminants: An overview, Advances in Materials Science and Engineering, **2014**, 2014, 1. DOI:10.1155/2014/715398.

[117] Sonja Schimmelpfennig, Bruno Glaser, One step forward toward characterization: Some important material properties to distinguish biochars, Journal of Environmental Quality, **2012**, 41, 1001. DOI:10.2134/jeq2011.0146.

[118] Caidi Yang, Jingjing Liu, Shenggao Lu, Pyrolysis temperature affects pore characteristics of rice straw and canola stalk biochars and biochar-amended soils, Geoderma, **2021**, 397, 115097. DOI:10.1016/j.geoderma.2021.115097.

[119] A. Shaaban, Sian-Meng Se, M. F. Dimin, Jariah M. Juoi, Mohd Haizal Mohd Husin, Nona Merry M. Mitan, Influence of heating temperature and holding time on biochars derived from rubber wood sawdust via slow pyrolysis, Journal of Analytical and Applied Pyrolysis, **2014**, 107, 31. DOI:10.1016/j.jaap.2014.01.021.

[120] Ammal Abukari, James Seutra Kaba, Evans Dawoe, Akwasi Adutwum Abunyewa, A comprehensive review of the effects of biochar on soil physicochemical properties and crop productivity, Waste Disposal & Sustainable Energy, **2022**, 4, 343. DOI:10.1007/s42768-022-00114-2.

[121] Si Chen, Chaoxian Qin, Teng Wang, Fangyuan Chen, Xuli Li, Haobo Hou, Min Zhou, Study on the adsorption of dyestuffs with different properties by sludge-rice husk biochar: Adsorption capacity, isotherm, kinetic, thermodynamics and mechanism, Journal of Molecular Liquids, **2019**, 285, 62. DOI:10.1016/j.molliq.2019.04.035.

[122] Nan Zhou, Honggang Chen, Junting Xi, Denghui Yao, Zhi Zhou, Yun Tian, Xiangyang Lu, Biochars with excellent Pb(II) adsorption property produced from fresh and dehydrated banana peels via hydrothermal carbonization, Bioresource Technology, **2017**, 232, 204. DOI:10.1016/j.biortech.2017.01.074.

[123] Françoise Rouquerol, Jean Rouquerol, Kenneth S. W. Sing, 2-thermodynamics of adsorption at the gas/solid interface, Adsorption by Powders and Porous Solids (Second Edition), **2014,** 25. DOI:10.1016/B978-0-08-097035-6.00002-4.

[124] Yingquan Chen, Haiping Yang, Xianhua Wang, Shihong Zhang, Hanping Chen, Biomass-based pyrolytic polygeneration system on cotton stalk pyrolysis: Influence of temperature, Bioresource Technology, **2012**, 107, 411. DOI:10.1016/j.biortech.2011.10.074.

[125] Nanthi Bolan, Son A. Hoang, Jingzi Beiyuan, Souradeep Gupta, Deyi Hou, Ajay Karakoti, Stephen Joseph, Sungyup Jung, Ki-Hyun Kim, M. B. Kirkham, Harn Wei Kua, Manish Kumar, Eilhann E. Kwon, Yong Sik Ok, Vishma Perera, Jörg Rinklebe, Sabry M. Shaheen, Binoy Sarkar, Ajit K. Sarmah, Bhupinder Pal Singh, Gurwinder Singh, Daniel C. W. Tsang, Kumar Vikrant, Meththika Vithanage, Ajayan Vinu, Hailong Wang, Hasintha Wijesekara, Yubo Yan, Sherif A. Younis, Lukas Van Zwieten, Multifunctional applications of biochar beyond carbon storage, International Materials Reviews, **2021,** 67, 150. DOI:10.1080/09506608.2021.1922047.

[126] Mahmood Laghari, Ravi Naidu, Bo Xiao, Zhiquan Hu, Muhammad Saffar Mirjat, Mian Hu, Muhammad Nawaz Kandhro, Zhihua Chen, Dabin Guo, Qamardudin Jogi, Zaidun Naji Abudi, Saima Fazal, Recent developments in biochar as an effective tool for agricultural soil management: A review, Journal of the Science of Food and Agriculture, **2016,** 96, 4840. DOI:10.1002/jsfa.7753.

[127] Riya Chatterjee, Baharak Sajjadi, Wei-Yin Chen, Daniell L. Mattern, Nathan Hammer, Vijayasankar Raman, Austin Dorris, Effect of pyrolysis temperature on physicochemical properties and acoustic-based amination of biochar for efficient CO_2 adsorption, Frontiers in Energy Research, **2020,** 8. DOI:10.3389/fenrg.2020.00085.

[128] Wan Azlina Wan Abdul Karim Ghani, Ayaz Mohd, Gabriel da Silva, Robert T. Bachmann, Yun H. Taufiq-Yap, Umer Rashid, Ala'a H. Al-Muhtaseb, Biochar production from waste rubber-wood-sawdust and its potential use in C sequestration: Chemical and physical characterization, Industrial Crops and Products, **2013,** 44, 18. DOI:10.1016/j.indcrop.2012.10.017.

[129] Anurita Selvarajoo, Dooshyantsingh Oochit, Effect of pyrolysis temperature on product yields of palm fibre and its biochar characteristics, Materials Science for Energy Technologies, **2020,** 3, 575. DOI:10.1016/j.mset.2020.06.003.

[130] Zainab Mahdi, Ali El Hanandeh, Qiming Yu, Influence of pyrolysis conditions on surface characteristics and methylene blue adsorption of biochar derived from date seed biomass, Waste and Biomass Valorization, **2017,** 8, 2061. DOI:10.1007/s12649-016-9714-y.

[131] Kwang Ho Kim, Jae-Young Kim, Tae-Su Cho, Joon Weon Choi, Influence of pyrolysis temperature on physicochemical properties of biochar obtained from the fast pyrolysis of pitch pine (Pinus rigida), Bioresource Technology, **2012,** 118, 158. DOI:10.1186/s40643-022-00618-z.

[132] Fungai N. D. Mukome, Xiaoming Zhang, Lucas C. R. Silva, Johan Six, Sanjai J. Parikh, Use of chemical and physical characteristics to investigate trends in biochar feedstocks, Journal of Agricultural and Food Chemistry, **2013,** 61, 2196. DOI:10.1021/jf3049142.

[133] Mahtab Ahmad, Sang Soo Lee, Anushka Upamali Rajapaksha, Meththika Vithanage, Ming Zhang, Ju Sik Cho, Sung-Eun Lee, Yong Sik Ok, Trichloroethylene adsorption by pine needle biochars produced at various pyrolysis temperatures, Bioresource Technology, **2013,** 143, 615. DOI:10.1016/j.biortech.2013.06.033.

[134] Rimena R. Domingues, Paulo F. Trugilho, Carlos A. Silva, Isabel Cristina NA de Melo, Leonidas C. A. Melo, Zuy M. Magriotis, Miguel A. Sanchez-Monedero, Properties of biochar derived from wood and high-nutrient biomasses with the aim of agronomic and environmental benefits, PLoS One, **2017,** 12, e0176884. DOI:10.1371/journal.pone.0176884.

[135] Wenqin Li, Qi Dang, Robert C. Brown, David Laird, Mark M. Wright, The impacts of biomass properties on pyrolysis yields, economic and environmental performance of the pyrolysis-bioenergy-biochar platform to carbon negative energy, Bioresource Technology, **2017,** 241, 959. DOI:10.1016/j.biortech.2017.06.049.

[136] Ronald J. Smernik, Jeffrey A. Baldock, J. Malcolm Oades, Andrew K. Whittaker, Determination of T1rhoH relaxation rates in charred and uncharred wood and consequences for NMR quantitation, Solid State Nuclear Magnetic Resonance, **2002,** 22, 50. DOI:10.1006/snmr. 2002.0064.

[137] A. Méndez, M. Terradillos, G. Gascó, Physicochemical and agronomic properties of biochar from sewage sludge pyrolysed at different temperatures, Journal of Analytical and Applied Pyrolysis, **2013,** 102, 124. DOI:10.1016/j.jaap.2013.03.006.

[138] Joyleene T. Yu, Amir Mehdi Dehkhoda, Naoko Ellis, Development of biochar-based catalyst for transesterification of canola oil, Energy & Fuels, **2011,** 25, 337. DOI:10.1021/ef100977d.

[139] Johannes Lehmann, John Gaunt, Marco Rondon, Bio-char sequestration in terrestrial ecosystems—A review, Mitigation and Adaptation Strategies for Global Change, **2006,** 11, 403. DOI:10.1007/s11027-005-9006-5.

[140] Shengsen Wang, Bin Gao, Andrew R. Zimmerman, Yuncong Li, Lena Ma, Willie G. Harris, Kati W. Migliaccio, Physicochemical and sorptive properties of biochars derived

from woody and herbaceous biomass, Chemosphere, **2015,** 134, 257. DOI:10.1016/j. chemosphere.2015.04.062.

[141] Huanliang Lu, Weihua Zhang, Yuxi Yang, Xiongfei Huang, Shizhong Wang, Rongliang Qiu, Relative distribution of Pb^{2+} sorption mechanisms by sludge-derived biochar, Water Research, **2012,** 46, 854. DOI:10.1016/j.watres.2011.11.058.

[142] Hongmei Jin, Sergio Capareda, Zhizhou Chang, Jun Gao, Yueding Xu, Jianying Zhang, Biochar pyrolytically produced from municipal solid wastes for aqueous As(V) removal: Adsorption property and its improvement with KOH activation, Bioresource Technology, **2014,** 169, 622. DOI:10.1016/j.biortech.2014.06.103.

[143] Rosie Wood, Ondrej Masek, Valentina Erastova, Biochars at the molecular level: Part 1— Insights into the molecular structures within biochars, ArXiv—Phys—Materials Science, **2023.** DOI:10.48550/arXiv.2303.09661.

[144] Peng Zhang, Wenyan Duan, Hongbo Peng, Bo Pan, Baoshan Xing, Functional biochar and its balanced design, ACS Environmental Au, **2022,** 2, 115. DOI:10.1021/acsenvironau.1c00032.

[145] Xinni Xiong, Iris K. M. Yu, Leichang Cao, Daniel C. W. Tsang, Shicheng Zhang, Yong Sik Ok, A review of biochar-based catalysts for chemical synthesis, biofuel production, and pollution control, Bioresource Technology, **2017,** 246, 254. DOI:10.1016/j.biortech.2017.06.163.

[146] Yunchao Li, Jingai Shao, Xianhua Wang, Yong Deng, Haiping Yang, Hanping Chen, Characterization of modified biochars derived from bamboo pyrolysis and their utilization for target component (furfural) adsorption, Energy & Fuels, **2014,** 28, 5119. DOI:10.1021/ef500725c.

[147] Patrick M. Godwin, Yuanfeng Pan, Huining Xiao, Muhammad T. Afzal, Progress in preparation and application of modified biochar for improving heavy metal ion removal from wastewater, Journal of Bioresources and Bioproducts, **2019,** 4, 31. DOI:10.21967/jbb.v4i1.180.

[148] Gabriel N. Kasozi, Andrew R. Zimmerman, Peter Nkedi-Kizza, Bin Gao, Catechol and humic acid sorption onto a range of laboratory-produced black carbons (biochars), Environmental Science & Technology, **2010,** 44, 6189. DOI:10.1021/es1014423.

[149] Shih-Hsin Ho, Shishu Zhu, Jo-Shu Chang, Recent advances in nanoscale-metal assisted biochar derived from waste biomass used for heavy metals removal, Bioresource Technology, **2017,** 246, 123. DOI:10.1016/j.biortech.2017.08.061.

[150] Shamim Gul, Joann K. Whalen, Ben W. Thomas, Vanita Sachdeva, Hongyuan Deng, Physico-chemical properties and microbial responses in biochar-amended soils: Mechanisms and future directions, Agriculture, Ecosystems & Environment, **2015,** 206, 46. DOI:10.1016/j.agee.2015.03.015.

[151] Weiping Song, Mingxin Guo, Quality variations of poultry litter biochar generated at different pyrolysis temperatures, Journal of Analytical and Applied Pyrolysis, **2012,** 94, 138. DOI:10.1016/j.jaap.2011.11.018.

[152] Ying Yao, Bin Gao, Ming Zhang, Mandu Inyang, Andrew R. Zimmerman, Effect of biochar amendment on sorption and leaching of nitrate, ammonium, and phosphate in a sandy soil, Chemosphere, **2012,** 89, 1467. DOI:10.1016/j.chemosphere.2012.06.002.

[153] Chumki Banik, Michael Lawrinenko, Santanu Bakshi, David A. Laird, Impact of pyrolysis temperature and feedstock on surface charge and functional group chemistry of biochars, Journal of Environmental Quality, **2018,** 47, 452. DOI:10.2134/jeq2017.11.0432.

[154] Jose L. Gomez-Eyles, Luke Beesley, Eduardo Moreno-Jimenez, Upal Ghosh, Tom Sizmur, The potential of biochar amendments to remediate contaminated soils, Biochar and Soil Biota, **2013,** 4, 100. DOI:10.13140/2.1.1074.9448.

[155] Lukas Van Zwieten, S. Kimber, S. Morris, K. Y. Chan, A. Downie, J. Rust, S. Joseph, Annette Cowie, Effects of biochar from slow pyrolysis of papermill waste on agronomic performance and soil fertility, Plant and Soil, **2010,** 327, 235. DOI:10.1007/s11104-009-0050-x.

[156] Jeffrey M. Novak, Warren J. Busscher, David L. Laird, Mohamed Ahmedna, Don W. Watts, Mohamed A. S. Niandou, Impact of biochar amendment on fertility of a southeastern coastal plain soil, Soil Science, **2009**, 174, 105. DOI:10.1097/SS.0b013e3181981d9a.

[157] Jiajun Fan, Chao Cai, Haifeng Chi, Brian J. Reid, Frédéric Coulon, Youchi Zhang, Yanwei Hou, Remediation of cadmium and lead polluted soil using thiol-modified biochar, Journal of Hazardous Materials, **2020**, 388, 122037. DOI:10.1016/j.jhazmat.2020.122037.

[158] Frederik Ronsse, Sven Van Hecke, Dane Dickinson, Wolter Prins, Production and characterization of slow pyrolysis biochar: Influence of feedstock type and pyrolysis conditions, GCB Bioenergy, **2013**, 5, 104. DOI:10.1111/gcbb.12018.

[159] Lichun Hsieh, Lei He, Mengya Zhang, Wanze Lv, Kun Yang, Meiping Tong, Addition of biochar as thin preamble layer into sand filtration columns could improve the microplastics removal from water, Water Research, **2022**, 221, 118783. DOI:10.1016/j.watres.2022.118783.

[160] Rajesh Chintala, Thomas E. Schumacher, Louis M. McDonald, David E. Clay, Douglas D. Malo, Sharon K. Papiernik, Sharon A. Clay, James L. Julson, Phosphorus sorption and availability from biochars and soil/biochar mixtures, Clean–Soil Air Water, **2014**, 42, 626. DOI:10.1002/clen.201300089.

[161] Ke Wang, Xiaoyuan Zhang, Cengceng Sun, Kaiqi Yang, Jiyong Zheng, Jihai Zhou, Biochar application alters soil structure but not soil hydraulic conductivity of an expansive clayey soil under field conditions, Journal of Soils and Sediments, **2021**, 21, 73. DOI:10.1007/s11368-020-02786-x.

[162] Hao Xia, Muhammad Riaz, Mengyang Zhang, Bo Liu, Zeinab El-Desouki, Cuncang Jiang, Biochar increases nitrogen use efficiency of maize by relieving aluminum toxicity and improving soil quality in acidic soil, Ecotoxicology and Environmental Safety, **2020**, 196, 110531. DOI:10.1016/j.ecoenv.2020.110531.

[163] Mangala Rai, T. G. Reeves, S. Pandey, L. Collette, Save and grow: A policymaker's guide to the sustainable intensification of smallholder crop production, Food and Agriculture Organization of the United Nations, **2011**. DOI:10.1017/S0014479711001049.

[164] Hanuman S. Jatav, Vishnu D. Rajput, Tatiana Minkina, Satish K. Singh, Sukirtee Chejara, Andrey Gorovtsov, Anatoly Barakhov, Tatiana Bauer, Svetlana Sushkova, Saglara Mandzhieva, Marina Burachevskaya, Valery P. Kalinitchenko, Sustainable approach and safe use of biochar and its possible consequences, Sustainability, **2021**, 13. DOI:10.3390/su131810362.

[165] Diego Tassinari, Maria luiza de Carvalho Andrade, Moacir de Souza Dias Junior, Ricardo Previdente Martins, Wellington Willian Rocha, Paula Sant'Anna Moreira Pais, Zélio Resende de Souza, Soil compaction caused by harvesting, skidding and wood processing in eucalyptus forests on coarse-textured tropical soils, Soil Use and Management, **2019**, 35, 400. DOI:10.1111/sum.12509.

[166] Deyi Hou, Sustainable remediation of contaminated soil and groundwater: Materials, processes, and assessment, **2019**, Elsevier.

[167] John Boardman, Karel Vandaele, Robert Evans, Ian D. L. Foster, Off-site impacts of soil erosion and runoff: Why connectivity is more important than erosion rates, Soil Use and Management, **2019**, 35, 245. DOI:10.1111/sum.12496.

[168] Hanuman Singh Jatav, Satish Kumar Singh, Surendra Singh Jatav, Vishnu D. Rajput, Manoj Parihar, Sonu Kumar Mahawer, Rajesh Kumar Singhal, Sukirtee, Applications of biochar for environmental safety, **2020**, IntechOpen.

[169] J. D. Mao, R. L. Johnson, J. Lehmann, D. C. Olk, E. G. Neves, M. L. Thompson, K. Schmidt-Rohr, Abundant and stable char residues in soils: Implications for soil fertility and carbon sequestration, Environmental Science & Technology, **2012**, 46, 9571. DOI:10.1007/978-94-007-5634-2_87.

[170] Lili Ye, Marta Camps-Arbestain, Qinhua Shen, Johannes Lehmann, Balwant Singh, Muhammad Sabir, Biochar effects on crop yields with and without fertilizer: A meta-analysis

of field studies using separate controls, Soil Use and Management, **2020,** 36, 2. DOI:10.1111/sum.12546.

[171] C. J. Atkinson, How good is the evidence that soil-applied biochar improves water-holding capacity? Soil Use and Management, **2018,** 34, 177. DOI:10.1111/sum.12413.

[172] Leigh D. Burrell, Franz Zehetner, Nicola Rampazzo, Bernhard Wimmer, Gerhard Soja, Long-term effects of biochar on soil physical properties, Geoderma, **2016,** 282, 96. DOI:10.1016/j.geoderma.2016.07.019.

[173] Humberto Blanco-Canqui, Biochar and soil physical properties, Soil Science Society of America Journal, **2017,** 81, 687. DOI:10.2136/sssaj2017.01.0017.

[174] Fatemeh Razzaghi, Peter Bilson Obour, Emmanuel Arthur, Does biochar improve soil water retention? A systematic review and meta-analysis, Geoderma, **2020,** 361, 114055. DOI:10.1016/j.geoderma.2019.114055.

[175] Johannes Lehmann, Matthias C. Rillig, Janice Thies, Caroline A. Masiello, William C. Hockaday, David Crowley, Biochar effects on soil biota—A review, Soil Biology and Biochemistry, **2011,** 43, 1812. DOI:10.1016/j.soilbio.2011.04.022.

[176] Simeng Li, Vanessa Barreto, Runwei Li, Gang Chen, Yuch P. Hsieh, Nitrogen retention of biochar derived from different feedstocks at variable pyrolysis temperatures, Journal of Analytical and Applied Pyrolysis, **2018,** 133, 136. DOI:10.1016/j.jaap.2018.04.010.

[177] E. Amoakwah, K. A. Frimpong, D. Okae-Anti, E. Arthur, Soil water retention, air flow and pore structure characteristics after corn cob biochar application to a tropical sandy loam, Geoderma, **2017,** 307, 189. DOI:10.1016/j.geoderma.2017.08.025.

[178] N. Saffari, M. A. Hajabbasi, H. Shirani, M. R. Mosaddeghi, G. Owens, Influence of corn residue biochar on water retention and penetration resistance in a calcareous sandy loam soil, Geoderma, **2021,** 383, 114734. DOI:10.1016/j.geoderma.2020.114734.

[179] Tanveer Ali Sial, Zhilong Lan, Muhammad Numan Khan, Ying Zhao, Farhana Kumbhar, Jiao Liu, Afeng Zhang, Robert Lee Hill, Altaf Hussain Lahori, Mehurnisa Memon, Evaluation of orange peel waste and its biochar on greenhouse gas emissions and soil biochemical properties within a loess soil, Waste Management, **2019,** 87, 125. DOI:10.1016/j.wasman.2019.01.042.

[180] Edith Juno, Inés Ibáñez, Biochar application and soil transfer in tree restoration: A meta-analysis and field experiment, Ecological Restoration, **2021,** 39, 158. DOI:10.3368/er.39.3.158.

[181] Aruna Olasekan Adekiya, Taiwo Michael Agbede, Adeniyi Olayanju, Wutem Sunny Ejue, Timothy A. Adekanye, Titilayo Tolulope Adenusi, Jerry Femi Ayeni, Effect of biochar on soil properties, soil loss, and cocoyam yield on a tropical sandy loam Alfisol, The Scientific World Journal, **2020,** 2020. DOI:10.1155/2020/9391630.

[182] Liping Lou, Binbin Wu, Lina Wang, Ling Luo, Xinhua Xu, Jiaai Hou, Bei Xun, Baolan Hu, Yingxu Chen, Sorption and ecotoxicity of pentachlorophenol polluted sediment amended with rice-straw derived biochar, Bioresource Technology, **2011,** 102, 4036. DOI:10.1016/j.biortech.2010.12.010.

[183] Fatima Sopeña, Kirk Semple, Saran Sohi, Gary Bending, Assessing the chemical and biological accessibility of the herbicide isoproturon in soil amended with biochar, Chemosphere, **2012,** 88, 77. DOI:10.1016/j.chemosphere.2012.02.066.

[184] Zhongxin Tan, Yuanhang Wang, Limei Zhang, Qiaoyun Huang, Study of the mechanism of remediation of Cd-contaminated soil by novel biochars, Environmental Science and Pollution Research, **2017,** 24, 24844. DOI:10.1007/s11356-017-0109-9.

[185] A. Venegas, A. Rigol, M. Vidal, Changes in heavy metal extractability from contaminated soils remediated with organic waste or biochar, Geoderma, **2016,** 279, 132. DOI:10.1016/j.geoderma.2016.06.010.

[186] Rongjun Bian, De Chen, Xiaoyu Liu, Liqiang Cui, Lianqing Li, Genxing Pan, Dan Xie, Jinwei Zheng, Xuhui Zhang, Jufeng Zheng, Biochar soil amendment as a solution to

prevent Cd-tainted rice from China: Results from a cross-site field experiment, Ecological Engineering, **2013,** 58, 378. DOI:10.1016/j.ecoleng.2013.07.031.

[187] Chaoyi Luo, Jingjing Yang, Wen Chen, Fengpeng Han, Effect of biochar on soil proper-ties on the Loess Plateau: Results from field experiments, Geoderma, **2020,** 369, 114323. DOI:10.1016/j.geoderma.2020.114323.

[188] Yongshan Chen, Marta Camps-Arbestain, Qinhua Shen, Balwant Singh, Maria Luz Cayuela, The long-term role of organic amendments in building soil nutrient fertility: A meta-analysis and review, Nutrient Cycling in Agroecosystems, **2018,** 111, 103. DOI:10.1007/s10705-017-9903-5.

[189] J. R. Quilty, S. R. Cattle, Use and understanding of organic amendments in Australian agri-culture: A review, Soil Research, **2011,** 49, 1. DOI:10.1071/SR10059.

[190] Zichuan Li, Zhaoliang Song, Bhupinder Pal Singh, Hailong Wang, The impact of crop residue biochars on silicon and nutrient cycles in croplands, Science of the Total Environment, **2019,** 659, 673. DOI:10.1016/j.scitotenv.2018.12.381.

[191] Marta Camps-Arbestain, Qinhua Shen, Tao Wang, Lukas van Zwieten, Jeff Novak, 10 Available nutrients in biochar, Biochar: A Guide to Analytical Methods, **2017,** 109.

[192] Md Zahangir Hossain, Md Mezbaul Bahar, Binoy Sarkar, Scott Wilfred Donne, Young Sik Ok, Kumuduni Niroshika Palansooriya, Mary Beth Kirkham, Saikat Chowdhury, Nanthi Bolan, Biochar and its importance on nutrient dynamics in soil and plant, Biochar, **2020,** 2, 379. DOI:10.1007/s42773-020-00065-z.

[193] Ali El-Naggar, Ahmed Hamdy El-Naggar, Sabry M. Shaheen, Binoy Sarkar, Scott X. Chang, Daniel C. W. Tsang, Jörg Rinklebe, Yong Sik Ok, Biochar composition-dependent impacts on soil nutrient release, carbon mineralization, and potential environmental risk: A review, Journal of Environmental Management, **2019,** 241, 458. DOI:10.1016/j.jenvman.2019.02.044.

[194] Asli Toptas Tag, Gozde Duman, Suat Ucar, Jale Yanik, Effects of feedstock type and pyrolysis temperature on potential applications of biochar, Journal of Analytical and Applied Pyrolysis, **2016,** 120, 200. DOI:10.1016/j.jaap.2016.05.006.

[195] Thomas H. DeLuca, Michael J. Gundale, M. Derek MacKenzie, Davey L. Jones, Biochar effects on soil nutrient transformations, Biochar for Environmental Management, **2015,** 2, 421. DOI:10.1007/s42773-020-00065-z.

[196] Natália Aragão de Figueredo, Liovando Marciano da Costa, Leônidas Carrijo Azevedo Melo, Evair Antônio Siebeneichlerd, Jairo Tronto, Characterization of biochars from dif-ferent sources and evaluation of release of nutrients and contaminants, Revista Ciência Agronômica, **2017,** 48, 3. DOI:10.5935/1806-6690.20170046.

[197] Rosa Marchetti, Fabio Castelli, Biochar from swine solids and digestate influence nutrient dynamics and carbon dioxide release in soil, Journal of Environmental Quality, **2013,** 42, 893. DOI:10.2134/jeq2012.0352.

[198] Zhanghong Wang, Haiyan Guo, Fei Shen, Gang Yang, Yanzong Zhang, Yongmei Zeng, Lilin Wang, Hong Xiao, Shihuai Deng, Biochar produced from oak sawdust by Lanthanum (La)-involved pyrolysis for adsorption of ammonium (NH4+), nitrate (NO3−), and phosphate (PO43−), Chemosphere, **2015,** 119, 646. DOI:10.1016/j.chemosphere.2014.07.084.

[199] Gang Xu, Junna Sun, Hongbo Shao, Scott X. Chang, Biochar had effects on phosphorus sorp-tion and desorption in three soils with differing acidity, Ecological Engineering, **2014,** 62, 54. DOI:10.1016/j.ecoleng.2013.10.027.

[200] Nikolas Hagemann, Stephen Joseph, Hans-Peter Schmidt, Claudia I. Kammann, Johannes Harter, Thomas Borch, Robert B. Young, Krisztina Varga, Sarasadat Taherymoosavi, K. Wade Elliott, Organic coating on biochar explains its nutrient retention and stimulation of soil fertility, Nature Communications, **2017,** 8, 1089. DOI:10.1038/s41467-017-01123-0.

[201] Claudia I. Kammann, Hans-Peter Schmidt, Nicole Messerschmidt, Sebastian Linsel, Diedrich Steffens, Christoph Müller, Hans-Werner Koyro, Pellegrino Conte, Stephen Joseph,

Plant growth improvement mediated by nitrate capture in co-composted biochar, Scientific Reports, **2015,** 5, 11080. DOI:10.1038/srep11080.

[202] Na Geng, Xirui Kang, Xiaoxiao Yan, Na Yin, Hui Wang, Hong Pan, Quangang Yang, Yanhong Lou, Yuping Zhuge, Biochar mitigation of soil acidification and carbon sequestration is influenced by materials and temperature, Ecotoxicology and Environmental Safety, **2022,** 232, 113241. DOI:10.1016/j.ecoenv.2022.113241.

[203] K. Dale Ritchey, J. Diane Snuffer, Limestone, gypsum, and magnesium oxide influence restoration of an abandoned Appalachian pasture, Agronomy Journal, **2002,** 94, 830. DOI:10.2134/agronj2002.8300.

[204] Amir Hass, Javier M. Gonzalez, Isabel M. Lima, Harry W. Godwin, Jonathan J. Halvorson, Douglas G. Boyer, Chicken manure biochar as liming and nutrient source for acid Appalachian soil, Journal of Environmental Quality, **2012,** 41, 1096. DOI:10.2134/jeq2011.0124.

[205] Mahmood Laghari, Muhammad Saffar Mirjat, Zhiquan Hu, Saima Fazal, Bo Xiao, Mian Hu, Zhihua Chen, Dabin Guo, Effects of biochar application rate on sandy desert soil properties and sorghum growth, Catena, **2015,** 135, 313. DOI:10.1016/j.catena.2015.08.013.

[206] Afeng Zhang, Rongjun Bian, Genxing Pan, Liqiang Cui, Qaiser Hussain, Lianqing Li, Jinwei Zheng, Jufeng Zheng, Xuhui Zhang, Xiaojun Han, Xinyan Yu, Effects of biochar amendment on soil quality, crop yield and greenhouse gas emission in a Chinese rice paddy: A field study of 2 consecutive rice growing cycles, Field Crops Research, **2012,** 127, 153. DOI:10.1016/j.fcr.2011.11.020.

[207] Simon Jeffery, Diego Abalos, Marija Prodana, Ana Catarina Bastos, Jan Willem Van Groenigen, Bruce A. Hungate, Frank Verheijen, Biochar boosts tropical but not temperate crop yields, Environmental Research Letters, **2017,** 12, 053001. DOI:10.1088/1748-9326/aa67bd.

[208] Julie Major, Marco Rondon, Diego Molina, Susan J. Riha, Johannes Lehmann, Maize yield and nutrition during 4 years after biochar application to a Colombian savanna oxisol, Plant and Soil, **2010,** 333, 117. DOI:10.1007/s11104-010-0327-0.

[209] Naba Raj Pandit, Jan Mulder, Sarah E. Hale, Andrew R. Zimmerman, Bishnu Hari Pandit, Gerard Cornelissen, Multi-year double cropping biochar field trials in Nepal: Finding the optimal biochar dose through agronomic trials and cost-benefit analysis, Science of the Total Environment, **2018,** 637, 1333. DOI:10.1016/j.scitotenv.2018.05.107.

[210] Ghulam Haider, Diedrich Steffens, Gerald Moser, Christoph Müller, Claudia I. Kammann, Biochar reduced nitrate leaching and improved soil moisture content without yield improvements in a four-year field study, Agriculture, Ecosystems & Environment, **2017,** 237, 80. DOI:10.1016/j.agee.2016.12.019.

[211] Ying Ding, Yuxue Liu, Weixiang Wu, Dezhi Shi, Min Yang, Zheke Zhong, Evaluation of biochar effects on nitrogen retention and leaching in multi-layered soil columns, Water, Air, & Soil Pollution, **2010,** 213, 47. DOI:10.1007/s11270-010-0366-4.

[212] Anja Sänger, Katharina Reibe, Jan Mumme, Martin Kaupenjohann, Frank Ellmer, Christina-Luise Roß, Andreas Meyer-Aurich, Biochar application to sandy soil: Effects of different biochars and N fertilization on crop yields in a 3-year field experiment, Archives of Agronomy and Soil Science, **2017,** 63, 213. DOI:10.1080/03650340.2016.1223289.

[213] Lei Wang, Lianqing Li, Kun Cheng, Chunying Ji, Qian Yue, Rongjun Bian, Genxing Pan, An assessment of emergy, energy, and cost-benefits of grain production over 6 years following a biochar amendment in a rice paddy from China, Environmental Science and Pollution Research, **2018,** 25, 9683. DOI:10.1007/s11356-018-1245-6.

[214] Wenliang Wei, Huaqing Yang, Mingsheng Fan, Haiqing Chen, Dayong Guo, Jian Cao, Yakov Kuzyakov, Corrigendum to "Biochar effects on crop yields and nitrogen loss depending on fertilization", Science of the Total Environment, **2020,** 705, 135991. DOI:10.1016/j.scitotenv.2019.135991.

[215] Sean C. Thomas, Nigel Gale, Biochar and forest restoration: A review and meta-analysis of tree growth responses, New Forests, **2015,** 46, 931. DOI:10.1007/s11056-015-9491-7.

[216] Atanu Mukherjee, Rattan Lal, Andrew R. Zimmerman, Impacts of 1.5-year field aging on biochar, humic acid, and water treatment residual amended soil, Soil Science, **2014,** 179. DOI:10.1097/ss.0000000000000076.

[217] S. Jeffery, F. G. A. Verheijen, M. van der Velde, A. C. Bastos, A quantitative review of the effects of biochar application to soils on crop productivity using meta-analysis, Agriculture, Ecosystems & Environment, **2011,** 144, 175. DOI:10.1016/j.agee.2011.08.015.

[218] Dali Song, Jiwei Tang, Xiangyin Xi, Shuiqing Zhang, Guoqing Liang, Wei Zhou, Xiubin Wang, Responses of soil nutrients and microbial activities to additions of maize straw biochar and chemical fertilization in a calcareous soil, European Journal of Soil Biology, **2018,** 84, 1. DOI:10.1016/j.ejsobi.2017.11.003.

[219] Zakaria M. Solaiman, Hossain M. Anawar, Application of biochars for soil constraints: Challenges and solutions, Pedosphere, **2015,** 25, 631. DOI:10.1016/S1002-0160(15)30044-8.

[220] S. D. Joseph, M. Camps-Arbestain, Yun Lin, P. Munroe, C. H. Chia, J. Hook, L. van Zwieten, S. Kimber, A. Cowie, B. P. Singh, J. Lehmann, N. Foidl, R. J. Smernik, J. E. Amonette, An investigation into the reactions of biochar in soil, Soil Research, **2010,** 48, 501. DOI:10.1071/SR10009.

[221] Yang Ding, Yunguo Liu, Shaobo Liu, Huang Xixian, Zhongwu Li, Xiaofei Tan, Guangming Zeng, Lu Zhou, Potential benefits of biochar in agricultural soils: A review, Pedosphere, **2017,** 27, 645. DOI:10.1016/S1002-0160(17)60375-8.

[222] Qiong Hou, Ting Zuo, Jian Wang, Shan Huang, Xiaojun Wang, Longren Yao, Wuzhong Ni, Responses of nitrification and bacterial community in three size aggregates of paddy soil to both of initial fertility and biochar addition, Applied Soil Ecology, **2021,** 166, 104004. DOI:10.1016/j.apsoil.2021.104004.

[223] Jared L. DeForest, Rael K. Otuya, Soil nitrification increases with elevated phosphorus or soil pH in an acidic mixed mesophytic deciduous forest, Soil Biology and Biochemistry, **2020,** 142, 107716. DOI:10.1016/j.soilbio.2020.107716.

[224] Chengyu Wang, Xue Zhou, Dan Guo, Jianghua Zhao, Li Yan, Guozhong Feng, Qiang Gao, Han Yu, Lanpo Zhao, Soil pH is the primary factor driving the distribution and function of microorganisms in farmland soils in northeastern China, Annals of Microbiology, **2019,** 69, 1461. DOI:10.1007/s13213-019-01529-9.

[225] Vandit Vijay, Sowmya Shreedhar, Komalkant Adlak, Sachin Payyanad, Vandana Sreedharan, Girigan Gopi, Tessa Sophia van der Voort, P. Malarvizhi, Susan Yi, Julia Gebert, Review of large-scale biochar field-trials for soil amendment and the observed influences on crop yield variations, Frontiers in Energy Research, **2021,** 9, 710766. DOI:10.3389/fenrg.2021.710766.

[226] Xiaohong Yin, Jiana Chen, Long Fan, Zui Tao, Min Huang, Yingbin Zou, Nitrospira bacteria in paddy soil reduced by biochar application, Agrosystems, Geosciences & Environment, **2020,** 3, e20009. DOI:10.1002/agg2.20009.

[227] Yufang Shen, Lixia Zhu, Hongyan Cheng, Shanchao Yue, Shiqing Li, Effects of biochar application on CO_2 emissions from a cultivated soil under semiarid climate conditions in Northwest China, Sustainability, **2017,** 9, 1482. DOI:10.3390/su9081482.

[228] Jingjing Chen, Hyunjin Kim, Gayoung Yoo, Effects of biochar addition on CO_2 and N_2O emissions following fertilizer application to a cultivated grassland soil, PLoS One, **2015,** 10, e0126841. DOI:10.1371/journal.pone.0126841.

[229] A. V. Gorovtsov, T. M. Minkina, S. S. Mandzhieva, L. V. Perelomov, G. Soja, I. V. Zamulina, V. D. Rajput, S. N. Sushkova, D. Mohan, J. Yao, The mechanisms of biochar interactions with microorganisms in soil, Environmental Geochemistry and Health, **2020,** 42, 2495. DOI:10.1007/s10653-019-00412-5.

[230] M. Kuśmierz, P. Oleszczuk, P. Kraska, E. Pałys, S. Andruszczak, Persistence of Polycyclic Aromatic Hydrocarbons (PAHs) in biochar-amended soil, Chemosphere, **2016,** 146, 272. DOI:10.1016/j.chemosphere.2015.12.010.

[231] M. Keiluweit, P. S. Nico, M. G. Johnson, M. Kleber, Dynamic molecular structure of plant biomass-derived black carbon (biochar), Environmental Science & Technology, **2010,** 44, 1247. DOI:10.1021/es9031419.

[232] Guixiang Zhang, Xiaofang Guo, Yuen Zhu, Xitao Liu, Zhiwang Han, Ke Sun, Li Ji, Qiusheng He, Lanfang Han, The effects of different biochars on microbial quantity, microbial community shift, enzyme activity, and biodegradation of polycyclic aromatic hydrocarbons in soil, Geoderma, **2018,** 328, 100. DOI:10.1016/j.geoderma.2018.05.009.

[233] Dinesh Mohan, Charles U. Pittman, Jr., Philip H. Steele, Pyrolysis of wood/biomass for bio-oil: A critical review, Energy & Fuels, **2006,** 20, 848. DOI:10.1021/ef0502397.

[234] Qi Wang, Ling Chen, Linyan He, Xiafang Sheng, Increased biomass and reduced heavy metal accumulation of edible tissues of vegetable crops in the presence of plant growth-promoting Neorhizobium huautlense T1-17 and biochar, Agriculture, Ecosystems & Environment, **2016,** 228, 9. DOI:10.1016/j.agee.2016.05.006.

[235] Richard S. Quilliam, Karina A. Marsden, Christoph Gertler, Johannes Rousk, Thomas H. DeLuca, Davey L. Jones, Nutrient dynamics, microbial growth and weed emergence in biochar amended soil are influenced by time since application and reapplication rate, Agriculture, Ecosystems & Environment, **2012,** 158, 192. DOI:10.1016/j.agee.2012.06.011.

[236] M. Pukalchik, F. Mercl, V. Terekhova, P. Tlustoš, Biochar, wood ash and humic substances mitigating trace elements stress in contaminated sandy loam soil: Evidence from an integrative approach, Chemosphere, **2018,** 203, 228. DOI:10.1016/j.chemosphere.2018.03.181.

[237] G. Fang, J. Gao, C. Liu, D. D. Dionysiou, Y. Wang, D. Zhou, Key role of persistent free radicals in hydrogen peroxide activation by biochar: Implications to organic contaminant degradation, Environmental Science & Technology, **2014,** 48, 1902. DOI:10.1021/es4048126.

[238] Daquan Sun, Lauren Hale, David Crowley, Nutrient supplementation of pinewood biochar for use as a bacterial inoculum carrier, Biology and Fertility of Soils, **2016,** 52, 515. DOI:10.1007/s00374-016-1093-9.

[239] Lydia Paetsch, Carsten W. Mueller, Ingrid Kögel-Knabner, Margit von Lützow, Cyril Girardin, Cornelia Rumpel, Effect of in-situ aged and fresh biochar on soil hydraulic conditions and microbial C use under drought conditions, Scientific Reports, **2018,** 8, 6852. DOI:10.1038/s41598-018-25039-x.

[240] Chenfei Liang, Xiaolin Zhu, Shenglei Fu, Ana Méndez, Gabriel Gascó, Jorge Paz-Ferreiro, Biochar alters the resistance and resilience to drought in a tropical soil, Environmental Research Letters, **2014,** 9, 064013. DOI:10.1088/1748-9326/9/6/064013.

[241] A. D. Igalavithana, S. E. Lee, Y. H. Lee, D. C. W. Tsang, J. Rinklebe, E. E. Kwon, Y. S. Ok, Heavy metal immobilization and microbial community abundance by vegetable waste and pine cone biochar of agricultural soils, Chemosphere, **2017,** 174, 593. DOI:10.1016/j.chemosphere.2017.01.148.

[242] Amit K. Jaiswal, Yigal Elad, Indira Paudel, Ellen R. Graber, Eddie Cytryn, Omer Frenkel, Linking the belowground microbial composition, diversity and activity to soilborne disease suppression and growth promotion of tomato amended with biochar, Scientific Reports, **2017,** 7, 44382. DOI:10.1038/srep44382.

[243] Xueyong Zhou, Zhe Yang, Huifen Liu, Xianzhi Lu, Jianchao Hao, Effect of soil organic matter on adsorption and insecticidal activity of toxins from Bacillus thuringiensis, Pedosphere, **2018,** 28, 341. DOI:10.1016/S1002-0160(18)60011-6.

[244] Xiaomin Zhu, Baoliang Chen, Lizhong Zhu, Baoshan Xing, Effects and mechanisms of biochar-microbe interactions in soil improvement and pollution remediation: A review, Environmental Pollution, **2017,** 227, 98. DOI:10.1016/j.envpol.2017.04.032.

[245] Ifeoma G. Edeh, Ondřej Mašek, Wolfram Buss, A meta-analysis on biochar's effects on soil water properties—New insights and future research challenges, Science of the Total Environment, **2020,** 714, 136857. DOI:10.1016/j.scitotenv.2020.136857.

[246] Morris Oduor Omondi, Xin Xia, Alphonse Nahayo, Xiaoyu Liu, Punhoon Khan Korai, Genxing Pan, Quantification of biochar effects on soil hydrological properties using meta-analysis of literature data, Geoderma, **2016,** 274, 28. DOI:10.1016/j.geoderma.2016.03.029.

[247] Kyle K. Shimabuku, Joshua P. Kearns, Juan E. Martinez, Ryan B. Mahoney, Laura Moreno-Vasquez, R. Scott Summers, Biochar sorbents for sulfamethoxazole removal from surface water, stormwater, and wastewater effluent, Water Research, **2016,** 96, 236. DOI:10.1016/j.watres.2016.03.049.

[248] Dominic Woolf, James E. Amonette, F. Alayne Street-Perrott, Johannes Lehmann, Stephen Joseph, Sustainable biochar to mitigate global climate change, Nature Communications, **2010,** 1, 56. DOI:10.1038/ncomms1053.

[249] Johannes Lehmann, Bio-energy in the black, Frontiers in Ecology and the Environment, **2007,** 5, 381. DOI:10.1890/1540-9295(2007)5[381:BITB]2.0.CO;2.

[250] Rattan Lal, Digging deeper: A holistic perspective of factors affecting soil organic carbon sequestration in agroecosystems, Global Change Biology, **2018,** 24, 3285. DOI:10.1111/gcb.14054.

[251] Bernardo Maestrini, Paolo Nannipieri, Samuel Abiven, A meta-analysis on pyrogenic organic matter induced priming effect, GCB Bioenergy, **2015,** 7, 577. DOI:10.1111/gcbb.12194.

[252] Yakov Kuzyakov, J. K. Friedel, Karl Stahr, Review of mechanisms and quantification of priming effects, Soil Biology and Biochemistry, **2000,** 32, 1485. DOI:10.1016/S0038-0717(00)00084-5.

[253] Lukas van Zwieten, The long-term role of organic amendments in addressing soil constraints to production, Nutrient Cycling in Agroecosystems, **2018,** 111, 99. DOI:10.1007/s10705-018-9934-6.

[254] Kathryn E. White, Eric B. Brennan, Michel A. Cavigelli, Richard F. Smith, Winter cover crops increase readily decomposable soil carbon, but compost drives total soil carbon during eight years of intensive, organic vegetable production in California, PLoS One, **2020,** 15, e0228677. DOI:10.1371/journal.pone.0228677.

[255] Shu-Yuan Pan, Cheng-Di Dong, Jenn-Fang Su, Po-Yen Wang, Chiu-Wen Chen, Jo-Shu Chang, Hyunook Kim, Chin-Pao Huang, Chang-Mao Hung, The role of biochar in regulating the carbon, phosphorus, and nitrogen cycles exemplified by soil systems, Sustainability, **2021,** 13, 5612. DOI:10.3390/su13105612.

[256] Ellen McHarg, Elena Mengo, Lisa Benson, Jody Daniel, Andre Joseph-Witzig, Paulette Posen, Tiziana Luisetti, Valuing the contribution of blue carbon to small island developing states' climate change commitments and Covid-19 recovery, Environmental Science & Policy, **2022,** 132, 13. DOI:10.1016/j.envsci.2022.02.009.

[257] Dominic Woolf, Johannes Lehmann, Modelling the long-term response to positive and negative priming of soil organic carbon by black carbon, Biogeochemistry, **2012,** 111, 83. DOI:10.1007/s10533-012-9764-6.

[258] Kalu Samuel Ukanwa, Kumar Patchigolla, Ruben Sakrabani, Edward Anthony, Sachin Mandavgane, A review of chemicals to produce activated carbon from agricultural waste biomass, Sustainability, **2019,** 11, 6204. DOI:10.3390/su11226204.

[259] Michael W. I. Schmidt, Angela G. Noack, Black carbon in soils and sediments: Analysis, distribution, implications, and current challenges, Global Biogeochemical Cycles, **2000,** 14, 777. DOI:10.1029/1999GB001208.

[260] Jinyang Wang, Zhengqin Xiong, Yakov Kuzyakov, Biochar stability in soil: Meta-analysis of decomposition and priming effects, GCB Bioenergy, **2016,** 8, 512. DOI:10.1111/gcbb.12266.

[261] Paul Balcombe, Jamie F. Speirs, Nigel P. Brandon, Adam D. Hawkes, Methane emissions: Choosing the right climate metric and time horizon, Environmental Science: Processes & Impacts, **2018,** 20, 1323. DOI:10.1039/C8EM00414E.

[262] Lars Biernat, Friedhelm Taube, Ralf Loges, Christof Kluß, Thorsten Reinsch, Nitrous oxide emissions and methane uptake from organic and conventionally managed arable crop

rotations on farms in Northwest Germany, Sustainability, **2020,** 12, 3240. DOI:10.3390/su12083240.

[263] Dengxiao Zhang, Genxing Pan, Gang Wu, Grace Wanjiru Kibue, Lianqing Li, Xuhui Zhang, Jinwei Zheng, Jufeng Zheng, Kun Cheng, Stephen Joseph, Biochar helps enhance maize productivity and reduce greenhouse gas emissions under balanced fertilization in a rainfed low fertility inceptisol, Chemosphere, **2016,** 142, 106. DOI:10.1016/j.chemosphere.2015.04.088.

[264] Yuze Song, Yongfu Li, Yanjiang Cai, Shenglei Fu, Yu Luo, Hailong Wang, Chenfei Liang, Ziwen Lin, Shuaidong Hu, Yongchun Li, Biochar decreases soil N2O emissions in Moso bamboo plantations through decreasing labile N concentrations, N-cycling enzyme activities and nitrification/denitrification rates, Geoderma, **2019,** 348, 135. DOI:10.1016/j.geoderma.2019.04.025.

[265] Mohammad I. Al-Wabel, Qaiser Hussain, Adel R. A. Usman, Mahtab Ahmad, Adel Abduljabbar, Abdulazeem S. Sallam, Yong Sik Ok, Impact of biochar properties on soil conditions and agricultural sustainability: A review, Land Degradation & Development, **2018,** 29, 2124. DOI:10.1002/ldr.2829.

[266] John L. Gaunt, Johannes Lehmann, Energy balance and emissions associated with biochar sequestration and pyrolysis bioenergy production, Environmental Science & Technology, **2008,** 42, 4152. DOI:10.1021/es071361i.

[267] Suzette P. Galinato, Jonathan K. Yoder, David Granatstein, The economic value of biochar in crop production and carbon sequestration, Energy Policy, **2011,** 39, 6344. DOI:10.1016/j.enpol.2011.07.035.

[268] Tristram O. West, Allen C. McBride, The contribution of agricultural lime to carbon dioxide emissions in the United States: Dissolution, transport, and net emissions, Agriculture, Ecosystems & Environment, **2005,** 108, 145. DOI:10.1016/j.agee.2005.01.002.

[269] Marion Huber-Humer, Julia Gebert, Helene Hilger, Biotic systems to mitigate landfill methane emissions, Waste Management & Research, **2008,** 26, 33. DOI:10.1177/0734242X070879.

[270] Charlotte Scheutz, Peter Kjeldsen, Jean E. Bogner, Alex De Visscher, Julia Gebert, Helene A. Hilger, Marion Huber-Humer, Kurt Spokas, Microbial methane oxidation processes and technologies for mitigation of landfill gas emissions, Waste Management & Research, **2009,** 27, 409. DOI:10.1177/0734242X09339325.

[271] C. N. Mulligan, R. N. Yong, B. F. Gibbs, Remediation technologies for metal-contaminated soils and groundwater: An evaluation, Engineering Geology, **2001,** 60, 193. DOI:10.1016/S0013-7952(00)00101-0.

[272] Hiba M. Alkharabsheh, Mahmoud F. Seleiman, Martin Leonardo Battaglia, Ashwag Shami, Rewaa S. Jalal, Bushra Ahmed Alhammad, Khalid F. Almutairi, Adel M. Al-Saif, Biochar and its broad impacts in soil quality and fertility, nutrient leaching and crop productivity: A review, Agronomy, **2021,** 11, 993. DOI:10.3390/agronomy11050993.

[273] Michał Kołtowski, Isabel Hilber, Thomas D. Bucheli, Patryk Oleszczuk, Effect of steam activated biochar application to industrially contaminated soils on bioavailability of polycyclic aromatic hydrocarbons and ecotoxicity of soils, Science of the Total Environment, **2016,** 566, 1023. DOI:10.1016/j.scitotenv.2016.05.114.

[274] Altaf Hussain Lahori, Zhanyu Guo, Zengqiang Zhang, Ronghua Li, Amanullah Mahar, Mukesh Kumar Awasthi, Feng Shen, Tanveer Ali Sial, Farhana Kumbhar, Ping Wang, Shuncheng Jiang, Use of biochar as an amendment for remediation of heavy metal-contaminated soils: Prospects and challenges, Pedosphere, **2017,** 27, 991. DOI:10.1016/S1002-0160(17)60490-9.

[275] Mahtab Ahmad, Anushka Upamali Rajapaksha, Jung Eun Lim, Ming Zhang, Nanthi Bolan, Dinesh Mohan, Meththika Vithanage, Sang Soo Lee, Yong Sik Ok, Biochar as a sorbent for contaminant management in soil and water: A review, Chemosphere, **2014,** 99, 19. DOI:10.1016/j.chemosphere.2013.10.071.

[276] Saravanan Rajendran, T. A. K. Priya, Kuan Shiong Khoo, Tuan K. A. Hoang, Hui-Suan Ng, Heli Siti Halimatul Munawaroh, Ceren Karaman, Yasin Orooji, Pau Loke Show, A critical review on various remediation approaches for heavy metal contaminants removal from contaminated soils, Chemosphere, **2022,** 287, 132369. DOI:10.1016/j. chemosphere.2021.132369.

[277] Hai Lin, Ziwei Wang, Chenjing Liu, Yingbo Dong, Technologies for removing heavy metal from contaminated soils on farmland: A review, Chemosphere, **2022,** 305, 135457. DOI:10.1016/j.chemosphere.2022.135457.

[278] P. C. Abhilash, N. Singh, Pesticide use and application: An Indian scenario, Journal of Hazardous Materials, **2009,** 165, 1. DOI:10.1016/j.jhazmat.2008.10.061.

[279] P. J. John, N. Bakore, P. Bhatnagar, Assessment of organochlorine pesticide residue levels in dairy milk and buffalo milk from Jaipur City, Rajasthan, India, Environment International, **2001,** 26, 231. DOI:10.1016/s0160-4120(00)00111-2.

[280] C. P. Kaushik, H. R. Sharma, A. Kaushik, Organochlorine pesticide residues in drinking water in the rural areas of Haryana, Environmental Monitoring and Assessment, **2012,** 184, 103. DOI:10.1007/s10661-011-1950-9.

[281] K. Mishra, R. C. Sharma, S. Kumar, Organochlorine pollutants in human blood and their relation with age, gender and habitat from North-east India, Chemosphere, **2011,** 85, 454. DOI:10.1016/j.chemosphere.2011.07.074.

[282] Eric F. Zama, Brian J. Reid, Hans Peter H. Arp, Guoxin Sun, Haiyan Yuan, Yongguan Zhu, Advances in research on the use of biochar in soil for remediation: A review, Journal of Soils and Sediments, **2018,** 18, 2433. DOI:10.1007/s11368-018-2000-9.

[283] Saurabh Mishra, Liu Cheng, Abhijit Maiti, The utilization of agro-biomass/byproducts for effective bio-removal of dyes from dyeing wastewater: A comprehensive review, Journal of Environmental Chemical Engineering, **2021,** 9, 104901. DOI:10.1016/j.jece.2020.104901.

[284] Segun Michael Abegunde, Kayode Solomon Idowu, Olorunsola Morayo Adejuwon, Tinuade Adeyemi-Adejolu, A review on the influence of chemical modification on the performance of adsorbents, Resources, Environment and Sustainability, **2020,** 1, 100001. DOI:10.1016/j. resenv.2020.100001.

[285] T. Sizmur, T. Fresno, G. Akgül, H. Frost, E. Moreno-Jiménez, Biochar modification to enhance sorption of inorganics from water, Bioresource Technology, **2017,** 246, 34. DOI:10.1016/j. biortech.2017.07.082.

[286] Arun Lal Srivastav, Agrochemicals detection, treatment and remediation, **2020,** Elsevier.

[287] Lin Ye, Xia Zhao, Encai Bao, Jianshe Li, Zhirong Zou, Kai Cao, Bio-organic fertilizer with reduced rates of chemical fertilization improves soil fertility and enhances tomato yield and quality, Scientific Reports, **2020,** 10, 177. DOI:10.1038/s41598-019-56954-2.

[288] Sanchita Mandal, Binoy Sarkar, Nanthi Bolan, Jeff Novak, Yong Sik Ok, Lukas Van Zwieten, Bhupinder Pal Singh, M. B. Kirkham, Girish Choppala, Kurt Spokas, Designing advanced biochar products for maximizing greenhouse gas mitigation potential, Critical Reviews in Environmental Science and Technology, **2016,** 46, 1367. DOI:10.1080/10643389.2016.1239975.

[289] Ali El-Naggar, Sang Soo Lee, Yasser Mahmoud Awad, Xiao Yang, Changkook Ryu, Muhammad Rizwan, Jörg Rinklebe, Daniel C. W. Tsang, Yong Sik Ok, Influence of soil properties and feedstocks on biochar potential for carbon mineralization and improvement of infertile soils, Geoderma, **2018,** 332, 100. DOI:10.1016/j.geoderma.2018.06.017.

[290] Kai Ling Yu, Pau Loke Show, Hwai Chyuan Ong, Tau Chuan Ling, Wei-Hsin Chen, Mohamad Amran Mohd Salleh, Biochar production from microalgae cultivation through pyrolysis as a sustainable carbon sequestration and biorefinery approach, Clean Technologies and Environmental Policy, **2018,** 20, 2047. DOI:10.1007/s10098-018-1521-7.

[291] N. L. Panwar, Ashish Pawar, B. L. Salvi, Comprehensive review on production and utilization of biochar, SN Applied Sciences, **2019,** 1, 1. DOI:10.1007/s42452-019-0172-6.

[292] Muhammad Ayaz, Dalia Feizienė, Vita Tilvikienė, Kashif Akhtar, Urte Stulpinaitė, Rashid Iqbal, Biochar role in the sustainability of agriculture and environment, Sustainability, **2021,** 13. DOI:10.3390/su13031330.

[293] Wojciech M. Budzianowski, High-value low-volume bioproducts coupled to bioenergies with potential to enhance business development of sustainable biorefineries, Renewable and Sustainable Energy Reviews, **2017,** 70, 793. DOI:10.1016/j.rser.2016.11.260.

[294] Patricia D. Fuentes-Saguar, Alfredo J. Mainar-Causapé, Emanuele Ferrari, The role of bio-economy sectors and natural resources in EU economies: A social accounting matrix-based analysis approach, Sustainability, **2017,** 9, 2383. DOI:10.3390/su9122383.

[295] Ke Sun, Jie Jin, Marco Keiluweit, Markus Kleber, Ziying Wang, Zezhen Pan, Baoshan Xing, Polar and aliphatic domains regulate sorption of Phthalic Acid Esters (PAEs) to biochars, Bioresource Technology, **2012,** 118, 120. DOI:10.1016/j.biortech.2012.05.008.

[296] Mohammed Baalousha, Aggregation and disaggregation of iron oxide nanoparticles: Influence of particle concentration, pH and natural organic matter, Science of the Total Environment, **2009,** 407, 2093. DOI:10.1016/j.scitotenv.2008.11.022.

[297] Noel P. Gurwick, Lisa A. Moore, Charlene Kelly, Patricia Elias, A systematic review of biochar research, with a focus on its stability in situ and its promise as a climate mitigation strategy, PLoS One, **2013,** 8, e75932. DOI:10.1371/journal.pone.0075932.

[298] Andrew R. Zimmerman, Bin Gao, Mi-Youn Ahn, Positive and negative carbon mineralization priming effects among a variety of biochar-amended soils, Soil Biology and Biochemistry, **2011,** 43, 1169. DOI:10.1016/j.soilbio.2011.02.005.

[299] Manav Saxena, Sheli Maity, Sabyasachi Sarkar, Carbon nanoparticles in "biochar" boost wheat (Triticum aestivum) plant growth, RSC Advances, **2014,** 4, 39948. DOI:10.1039/C4RA06535B.

[300] Xinde Cao, Lena Ma, Bin Gao, Willie Harris, Dairy-manure derived biochar effectively sorbs lead and atrazine, Environmental Science & Technology, **2009,** 43, 3285. DOI:10.1021/es803092k.

[301] Beluri Kavitha, Pullagurala Venkata Laxma Reddy, Bojeong Kim, Sang Soo Lee, Sudhir Kumar Pandey, Ki-Hyun Kim, Benefits and limitations of biochar amendment in agricultural soils: A review, Journal of Environmental Management, **2018,** 227, 146. DOI:10.1016/j.jenvman.2018.08.082.

[302] Jan Matuštík, Tereza Hnátková, Vladimír Kočí, Life cycle assessment of biochar-to-soil systems: A review, Journal of Cleaner Production, **2020,** 259, 120998. DOI:10.1016/j.jclepro.2020.120998.

[303] Z. Hauschild Michael, K. Rosenbaum Ralph, Irving Olsen Stig, Life cycle assessment: Theory and practice, **2018,** Springer.

[304] Maga Ram Patel, Narayan Lal Panwar, Biochar from agricultural crop residues: Environmental, production, and life cycle assessment overview, Resources, Conservation & Recycling Advances, **2023,** 19, 200173. DOI:10.1016/j.rcradv.2023.200173.

CHAPTER 8

Other Potential Applications for Biomass-Derived Porous Carbon

Song Yang*

ABSTRACT

According to the previous introduction, it can be seen that biomass-derived porous carbon has an extremely wide range of applications. In recent years, advancements in technology have spurred increased exploration of porous carbon applications, garnering broader attention in new areas, such as electrolysis, nuclear waste removal, and seawater desalination/batteries. This chapter first introduces the mechanism and recent advances of porous carbon or heteroatom-doped porous carbon for oxygen reduction, hydrogen evolution, oxygen evolution, carbon dioxide reduction, and organic pollutant degradation. Second, the application of porous carbon in nuclear waste removal of iodine, uranium, and cesium elements is summarized. Lastly, the application of porous carbon in seawater desalination and batteries is emphasized.

8.1 APPLICATION STATUS OF POROUS CARBON IN THE FIELD OF ELECTROCATALYSIS

Electrocatalysts have long been essential for the progression of green energy conversion technologies[1]. Exploring appropriate electrocatalysts for diverse energy conversion reactions stands as a highly dynamic field of contemporary investigation. In crafting high-performance electrocatalysts, three fundamental factors require consideration: the physical and chemical attributes of the active site, the substrate's conductivity, and its porous nature[2,3]. The size and electronic structure primarily dictate the physical and chemical characteristics of the catalytic center[4].

Porous carbon materials are commonly used as catalyst supports. Compared to traditional catalyst supports, the pore structure of new nanoporous carbon materials is ordered

DOI: 10.1201/9781003520566-8

and controllable in size. After loading, the catalyst is confined within the pores and is not easily aggregated, which significantly improves the catalytic efficiency[5,6]. Moreover, porous carbon electrodes offer numerous advantages, such as high electronic conductivity, adjustable molecular structure, and robust resistance to acidic or alkaline conditions. Furthermore, their non-metallic properties mitigate the risk of metal ion release, thereby minimizing environmental impact[7]. The pivotal characteristics of porous carbon electrodes establish them as essential catalysts for various reactions, spanning from oxygen reduction (ORR) and hydrogen evolution (HER) to oxygen evolution (OER), carbon dioxide reduction (CRR), and the degradation of organic pollutants in wastewater[8–11].

8.1.1 Application of Porous Carbon in ORR Reaction

ORR, a cathodic reaction crucial for various energy applications like fuel cells, metal-air batteries, and chlor-alkali electrolysis, is intricate, involving four linked electron and proton transfers due to the robust O=O bond. Presently, platinum-based electrocatalysts are considered the most efficient for ORR. However, their practical utility is curtailed by high cost, poor stability, cross-effects, CO poisoning, and limited natural reserves. Numerous studies aim to develop superior non-precious ORR catalysts. In the past decade, several attempts have focused on creating non-metallic carbon catalysts for ORR. Certain carbon materials doped with diverse heteroatoms like N, B, S, or P have exhibited activity and longevity comparable to state-of-the-art Pt-based electrocatalysts[12–14].

A study conducted by A. Celzard et al.[15] aimed to optimize porous carbon electrocatalysts' oxygen reduction reaction (ORR) efficiency in alkaline media. The investigation utilized rotating ring disk electrodes (RRDE) or Koutecky Levich (KL) equations to evaluate the number of transferred electrons. Factors such as ionomers (Nafion), electrode carbon load, carbon grinding method, and upper potential selection during ORR measurement were found to be influential. These findings provide valuable insights into the appropriate procedures and assessment of ORR performance for non-metallic porous carbon materials.

Choi et al.[16] presented a carbon template doped with sulphur in zeolite configuration, demonstrating a high sulphur content of 17 wt% S and a unique carbon structure composed of a three-dimensional network of graphene nanoribbons with high curvature. This configuration stabilized platinum loading of up to 5 wt%, in the form of highly dispersed species which included atoms that were site-separated. Density functional theory (DFT) was employed to establish a Pt S4 model structure for two thiophenes and two sulfates connected to Pt. The sulphur in the porous carbon coordinates with platinum-single atom catalysts via a dual electron pathway, effectively inhibiting the four-electron pathway due to a significantly higher kinetic barrier. The sulphur in the porous carbon coordinates with platinum-single atom catalysts via a dual electron pathway, effectively inhibiting the four-electron pathway due to a significantly higher kinetic barrier. The sulphur in the porous carbon coordinates with platinum-single atom catalysts via a dual electron pathway, effectively inhibiting the four-electron pathway due to a significantly higher kinetic barrier. This leads to significant selectivity for ORR to generate hydrogen peroxide. Zhang et al.[17] utilized nitrogen and sulphur to jointly regulate the carbon charge group, achieving a precise and balanced distribution of charges for optimal performance (Figure 8.1).

When pyrrole-N coexists with a hexagonal nitrogen ring or thiophenes, it may disturb charge delocalization. Consequently, sulphur is preferentially released to the site of formation for the five-membered ring which in turn induces the precise formation of the six-membered nitrogen ring. The electron-withdrawing properties of six-membered nitrogen and thiophenes lead to positively charged adjacent carbon compared to single heteroatoms, resulting in a noteworthy increase in reduction activity. In addition, DFT calculations have revealed that N and S-ACTP display reduced energy barriers for *OOH fracture into *O. This promotes a smoother progression of the ORR reaction.

Metal nitrogen carbon (M-N-C, M = Fe, Co, etc.) catalysts are widely regarded as the most promising electrocatalysts for ORR and have attracted significant interest from researchers in recent years[18,19]. M-N-C catalysts, in particular, demonstrate ideal ORR performance, which is linked to the exclusive active sites of surface nitrogen and matching metal. Liu et al.[20] developed a model to predict Fe-N_4 site ORR activity using density functional theory calculations, focusing on frontier orbital energy levels and spatial structure. They demonstrated how vacancy defects at the Fe-N_4 site regulate ORR activity. Results indicated that Fe-N_4 ORR activity stems from hybridization between Fe $3dz^2$, 3dyz (3dxz), and O_2 3 orbitals. Fe-O bond length, spin state d-band center gap, Fe site magnetic moment, and $*O_2$ were identified as accurate predictors of Fe-N_4 site ORR activity. Furthermore, these descriptors revealed that ORR activity in the Fe-N_4 site is primarily localized within two distinct regions, closely associated with the height of the projected Fe 3d orbit in the Z direction. This study presents novel insights into the ORR activity of single-atom M-N-C catalysts.

Furthermore, there has been significant research on electrocatalysts based on porous carbon doped with dual or triple elements, such as Fe and N doped porous graphite

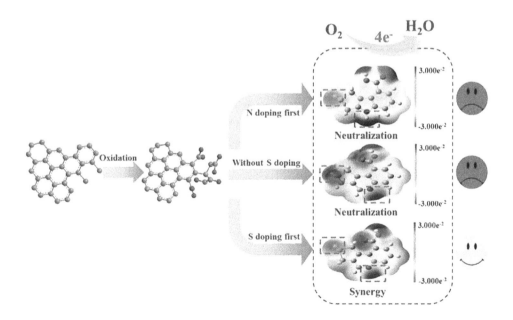

FIGURE 8.1 Charge distribution map of N and S co-doped carbon for ORR[17].

carbon[20], three-dimensional porous graphite carbon doped with Fe, Co, and N[21], and porous carbon doped with N, S, and Fe[22]. Therefore, an efficient and stable electrocatalyst for ORR can be reasonably designed by combining the porous structure of bio-based carbon with the electronic structure induced by heteroatom doping (N, S, Fe, Co, etc.) to induce lattice distortion and electron density changes.

8.1.2 Application of Porous Carbon in HER and OER Reactions

Hydrogen is a sustainable energy carrier, sought after as a substitute for fossil fuels. Hydrogen production through water splitting is a promising avenue, with electrocatalytic water splitting representing the reverse process of fuel cell reactions, notably the cathodic HER and anodic OER. Efficient catalysts are imperative to bolster reaction kinetics for both processes. Presently, precious metals like Pt and precious metal oxides such as IrO_2 demonstrate optimal performance for HER and OER, respectively, particularly in alkaline media. Nevertheless, their high cost and scarcity significantly constrain their applicability, and porous carbon materials have been discovered to possess unforeseen characteristics in HER and OER.

Carbon is generally inactive in most electrochemical reactions, such as HER and OER. In order to enhance the binding energy of reactants and reaction intermediates on carbon, surface modification is required to modify their structure. Up until now, most surface modification strategies have followed the strategy used in ORR. Incorporating heteroatoms into carbon matrices has proved to be a successful technique for modifying carbon's electronic properties, leading to the formation of active sites for HER or OER. As these dopants have different electron densities compared to carbon, they can influence the valence orbital energy levels of adjacent carbon atoms, alongside enabling the adsorption of hydrogen or oxygen atoms.

The range of elements doped into carbon catalysts spans from non-metals to transition metals. To further improve its efficacy, carbon is frequently doped with more than one element. Table 8.1 details common doped carbon catalysts and their HER overpotentials at 10 mA/cm². Notably, these doped elements can considerably enhance the HER or OER activity of carbon. Among them, N-doped carbon is considered one of the most extensively studied HER and OER materials. Lu et al.[23] employed N-doped graphite sheets (N-GP) as a research model to assess the hydrogenation of N and C atoms during HER, and the subsequent reconstruction of the carbon framework. This resulted in significant improvement in HER activity. The N dopant gradually underwent hydrogenation and dissolved almost completely as ammonia. Theoretical simulation calculations uncovered heteroatom-doped carbon as the active source for HER (Figure 8.2). Compared to HER, the detailed mechanism of OER is difficult to describe due to the involvement of complex electrochemical reactions and multi-electron transfer steps. In a study by Zhao et al.[24], Nitrogen-doped carbon exhibited an overpotential of 380 mV under 0.1 M KOH conditions and a mass load of 0.2 mg/cm² at 10 mA/cm². The OER activity predominantly depends on the abundance of pyridine or/and quaternary ammonium salt-related active sites. Additionally, different carbon structures like mesoporous carbon, graphene, carbon nanotubes, and carbon fibers demonstrate significant activity towards OER in alkaline environments via heteroatom doping. Table 8.2 displays some typical non-metallic doped carbon structures and their corresponding properties.

TABLE 8.1 Overpotential Review of Porous Carbon Structure HER

Catalysts	Overpotential at 10 mA/cm² (mV)
WS₂ nanolayers with P/N/O-doped graphene[25]	104
WS₂ anchored to hollow N-doped carbon nanofibers[26]	280
Mo₂C encapsulated by N/P-codoped carbon shells and N/P-codoped reduced graphene oxide[27]	24
MoSₓ layer on vertical N-doped carbon nanotube[28]	92
SnS on N-reduced graphene sheets[29]	130
MoSₓ clusters decorated N-doped graphene[30]	181
FeP nanoparticles on graphene sheets[31]	123

TABLE 8.2 Overview of OER Overpotential of Non-Metallic and Metal Doped Carbon-Based Materials

Catalysts	Overpotential at 10 mA/cm² (mV)
N-doped porous carbon@graphene[32]	412
N-doped mesoporous carbon nanosheet/carbon nanotube hybrid[33]	313
N-doped ordered mesoporous carbon/graphene framework[34]	341
N/P-codoped graphene/carbon nanosheets[35]	319
N/S-codoped carbon nanotubes[36]	351
Fe–N-codoped graphene[37]	402
Atomic dispersion of Fe–Nₓ species on N and S co-decorated hierarchical carbon layers[38]	397

8.1.3 Application of Porous Carbon in CO_2RR Reaction

The rapid consumption of fossil fuels results in a significant amount of CO_2 being produced in the atmosphere, which causes rapid environmental degradation. To minimize our dependence on fossil fuels, CO_2RR is a highly encouraging direction as it diminishes the accumulation of carbon dioxide and converts intermittent renewable electricity into energy-intensive fuels. However, implementing CO_2RR in aqueous media remains challenging due to issues with existing electrocatalysts, including poor selectivity, competitive HER side reactions, and low electrochemical stability. Overcoming the activation barrier and enhancing reaction kinetics can be achieved by increasing the overpotential; however, this yields an unfavorable competitive HER. Porous carbon with a rich geometric structure and controllable surface chemical properties has been developed as a carrier catalyst to address this issue. This catalyst enables efficient and highly selective CO_2RR.

Xue et al.[39] have reported on their latest research developments regarding the creation of various defect types such as intrinsic carbon defects, heteroatom doping defects, metal atom sites, and edge detection in carbon materials for CO_2RR. Their findings reveal the structure-activity relationships and catalytic mechanisms involved (Figure 8.3(a)). In particular, the existence of innate flaws, heteroatom doping, and metallic atom introduction primarily modify the nearby electronic structure of the carbon framework, leading

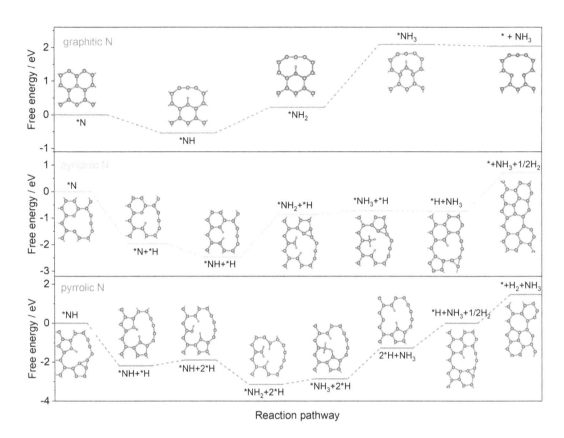

FIGURE 8.2 DFT calculation of possible hydrogenation pathways for graphite N, pyridine N, and pyrrole N[23].

to uneven charge distribution. By employing charge redistribution to refine the carbon matrix, one can enhance the density of reactive sites and the catalyst's catalytic efficacy. Different modified porous carbon catalysts were constructed according to Figure 8.3(b) to selectively use distinct reaction pathways for carbon monoxide (CO), methane (CH_4), formic acid (HCOOH), formaldehyde (HCHO), and methanol (CH_3OH) as C1 and poly-carbonate products (C^{2+}, ethanol (C_2H_5OH), ethane (C_2H_6), ethylene (C_2H_4), etc.). For instance, Wallace et al.[40] demonstrated that tin nanoparticles supported on nitrogen-doped carbon catalyst can efficiently facilitate the production of formates, with a Faradaic efficiency of 62% at an overpotential of 690 mV. Additionally, tin single atoms on carbon (SAC) catalyst can selectively transform CO_2 into CO with a high Faradaic efficiency of 91% at a relatively low overpotential of 490 mV. Recent studies conducted by He et al.[41] revealed that copper single atoms on carbon catalyst can generate nearly pure methanol on porous carbon nanofiber membranes with a liquid-phase Faradaic efficiency of 44% and a current density of 93 mA/cm². Based on DFT calculations, the catalytic center (Cu-N_4) has a free energy of 0.12 eV, indicating it is an important step. The adsorbed *CO intermediate can be further reduced to methanol at this stage. Additionally, the porous structure facili-tates mass transfer, significantly boosting the efficiency of CO_2RR.

(a)

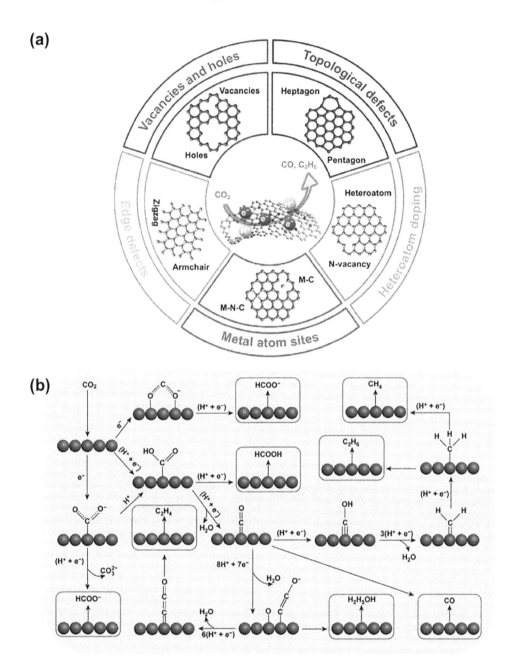

FIGURE 8.3   (a) Construction of different types of defects in carbon materials in ECR. (b) Potential ECR pathways for the formation of CO, HCOOH, C_2H_4, C_2H_6, and C_2H_5OH products[39].

8.1.4   Application of Porous Carbon in Electrocatalytic Oxidation Degradation of Organic Pollutants in Wastewater

With the growth of industry, the release of wastewater from diverse industries is increasingly prevalent, and industrial wastewater significantly affects both the watershed environment and the livelihood of residents. Industrial organic wastewater is characterized by its diverse composition, which varies depending on the industry. This wastewater is typically complex, containing hazardous heavy metals and persistent organic pollutants[42]. Excessive accumulation of these components in organisms can result in poisoning and cancer, presenting a hazard to both the environment and human health[43]. As a result, industrial organic wastewater pollution has become one of the most pressing environmental issues that needs to be addressed, given the increasingly severe risks and stringent regulations.

Porous carbon possesses a high specific surface area, a well-developed porous structure, and abundant oxygen-containing surface functional groups. These features facilitate the enrichment and adsorption of pollutants[44]. Porous carbon stands as the most extensively employed electrode material. Simultaneously, its high specific surface area and porosity enable both adsorption and electroadsorption during the electrochemical process, thus enhancing degradation efficiency even further. Zhang[45] and colleagues constructed biochar particle electrodes, Ti Sn Co/bamboo BC, to eliminate coking wastewater. Sun et al.[46] engineered powdered sludge carbon through pyrolysis at 800 °C that can efficiently eliminate acid orange 7 (AO7).

This section presents a comprehensive review of the current research status and applications of bio-based porous carbon in the field of electrocatalysis. Porous carbon exhibits tremendous potential as an electrocatalyst for diverse energy conversion applications due to its excellent heteroatom doping, high conductivity, high porosity, and low cost. By adjusting the heteroatoms or their configurations appropriately, the electronic structure of porous carbon can be vastly different, leading to varying electrochemical activity and selectivity depending on the needs. Owing to its numerous advantages as a carrier, it is widely believed that porous carbon carriers hold great potential in various energy-related sectors.

8.2 APPLICATION STATUS OF POROUS CARBON IN NUCLEAR WASTE REMOVAL

As a safe, reliable and clean energy source, nuclear energy has shown great advantages in alleviating the energy crisis and environmental problems, but the development of nuclear energy inevitably generates and releases radionuclides into the environmental system, posing a serious threat to ecological and environmental safety. From the perspective of environmental radiation safety, the enrichment and separation of radioactive pollutants and the decontamination and remediation of contaminated media can provide theoretical and technical support for the control of radioactive contamination risks and nuclear environmental safety, and is also of great significance for nuclear emergency response and the sustainable development of nuclear energy.

The methods of radionuclide enrichment and separation include precipitation, reverse osmosis, ion exchange, electrodialysis, solvent extraction, and evaporation and adsorption. Among them, adsorption method has been widely concerned and applied due to the advantages of simple operation, strong relevance, and a wide range of material sources, etc.[47]. However, traditional adsorbents (e.g., clay minerals[48] and zeolites[49]) are restricted from wide application due to the drawbacks such as low adsorption capacity, poor selectivity, and environmental friendliness, and the development of new environmentally friendly and highly efficient and stable adsorbent materials is becoming a hot research topic in the field of environmental radiochemistry. However, traditional adsorbents (e.g., clay minerals[48] and zeolites[49]) are limited by their low adsorption capacity, poor selectivity, poor environmental friendliness and other shortcomings. Among the common new adsorbents, carbon-based nanomaterials have been widely studied.

Organic porous materials have great potential for adsorption, and some functionalized materials for the adsorption and separation of uranium, copper, and iodine have been reported, but the adsorption performance of different materials cannot fully reflect the advantages and disadvantages of each material when measured under different pH, temperature, and pressure conditions. The conformational relationship between organic porous materials and radionuclides has a significant influence on the adsorption and separation effects, and the organic porous materials can be endowed with good specific functions through precise design and modification. Therefore, in order to scientifically evaluate the advanced nature of each material and develop better adsorbents, it is necessary to analyze the characteristics of each material in detail in terms of the conformational relationship. By combining different monomers and structures through software simulation, it is possible to simulate the design of the constitutive relationship and to evaluate the effect. Using a simple method, agricultural wastes can be made into activated carbon with different pore sizes[50]. We begin with an exploration of the conformational relationships and structural design of organic porous materials, progressing to discuss their application in radionuclide adsorption and separation. Additionally, we delve into the utilization of organic porous materials in treating radioactive nuclear wastes, examining aspects such as computer-assisted design, effect evaluation, and practical application. Our discussion centers around covalent organic frameworks as a representative example.

8.2.1 Application of Porous Carbon in the Adsorptive Separation of Iodine Elements

Energy sustainability stands as one of humanity's paramount scientific and technological challenges. With fossil fuel reserves dwindling and global energy demand surging, nuclear energy emerges as a safe and clean alternative capable of significantly reducing greenhouse gas emissions. Radioactive iodine, a byproduct of nuclear fission, is characterized by its high volatility and mobility[51]. These harmful iodine with long half-life accumulate in biological thyroid tissues through the food chain, affecting the normal metabolic process and posing a great danger to the human body[52]. ^{129}I and ^{131}I, the most abundant radioactive iodine nuclides, are usually found in the combination of I_2 with other hydrocarbons to produce organic substances such as methyl iodide (CH_3I). Additionally, iodine ions have been identified in wastewater discharged from industries such as biological, pharmaceutical, and food, leading to irreversible pollution of groundwater. Hence, effectively separating and treating radioactive iodine elements poses significant challenges in safeguarding the environment and human health.

Under different synthesis conditions, the organic porous materials can have tubular, spherical, solid, and hollow morphologies, and thus the adsorption of iodine is not limited to the intrinsic pore structure. Chen et al.[53] prepared CMPN-1-3, which had an adsorption capacity of 2080 mg/g of gaseous iodine, and the presence of iodine crystals on the inside and the surface of the tube was demonstrated by transmission electron microscopy.

Transmission electron microscopy showed the presence of iodine crystals inside and on the surface of the CMPN-3 tubes, and it was also found that the structure of the monomer and the solvent conditions had a significant effect on the morphology of the CMPN products. Ren et al.[54] synthesized two thiophene-containing co-compacted microspore polymers, SCMPs, with palladium as catalyst. SCMPs are aggregates composed of microspheres, and due to the three-dimensional honeycomb porous mesh structure, SCMP-I has an adsorption capacity of 3,450 mg/g of gaseous iodine. Yin et al.[55] reported the adsorption of iodine by heteroporous COF hollow microspheres SIOC-COF-7, which has an adsorption capacity of 4,810 mg/g of gaseous iodine, and found that iodine was mainly distributed in the inner cavities of the micropore and in the porous shells, and that the heterogeneous It was found that the iodine was mainly distributed in the microporous inner cavities and porous shells, and the heterogeneous aromatic ring, high ammonia content and ordered lattice structure contributed to the iodine enrichment. Lin et al.[56] synthesized a hydrogen-bonded covalent organic polymer, pha-HcOP-1, which was a spherical particle of ca. 20 nm using isonicotinic tripe as a bifunctional connector, and the adsorption capacity for gaseous iodine was 1310 mg/g, and the solution of the polymer was also characterized by the presence of a pore-rich, nt-co-alternate phenyl ring, and the functionalization of one-CO-one-NH-one building blocks. The adsorption capacity of iodine in solution was 833.33 mg/g.

Activated carbon (AC) has garnered significant attention as the primary component of iodine adsorption devices, notably iodine filter cartridges, owing to its high specific surface area and stability. Typically derived from organic raw materials through carbonization and activation processes, activated carbon undergoes chemical activation using substances like KOH, $ZnCl_2$, and H_3PO_4 to yield a porous structure and high specific surface area. However, most activated carbons utilized in iodine filters exist in granular form, exhibiting poor mechanical strength and processing properties. To further enhance the physicochemical attributes of activated carbon materials, doping with elements such as N and S has been introduced.

Flexible phenolic porous carbon fibers with high specific surface area of 1,500–2,200 m^2/g were prepared under KOH activation (Figures 8.4 and 8.5)[57]. The nitrogen-doped sample NDAC-4 exhibited impressive liquid and gaseous iodine adsorption capacities, reaching 2.12 g/g, a remarkable 3.4 times higher than that of commercially available activated carbon particles (0.84 g/g) and 5,530 mg/g. Moreover, NDAC-4 demonstrated effective adsorption of radioactive methyl iodine at 0.51 g/g, showcasing a significant 1.96 times increase compared to commercial unmodified viscose fibers (0.26 g/g).

Among various types of coal-based activated carbon, columnar coal-based activated carbon is[58] the most widely used. The process of making columnar coal-based activated carbon has two main steps: carbonization and activation. Carbonization involves heating activated carbon raw materials without oxygen at high temperatures. Activation enhances the carbon's pore structure through additional treatments, which can be either physical or chemical[59]. Physical activation involves using steam, CO_2, air, or other activators at high temperatures to activate char. Chemical activation, on the other hand, involves adding chemical reagents like $ZnCl_2$, H_3PO_4, KOH, and so forth[58,60]. Typically, for large-scale activated carbon production, physical activation is preferred due to its cost-effectiveness and environmental

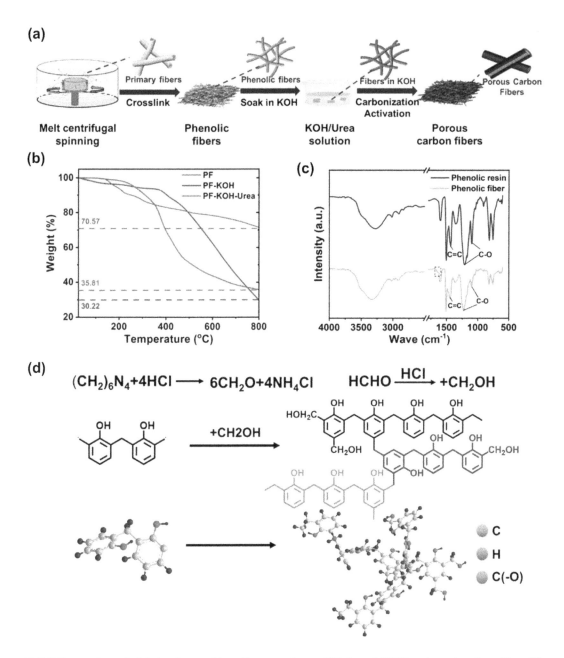

FIGURE 8.4 Sample fabrication and bonding reactions of N-doped KOH activated carbon fibers[57].

friendliness[61,62]. Currently, most studies on activated carbon focus on the pore structure and adsorption properties (Liu et al.[63], Jing et al.[64], Muthmann et al.[65], etc.).

8.2.2 Application of Porous Carbon in Adsorption Separation of Uranium Elements

The emergence of the self-assembly concept has made it possible to extend organic small molecules from zero-dimensional to one-, two- and even three-dimensional periodic

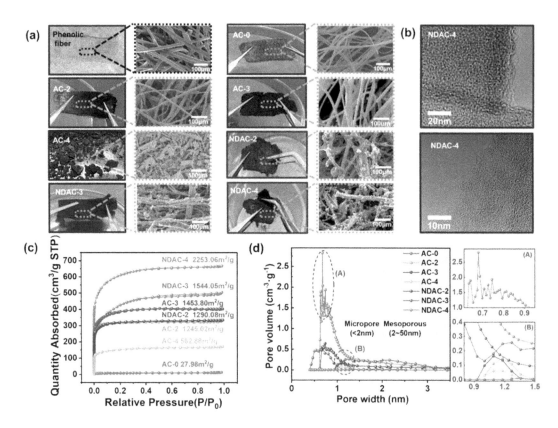

FIGURE 8.5 Surface structure and pore size distribution of N-doped KOH activated carbon fibers[57].

structures[66]. In this context, organic porous materials based on rigid organic small molecules connected and propped up with open pores have gained wide attention in the adsorption and separation of uranium due to their better stability and ease of structural modulation and functionalization. In order to make the organic porous adsorbents meet the application requirements of acid and alkali resistance, irradiation resistance, and reusability, strategies such as size sieving, introduction of hydrogen bonding, composite materials, reciprocal isomerization, and introduction of stabilizing monomers have been used for the construction of the organic porous adsorbents; in order to make the organic porous materials realize the high selectivity of the organic porous materials to uranium, the methods of introduction of heteroatoms, post-modification, coordinated coordination, introduction of auxiliary groups, and density functional calculations have been used for the design of organic In order to make organic porous adsorbents easy to prepare and process, methods such as hydrogen bonding, membrane making and spinning have been applied to organic porous adsorbents. These strategies provide a reference for the development of more advanced organic porous adsorbents.

In the early days, organic porous materials were mainly composed of elements such as B, C, N, O, and H. On the one hand, the lighter mass of the element makes the material have a larger specific surface area, on the other hand, the less variety of the element makes the material's structural diversity is not rich enough, therefore, it has been reported that other atoms are introduced into the organic porous materials in order to enrich the structural diversity and to broaden the scope of application. The hexachlorocyclotriphosphonitrile has a symmetric six-membered ring structure with good chemical stability, and the chlorine in the phosphorus-chlorine bond can be easily replaced, which can be used to synthesize organic porous materials. Zhang et al.[67] used hexachlorocyclotriphosphonitrile and p-phenylenediamine as the monomers to construct a two-dimensional ultramicroporous phosphonitrile covalent organic framework (MPCOF) with a "three-dimensional" structure. In a solution of 1 mol/L HNO_3 and 12 cations, the adsorption capacity of MPCOF for uranium reached 57 mg/g, and the uranium selectivity (uranium adsorption as a percentage of total cation adsorption) reached 64%. Photoelectron spectroscopy showed that there was no chemical interaction between MPCOF and uranium, and the selectivity of the material for uranium was mainly due to the size-sieving effect of the ultramicroporous structure. This strategy, which employs stable monomers and an acidity-independent size-screening mechanism to achieve selective adsorption of clocks, provides an idea for the adsorption and purification of highly acidic uranium-containing wastewater. Based on the same strategy, Zhang et al.[68] utilized the stability of phosphorus-oxygen bonding to construct two pore-size microporous covalent phosphonitrile frameworks (CPFs), CPF-D and CPF-T, using hexachlorocyclotriphosphate (HCTP), p-phenylenebenzofuran (PBF), and m-tritylenebenzofuran (m-TBF) as the monomers, and the adsorption capacity of uranium was maximum (62 mg/kg) at pH 4.0, with a maximum of 1 mg/kg (2 mg). The maximum capacity of uranium adsorption was achieved at pH=4.0 (62 mg/g), and the capacity of uranium adsorption was still 48 mg/g at 3 mol/L HNO_3: Compared with CPF-D (pore size of 1.82 nm), the pore size of CPF-T (0.95 nm) was closer to that of the hydrated uranyl ions (0.59–0.66 nm), which was more favorable for the water molecules in the hydrated uranyl ions to form hydrogen bonds with the ammonia oxygen atoms of the pore channels. CPF-D loses its ability to adsorb uranium due to the protonation of functional groups at high acidity, while CPF-T effectively prevents the protonation of functional groups due to the size-matching effect, resulting in the excellent uranium adsorption ability of CPF-T even at high acidity. In this work, hydrogen bonding was utilized to improve the chemical stability of the material, and the adsorption of uranium was achieved under the condition of 3 mol/L HNO_3, which showed the potential application of organic porous materials in the treatment of uranium-containing wastewater with high acidity.

Compared with covalent bonding, it is easier to construct organic porous materials by hydrogen bonding, and the materials are easier to be regenerated. Li et al.[69] synthesized a hydrogen bonded supramolecular organic framework (SOF), MA-TMA, by using melamine (MA) containing nitrogen atoms and homotrimethylene tricarboxylic acid (TMA) containing oxygen atoms. The adsorption capacity of MA-TMA for uranium was up to 324 mg/g at pH = 2.5 and in the presence of 12 cations, with 92% selectivity. The adsorption capacity of MA-TMA for uranium was 324 mg/g with 92% uranium selectivity, and the morphology of the adsorbed uranium was changed from nanoribbon structure to tens of micrometer agglomerated particles. After the interaction of MA-TMA with uranium, the hydrogen bond between MA and TMA was replaced by the coordination bond between uranyl ion and ammonia oxygen atom to form MA-UO+ -TMA. The uranium-selective organic porous materials constructed based on the concept of induced-matching ion recognition provide ideas for the design of highly selective uranium adsorbents.

The mesoporous silicon oxide SBA-15 was used as a template, and different masses of iron and carbon sources were introduced into the template pores as precursors by the nano-casting method. After the in-situ polymerization reaction, the surfaces of the mesoporous SBA-15 were then post-grafted and modified using[3-(trimethoxysilyl) propyl]urea (UPTS) and aminopropyltriethoxysilane (APS) organic reagents to obtain the functionalized charcoal-based magnetic materials (FCMMCs) with a well-graded and orderly mesoporous structure. Functionalized carbon based magnetic mesoporous materials (FCMMCs) were obtained by post-grafting modification of their surfaces with organic reagents. The structure of FCMMC was characterized by infrared spectroscopy (FT-IR) and N_2 adsorption-desorption, respectively. The effects of solution pH, initial uranium concentration, adsorbent dosage, and adsorption time on the adsorption of uranium by FCMMC were investigated. The results showed that: Both carbon and iron groups were loaded on the mesoporous silicon oxide matrix, and the FCMMC had a high specific surface area and a narrow pore size distribution. The adsorption kinetic model and adsorption isotherm model were analyzed, and the adsorption kinetic process of uranium on FCMMC conformed to the quasi-secondary kinetic model, and the adsorption isotherm conformed to the Langmuir isotherm model, and the maximum theoretical adsorption capacity was 128.69 mg/g. Meanwhile, the desorption and regeneration of FCMMC for 8 times by three different desorbents showed that the adsorption rate of uranium on FCMMC was over 80%, which indicates that FCMMC has good regeneration capacity. This indicates that FCMMC has good regeneration performance.

8.2.3 Application of Porous Carbon in Adsorption Separation of Cesium Elements

Nuclear energy emerges as an alternative to fossil fuels, but nuclear plant mishaps pose severe environmental threats. In 2011, the Fukushima nuclear disaster, triggered by a magnitude 9 earthquake, released substantial radionuclides like radioactive iodine (^{131}I), strontium (^{90}Sr), and cesium (^{137}Cs, ^{134}Cs). Approximately 6.3 bins of these isotopes contaminated soil, air, and water. Cs, with a half-life of 30.2 years, is particularly concerning in nuclear plant wastewater[70–73]. Absorbed cesium can cause harm to the human body and cells, resulting in functional defects and genetic disorders[74]. Several methods have been explored for removing Cs+ from wastewater, such as membrane distillation, chemical precipitation, evaporation, and adsorption[75]. Among the mentioned treatment methods, evaporation might lead to issues like corrosion, scaling, and foaming. Membrane filtration has drawbacks such as high cost, process complexity, membrane contamination, and low permeate flux. In contrast, adsorption stands out for its affordability, ease of operation, and eco-friendliness. It is widely regarded as one of the simplest and most efficient approaches for eliminating Cs+ from wastewater[71]. Numerous Cs adsorbents have been created for cesium ion removal, including clay, mineral oxides, zeolites, titanate-based materials, and microbial fuel cells[76]. However, many of these adsorbents are costly, have limited adsorption capacity, and involve complex production processes, making large-scale deployment impractical[77]. Moreover, residual adsorbents in wastewater can lead to secondary pollution[78]. Consequently, there's a growing focus on developing safe and efficient new types of adsorbents for removing radiocesium.

Prussian blue (PB) is a widely used inexpensive dark blue dye pigment with a face-centered cubic lattice structure that includes an open zeolite-like framework[79]. It can be easily synthesized by mixing Fe^{3+} and $[Fe(CN)_6]^{4-}$ in an aqueous solution, making it cost-effective and suitable for large-scale applications[80]. However, PB tends to aggregate due to its small particle size, hindering the exposure of adsorption sites and dispersion in solution. Moreover, its strong water solubility makes it challenging to separate from the waste solution. To address these issues, various carrier systems such as graphene oxide foams, carbon nanotubes, nonwoven fabrics, and sodium titanate nanoribbon membranes have been developed to increase the specific surface area of the adsorbent and prevent PB aggregation[71,73,81]. However, these carriers are not derived from natural sources, raising concerns about their biological and safety implications. Therefore, there is growing interest in developing "green economy" technologies for the immobilization and separation of PB[82,83].

Organic porous materials have shown great advantages and potentials in radionuclide adsorption and separation applications. Due to the wide variety of organic porous materials and their inconspicuous boundaries, this paper focuses on crystallographically ordered COFs and discusses the opportunities and challenges of organic porous materials in terms of functional group synergistic coordination and theoretical calculation-assisted design.

Currently, the influence of the functional ligand space configuration of adsorbent materials on the conformational relationship and adsorption properties has received much attention, but only a few reports have carried out theoretical computational studies on the problems related to the space configuration of functional ligands[84]. Computer-assisted design can play an important role in screening the target structures and realizing the design of high-performance ligands. Hay et al.[85] Starting from the size, charge, and electronegativity of the metal ions, through the empirical formulas to derive five parameters to define M-O stretching, M-O=C bending, and M-0=C one-X torsion, according to which 52 metal-coolamine complexes were.

The crystal structure data were fitted to obtain the molecular mechanics force field of the metal-coolamine complex (MM3(96)). Lumetta et al. applied the MM3 force field to theoretical calculations of the Eu(I) complex of propylene dicoolamine, and the results showed that the donor oxygen atom of the structure of propylene dicoolamine was poorly bonded to the trivalent region elements, and that the new ligands obtained by restructuring of the dicoolamine structure had much higher affinity for Eu(M). was greatly improved. Subsequently, Lumetta et al.[86] synthesized a computer-aided design of the new ligand, and the results of liquid-liquid extraction showed that the affinity of the new ligand for Eu(I) was increased by seven orders of magnitude.

To further utilize the role of computer-aided design, the molecular software HostDesigner has been developed in the field of coordination chemistry. The scoring algorithm contained in HostDesigner can rapidly screen a large number of candidate structures[86,87], and together with the molecular mechanics software PCModel or the GMMX module, a large number of potential target structures can be generated and screened, which is a good guide for the ab initio design of functional organic materials[87–89], and has attracted attention in the design of organic porous materials. Duncan et al. investigated the thermodynamic, kinetic, and structural factors for the synthesis of covalent organic frameworks, and in the process of constructing the covalent organic frameworks, HostDesigner was used to screen five candidate monomers from more than 8,000 molecular fragments. However, HostDesigner could not resolve the entropy, kinetics, template, and structural factors. cannot solve the problems of entropy, kinetics, template action, reaction conditions (solvent, temperature, concentration and catalyst, etc.), the degree of interpenetration of dia topological frameworks cannot be predicted and the number of atoms of the screenable fragments is not more than 200[90], Therefore, there is still a long way to go before assisting in the design of the complete structure of organic porous materials. Currently, computer-assisted tools are widely used to predict the gas adsorption properties of organic porous materials. Babarao et al. studied the adsorption of carbon dioxide on three-dimensional covalent organic frameworks, in which carbon dioxide adsorption was simulated by using the giant canonical Monte Carlo (GCMC) method, which is widely used in gas adsorption simulation software.

The reports on the use of computer-aided design for organic porous materials are fewer, protein adsorption, Zhou et al.[91] with the help of uranium, a large-scale screening algorithm, selected the best solution from a large number of simulated protein structures by choosing the coordination mode, selecting the scaffolds to establish the oxygen and hydrogen libraries, searching for the uranium sites, scoring and sorting, and selecting the top solution, etc. The adsorption experiments showed that the selectivity of this protein for uranium was 10,000 times higher than that for other metal ions. The adsorption experiments showed that the selectivity of this protein for uranium is 10,000 times higher than that for other metal ions. This work demonstrates the potential of computer-aided design, which provides a better idea for the screening and design of functional organic porous materials. Since COFs are characterized by crystalline ordering, which is suitable for precise regulation of functional group positions, the use of computer-aided design to obtain synergistically functionalized COFs may be a strategy for achieving high selectivity in organic porous materials in the future.

8.3 POROUS CARBON IN SEAWATER-RELATED APPLICATIONS

In the context of the continuous growth of the global population, the increasing scarcity of freshwater resources, and the rising demand for industrial and agricultural water, 6 billion people worldwide are facing a serious problem of freshwater resource shortage. However, the use of porous carbon in the process of seawater desalination provides an effective solution to this problem. At present, a variety of technological means are being widely used, such as interfacial solar seawater desalination technology and adsorption cooling and seawater desalination (ACD) systems. In addition, to address the energy shortage caused by population growth, porous carbon materials are also applied in the field of seawater batteries. The emergence of these technologies provides new ideas and approaches for humans to solve the problems of water resource shortage and energy crisis.

8.3.1 Application of Porous Carbon Materials in Seawater Desalination

Seawater desalination methods include thermal processes, membrane processes, and other techniques. Thermal processes involve methods like multi-stage flash distillation and vapor compression distillation, while membrane processes include reverse osmosis and electrodialysis[92]. Other methods include ion exchange, hydration, and solar seawater desalination. Reverse osmosis and multi-stage flash distillation are the most commonly used methods, with reverse osmosis being the most widespread, accounting for over 50% of installed capacity[93,94].

8.3.1.1 Solar Energy Desalination of Seawater

With the development of technology, solar thermal evaporation, and humid air desalination technology have an increasing impact on remote areas or very small communities, especially in developing countries[95,96]. Solar energy is used more and more for seawater desalination, and the design of using solar energy mainly involves around three steps: (1) absorbing solar energy and effectively converting it into heat; (2) allowing steam to evaporate in a rich porous structure; (3) having a floating evaporation structure to maximize evaporation efficiency while continuously providing liquid with reduced energy loss. Delaziz et al.[97] tested solar desalination combined with reusable activated carbon, achieving a minimum cost of $0.01090 per liter. El-Said et al.[98,99] experimented with a low-power, fast ultrasonic humidifier for solar desalination, yielding a distillate cost of $0.0112 per liter. El-Said and Abdelaziz[100] enhanced a solar still with HFUA to boost evaporation efficiency and assessed its economic feasibility. Peng et al.[101] developed a compact flat-plate solar still with a maximum daily output of 7 kg/m2. Kandel et al.[102] analyzed the economic performance of a solar distiller desalination system integrated with copper chips, nanofluids, and nano-based PCM, showing significant efficiency improvements compared to conventional systems.

Peng et al.[103] and Sharshir et al.[104] investigated how porous materials, like floating coal, cotton fabric, and carbon black nanoparticles, impact solar distillation systems. Sharshir et al.[105] tested a tubular solar still combined with mushrooms and nanofluid carbon coatings. Results revealed a 59.05% increase in yield and a 33.85% reduction in production cost compared to conventional systems.

Recent studies have explored solar desalination and HDH technology, suggesting advancements in solar-driven desalination. These methods utilize wet cores, cooling covers, reflectors, and nano-enhanced energy storage materials[106,107]. Abdelaziz et al.[108] reviewed many other studies on this topic. The current state-of-the-art hybrid solar desalination technology is introduced, as well as future research and development directions in this field.

Porous carbon is used for light-based conversion materials, such as graphite[109-112], graphite[113], and carbon fiber nanotubes, which have good absorption capabilities for sunlight[114]. Graphene-based photothermal conversion materials. For example, Liu et al. proposed a graphene multilevel pore photothermal conversion material[115], Graphene foam with a multilevel pore structure was prepared by plasma-enhanced chemical vapor deposition. This material has enhanced broad-band light absorption characteristics, its photothermal conversion efficiency is as high as 93.4%. Li et al. proposed a foldable graphene oxide film photothermal conversion material[111] that does not require any support system as an efficient photothermal conversion material. The solar-to-steam conversion efficiency of the material under sunlight is 80%.

Graphite carbon fiber-based materials have been used in photothermal conversion. For example, Ghasemi et al.[113] reported a double-layer structure composed of a carbon foam support layer and a graphite layer. The bottom carbon foam support layer is thermally insulating and has a smaller pore size for liquid Transport, while the top peeled graphite layer has a larger pore size for completing photothermal conversion and steam escape. This structure concentrates thermal energy and fluid at the location required for phase change and minimizes the dissipated energy. Experiments have shown that this material can achieve a photothermal conversion efficiency of 85% at an optical power density of 10 kW/m². Zhu et al.[116] introduced a honeycomb carbon fiber sponge photothermal conversion material. The sponge structure can automatically confine water in the photothermal conversion active center, and its photothermal conversion efficiency is 2.5 times that of the original material. Carbon nanotube photothermal conversion material. Based on the concept of interfacial heating, Wang et al. prepared a photothermal conversion material composed of a top layer of self-floating hydrophobic carbon nanotube film and a bottom layer of hydrophilic macroporous silica substrate. The carbon nanotube film on the top can collect and convert almost all incident light into heat for interfacial vapor generation, and the macroporous silica substrate on the bottom can transport liquid to act as a mechanical support and thermal barrier. role. The material enables energy-efficient solar-driven water evaporation with a photothermal conversion efficiency of 82%.

8.3.1.2 Desalination of Seawater by Solar Interfacial Evaporation

Solar interfacial evaporation has the advantages of low cost and environmental friendliness. It is a clean and portable interfacial solar seawater desalination technology. This technology uses photothermal materials floating at the air-liquid interface to capture solar energy and convert it into thermal energy so that the thermal energy is localized at the water evaporation interface. It can not only reduce the heat loss caused by the photothermal material's overall water heating, but it can also provide a larger surface area and quickly release steam to achieve higher evaporation efficiency (can reach more than 80%)[117–119]. Additionally, this technology simply floats a porous solar-absorbing material on the surface of the water, eliminating the need for complex pressure controls and expensive infrastructure. Based on these advantages, scientists in recent years have devoted themselves to developing and optimizing the structure and performance of interfacial evaporation materials to achieve green and efficient solar seawater desalination.

Carbon nanotubes can be directly used in the desalination process because the cavity inside the carbon nanotube provides a great possibility for the desalination process. The high aspect ratio of the carbon nanotube, the smooth hydrophobic wall, and the internal pore size can provide water flow. A nearly frictionless channel, this structure enables carbon nanotubes to transport water molecules ultra-efficiently[96,120]. The transport speed of water molecules in carbon nanotubes is several orders of magnitude higher than theoretically calculated[121].

The size of the inner diameter of the carbon nanotube is equivalent to the critical size of the capillary behavior of the size exclusion effect pipe. When the inner diameter of the carbon nanotube is in the range of 0.34–0.39 nm (the separation point size of Na and Cl), water molecules can pass through the carbon nanotube, while larger sodium ions and chloride ions are excluded[122]. Therefore, it is kinetically feasible to improve the water flux and salt rejection rate of carbon nanotubes, and this has potential application value. Carbon nanotubes also have special physical and chemical properties such as antibacterial and stain resistance[123], chlorine resistance[124], and oxidation resistance[125], which can improve the stability of carbon nanotube-based materials in long-term applications and extend their service life. In addition, studies have shown that carbon nanotubes can not only provide channels for water molecules and increase the effective mass transfer surface area of membrane materials based on carbon nanotubes but can also absorb water vapor and enhance the permeability and permselectivity of the membrane[126]. These properties are critical to the desalination process.

High energy consumption and low thermal efficiency are problems that need to be solved in the traditional seawater desalination process. The anti-fouling and stability of composite membranes based on carbon nanotubes are better than those of the original membranes. The development of new composite materials based on carbon nanotubes can solve the above two problems. It is of great significance to broaden the application scope of seawater desalination technology and reduce operating costs. For example, Mitra et al.[127] doped untreated carbon nanotubes into PTFE, PP, and PVDF respectively to prepare membrane materials with a flux increased by 51.5% compared with the original membrane, and a salt rejection rate greater than 99.9%. It is reported that a composite film made of polyvinyl alcohol and carbon nanotubes can generate heat under electrical conditions. This heat can make the composite film a high-ion concentration operation that can be used in the film evaporation process. Corrosion-resistant direct contact heat exchange surfaces under normal conditions.

Composite materials based on carbon nanotubes have better mechanical strength and light absorption capabilities. The vertically arranged carbon nanotube array can absorb 99.97% of the energy of direct sunlight. Therefore, compared with expanding the spectral absorption range of carbon-based materials, the mechanical strength of the carbon nanotube array can be improved by compounding with other materials, and reducing the carbon-based Light reflection on the material surface is the key to improving the light absorption capacity of carbon nanotube-based photothermal conversion materials. There are reports in the literature of a double-layer material based on carbon nanotubes and prepared by a filtration method. The material includes a porous carbon nanotube top layer and silica substrate with adjustable thickness and has a high solar evaporation rate and energy conversion efficiency[128].

However, attention has been rather limited to the evaporation performance of brine (especially high-concentration brine), resulting in a considerable gap between the current state of the art and practical applications. The main challenge facing interfacial solar desalination technology is the crystallization of salt on the evaporation surface. Sedimentation, including the crystallization of soluble salts above saturation concentrations and the accumulation of small amounts of soluble salts[129]. The deposited salt will block the capillary water transmission channel, reduce the light capture ability of the evaporation surface, lead to an imbalance between water supply and demand, and lead to a reduction in the evaporation efficiency of brine[130,131]. Therefore, resisting salt deposition and maintaining sustained high efficiency of solar desalination remains a huge challenge for the practical application of this technology.

8.3.1.3 Graphene Oxide Membrane Used in Seawater Desalination

Graphene oxide membranes, made from partially oxidized stacked graphene sheets, offer thin, efficient membranes for precise ion and molecular screening in water[132–134]. They have diverse applications in seawater desalination, gas separation, biosensors, batteries, and supercapacitors. Unlike carbon nanotube films with fixed pore sizes, graphene oxide films have variable pore sizes between the sheets. In addition, reducing the interlayer spacing enough to exclude small ions while preventing swelling of graphene oxide films in water poses a challenge[135–139]. There have been previous attempts to adjust the spacing between layers. For example, it can be widened to increase the permeability of graphene oxide membranes (GOMs) by inserting large nanomaterials[135,136] and by cross-linking large and rigid molecules[137]. Lowering GOMs results in a drastic reduction in interlayer spacing but makes them highly impermeable to all gases, liquids, and corrosive chemicals.

8.3.2 Application of Porous Carbon in Seawater Batteries

The seawater battery was first designed by Bell Laboratories in the United States during World War II and developed by General Electric Company. It is mainly used as a torpedo battery pack. The battery uses magnesium as the anode, silver chloride as the cathode, and seawater as the electrolyte. The battery is a magnesium-silver chloride series battery, which is a one-time activated battery. Its main performance characteristics are: high specific energy, up to 88 W·h/kg; long battery storage life, up to 5 years; discharge voltage Smooth, safe, and reliable; fast activation, activation time is only 2 s (10 s for silver-zinc battery). The disadvantages are that the design structure is complex, the cost is expensive, and the battery performance is easily affected by factors such as seawater temperature and concentration. My country's research on seawater batteries began in the 1990s. In 1991, Chinese researchers successfully developed an aluminum-air seawater battery. This seawater battery uses aluminum as the anode active material, oxygen in the air as the cathode active material, and seawater as the electrolyte, which relies on the electrochemical reaction occurring on the two poles to generate electric current, and has been successfully used as a navigation mark light. Although seawater batteries started early, few seawater batteries are used by civilians. Currently, seawater batteries are mainly used in the military field. There are three main concerns for this: (1) The seawater environment is complex and changeable, and factors such as temperature, concentration, flow rate, and dissolved oxygen have a certain impact on battery discharge performance. (2) Although seawater batteries do not need to carry electrolytes, most seawater batteries require an electrolyte control system to ensure the renewal of electrolytes and the elimination of discharge products. The quality of the electrolyte control system directly determines the discharge performance of the battery. (3) The cathode active materials should be resistive to seawater.

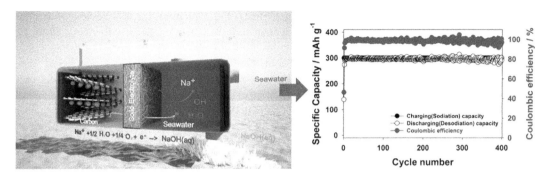

FIGURE 8.6   Rechargeable seawater battery[140].

The seawater battery, a recent innovation in energy storage, utilizes seawater as a vast sodium reservoir for its positive electrode, making it environmentally friendly and avoiding the need for expensive and toxic elements like nickel and cobalt found in conventional lithium-ion batteries. While the system shows promise, challenges arise from the choice of negative active materials. In a study by Kim et al.[140], starch-derived hard carbon was investigated to address this issue (Figure 8.6). Compared to commercial hard carbon, starch hard carbon demonstrated superior reversible capacity, current rate capability, and cycling ability due to its improved disordered structure. The material achieved a high maximum power density of 700 W/kg (based on hard carbon weight) when discharged at 900 mA/g, while remaining active at 2700 mA/g. These findings mark a significant advancement towards the practical use of sodium-based seawater battery technology.

8.4 CONCLUSION

This chapter introduces the potential applications of porous carbon from multiple perspectives. Porous carbon materials have a regular and ordered structure, high porosity, and high conductivity. As a catalyst carrier, they can significantly improve catalytic efficiency in electrocatalysis. Due to its high adsorption capacity, stable properties, and environmental friendliness, it has many applications in the removal of nuclear waste such as iodine, uranium, and cesium. At the same time, porous carbon materials have also made great progress in the application of seawater desalination technology and batteries, providing new ideas and ways for humanity to solve water resource shortages and energy crises. Therefore, continuous exploration of the application of porous carbon is of great significance.

NOTE
* Corresponding author.

REFERENCES

[1] Camila Rivera-Cárcamo, Philippe Serp, Cover feature: Single atom catalysts on carbon-based materials, ChemCatChem, **2018,** 10, 5056. DOI:10.1002/cctc.201801766.
[2] Hong Wang, Yue Shao, Shilin Mei, Yan Lu, Miao Zhang, Jian-ke Sun, Krzysztof Matyjaszewski, Markus Antonietti, Jiayin Yuan, Polymer-derived heteroatom-doped porous carbon materials, Chemical Reviews, **2020,** 120, 9363. DOI:10.1021/acs.chemrev.0c00080.

[3] Jiajia Song, Chao Wei, Zhen Feng Huang, Chuntai Liu, Zhichuan J. Xu, A review on funda-mentals for designing oxygen evolution electrocatalysts, Chemical Society Reviews, **2020, 49**. DOI:10.1039/c9cs00607a.

[4] Zhihong Bi, Qingqiang Kong, Yufang Cao, Guohua Sun, Fangyuan Su, Xianxian Wei, Xiaoming Li, Aziz Ahmad, Lijing Xie, Cheng-Meng Chen, Biomass-derived porous car-bon materials with different dimensions for supercapacitor electrodes: A review, Journal of Materials Chemistry A, **2019, 7**, 16028. DOI:10.1039/C9TA04436A.

[5] Anand S. Burange, Manoj B. Gawande, Frank L. Y. Lam, Radha V. Jayaram, Rafael Luque, Heterogeneously catalyzed strategies for the deconstruction of high density polyeth-ylene: Plastic waste valorisation to fuels, Green Chemistry, **2015**, 17, 146. DOI:10.1039/C4GC01760A.

[6] Zhengkun Yang, Yu Wang, Mengzhao Zhu, Zhijun Li, Wenxing Chen, Weichen Wei, Tongwei Yuan, Yunteng Qu, Qian Xu, Changming Zhao, Xin Wang, Peng Li, Yafei Li, Yuen Wu, Yadong Li, Boosting oxygen reduction catalysis with Fe-N$_4$ sites decorated porous car-bons toward fuel cells, ACS Catalysis, **2019, 9**, 2158. DOI:10.1021/acscatal.8b04381.

[7] Gurwinder Singh, Ajanya Maria Ruban, Xun Geng, Ajayan Vinu, Recognizing the poten-tial of K-salts, apart from KOH, for generating porous carbons using chemical activation, Chemical Engineering Journal, **2023, 451**, 139045. DOI:10.1016/j.cej.2022.139045.

[8] Yue Qiao, Yuanyuan Zhang, Shuhui Xia, Chaolong Wei, Yuehui Chen, Shuo Chen, Jianhua Yan, Stabilizing high density Cu active sites with ZrO$_2$ quantum dots as chemical ligand in N-doped porous carbon nanofibers for efficient ORR, Small, **2023, 19**, 2206823. DOI:10.1002/smll.202206823.

[9] Jiaqi Xu, Mengxiao Zhong, Na Song, Ce Wang, Xiaofeng Lu, General synthesis of Pt and Ni co-doped porous carbon nanofibers to boost HER performance in both acidic and alkaline solutions, Chinese Chemical Letters, **2023, 34**, 107359. DOI:10.1016/j.cclet.2022.03.082.

[10] Chuanlai Jiao, Zian Xu, Jingze Shao, Yu Xia, Jochi Tseng, Guangyuan Ren, Nianji Zhang, Pengfei Liu, Chongxuan Liu, Guangshe Li, Shi Chen, Shaoqing Chen, Hsing-Lin Wang, High-density atomic Fe-N$_4$/C in tubular, biomass-derived, nitrogen-rich porous carbon as air-electrodes for flexible Zn-air batteries, Advanced Functional Materials, **2023, 33**, 2213897. DOI:10.1002/adfm.202213897.

[11] Yanjia Cui, Yonghui Cheng, Caili Yang, Yingshi Su, Defu Yao, Biping Liufu, Jialei Li, Yiwen Fang, Suyao Liu, Ziyi Zhong, Xiaoming Wang, Yibing Song, Zhen Li, High-performance electrocatalytic CO$_2$ reduction for CO generation using hydrophobic porous carbon sup-ported Au, ACS Sustainable Chemistry & Engineering, **2023, 11**, 11229. DOI:10.1021/acssuschemeng.3c02291.

[12] Javier Quílez-Bermejo, Emilia Morallón, Diego Cazorla-Amorós, Metal-free heteroatom-doped carbon-based catalysts for ORR: A critical assessment about the role of heteroatoms, Carbon, **2020, 165**, 434. DOI:10.1016/j.carbon.2020.04.068.

[13] Chuangang Hu, Liming Dai, Carbon-based metal-free catalysts for electrocatalysis beyond the ORR, Angewandte Chemie International Edition, **2016**, 55, 11736. DOI:10.1002/anie.201509982.

[14] Ruifeng Zhou, Yao Zheng, Mietek Jaroniec, Shizhang Qiao, Determination of the elec-tron transfer number for the oxygen reduction reaction: From theory to experiment, ACS Catalysis, **2016, 6**, 4720. DOI:10.1021/acscatal.6b01581.

[15] L. Bouleau, S. Pérez-Rodríguez, J. Quílez-Bermejo, M. T. Izquierdo, F. Xu, V. Fierro, A. Celzard, Best practices for ORR performance evaluation of metal-free porous carbon electrocatalysts, Carbon, **2022, 189**, 349. DOI:10.1016/j.carbon.2021.12.078.

[16] Chang Hyuck Choi, Minho Kim, Han Chang Kwon, Sung June Cho, Seongho Yun, Hee-Tak Kim, Karl J. J. Mayrhofer, Hyungjun Kim, Minkee Choi, Tuning selectivity of electrochemi-cal reactions by atomically dispersed platinum catalyst, Nature Communications, **2016**, 7, 10922. DOI:10.1038/ncomms10922.

[17] Tong Zhang, Huanhuan Wang, Jintao Zhang, Jing Ma, Zhi Wang, Junhao Liu, Xuzhong Gong, Carbon charge population and oxygen molecular transport regulated by program-doping for highly efficient 4e-ORR, Chemical Engineering Journal, **2022**, 444, 136560. DOI:10.1016/j. cej.2022.136560.

[18] Kai Huang, Le Zhang, Ting Xu, Hehe Wei, Ruoyu Zhang, Xiaoyuan Zhang, Binghui Ge, Ming Lei, Jing-Yuan Ma, Li-Min Liu, Hui Wu, −60 °C solution synthesis of atomically dispersed cobalt electrocatalyst with superior performance, Nature Communications, **2019**, 10, 606. DOI:10.1038/s41467-019-08484-8.

[19] Wengang Liu, Yinjuan Chen, Haifeng Qi, Leilei Zhang, Wensheng Yan, Xiaoyan Liu, Xiaofeng Yang, Shu Miao, Wentao Wang, Chenguang Liu, Aiqin Wang, Jun Li, Tao Zhang, A durable nickel single-atom catalyst for hydrogenation reactions and cellulose valorization under harsh conditions, Angewandte Chemie International Edition, **2018**, 57, 7071. DOI:10.1002/anie. 201802231.

[20] Kang Liu, Junwei Fu, Yiyang Lin, Tao Luo, Ganghai Ni, Hongmei Li, Zhang Lin, Min Liu, Insights into the activity of single-atom Fe-N-C catalysts for oxygen reduction reaction, Nature Communications, **2022**, 13, 2075. DOI:10.1038/s41467-022-29797-1.

[21] Dahuan Li, Yongfang Qu, Xuejun Liu, Cuiping Zhai, Yong Liu, Preparation of three-dimensional Fe-N co-doped open-porous carbon networks as an efficient ORR electrocatalyst in both alkaline and acidic media, International Journal of Hydrogen Energy, **2021**, 46, 18364. DOI:10.1016/j.ijhydene.2021.03.008.

[22] Koroush Sasan, Aiguo Kong, Yuan Wang, Mao Chengyu, Quanguo Zhai, Pingyun Feng, From hemoglobin to porous N-S-Fe-doped carbon for efficient oxygen electroreduction, The Journal of Physical Chemistry C, **2015**, 119, 13545. DOI:10.1021/acs.jpcc.5b04017.

[23] Shanshan Lu, Chuanqi Cheng, Yanmei Shi, Yongmeng Wu, Zhipu Zhang, Bin Zhang, Unveiling the structural transformation and activity origin of heteroatom-doped carbons for hydrogen evolution, Proceedings of the National Academy of Sciences, **2023**, 120, e2300549120. DOI:10.1073/pnas.2300549120.

[24] Man Zhao, Xiaoru Cheng, He Xiao, Jianru Gao, Shoufeng Xue, Xiaoxia Wang, Haishun Wu, Jianfeng Jia, Nianjun Yang, Cobalt-iron oxide/black phosphorus nanosheet heterostructure: Electrosynthesis and performance of (photo-)electrocatalytic oxygen evolution, Nano Research, **2023**, 16, 6057. DOI:10.1007/s12274-022-4676-9.

[25] Jingjing Duan, Sheng Chen, Mietek Jaroniec, Shi Zhang Qiao, Heteroatom-doped graphene-based materials for energy-relevant electrocatalytic processes, ACS Catalysis, **2015**, 5, 5207. DOI:10.1021/acscatal.5b00991.

[26] Sunmoon Yu, Jaehoon Kim, Ki Ro Yoon, Ji-Won Jung, Jihun Oh, Il-Doo Kim, Rational design of efficient electrocatalysts for hydrogen evolution reaction: Single layers of WS$_2$ nanoplates anchored to hollow nitrogen-doped carbon nanofibers, ACS Applied Materials & Interfaces, **2015**, 7, 28116. DOI:10.1021/acsami.5b09447.

[27] Ji-Sen Li, Yu Wang, Chun-Hui Liu, Shun-Li Li, Yu-Guang Wang, Long-Zhang Dong, Zhi-Hui Dai, Ya-Fei Li, Ya-Qian Lan, Coupled molybdenum carbide and reduced graphene oxide electrocatalysts for efficient hydrogen evolution, Nature Communications, **2016**, 7. DOI:10.1038/ ncomms11204.

[28] Dong Jun Li, Uday Narayan Maiti, Joonwon Lim, Dong Sung Choi, Won Jun Lee, Youngtak Oh, Gil Yong Lee, Sang Ouk Kim, Molybdenum sulfide/N-doped CNT forest hybrid catalysts for high-performance hydrogen evolution reaction, Nano Letters, **2014**, 14, 1228. DOI:10.1021/nl404108a.

[29] S. S. Shinde, Abdul Sami, Dong-Hyung Kim, Jung-Ho Lee, Nanostructured SnS-N-doped graphene as an advanced electrocatalyst for the hydrogen evolution reaction, Chemical Communications, **2015**, 51, 15716. DOI:10.1039/c5cc05644f.

[30] Sheng Chen, Jingjing Duan, Youhong Tang, Bo Jin, Shi Zhang Qiao, Molybdenum sulfide clusters-nitrogen-doped graphene hybrid hydrogel film as an efficient three-dimensional hydrogen evolution electrocatalyst, Nano Energy, **2015**, 11, 11. DOI:10.1016/j.nanoen.2014.09.022.

[31] Zhe Zhang, Baoping Lu, Jinhui Hao, Wenshu Yang, Jilin Tang, FeP nanoparticles grown on graphene sheets as highly active non-precious-metal electrocatalysts for hydrogen evolution reaction, Chemical Communications, **2014**, 50, 11554. DOI:10.1039/c4cc05285d.

[32] Shengwen Liu, Haimin Zhang, Qian Zhao, Xian Zhang, Rongrong Liu, Xiao Ge, Guozhong Wang, Huijun Zhao, Weiping Cai, Metal-organic framework derived nitrogen-doped porous carbon@graphene sandwich-like structured composites as bifunctional electrocatalysts for oxygen reduction and evolution reactions, Carbon, **2016**, 106, 74. DOI:10.1016/j.carbon.2016.05.021.

[33] Xinzhe Li, Yiyun Fang, Shiling Zhao, Juntian Wu, Feng Li, Min Tian, Xuefeng Long, Jun Jin, Jiantai Ma, Nitrogen-doped mesoporous carbon nanosheet/carbon nanotube hybrids as metal-free bi-functional electrocatalysts for water oxidation and oxygen reduction, Journal of Materials Chemistry A, **2016**, 4, 13133. DOI:10.1039/c6ta04187f.

[34] Changlin Zhang, Biwei Wang, Xiaochen Shen, Jiawei Liu, Xiangkai Kong, Steven S. C. Chuang, Dong Yang, Angang Dong, Zhenmeng Peng, A nitrogen-doped ordered mesoporous carbon/graphene framework as bifunctional electrocatalyst for oxygen reduction and evolution reactions, Nano Energy, **2016**, 30, 503. DOI:10.1016/j.nanoen.2016.10.051.

[35] Rong Li, Zidong Wei, Xinglong Gou, Nitrogen and phosphorus dual-doped graphene/carbon nanosheets as bifunctional electrocatalysts for oxygen reduction and evolution, ACS Catalysis, **2015**, 5, 4133. DOI:10.1021/acscatal.5b00601.

[36] Konggang Qu, Yao Zheng, Yan Jiao, Xianxi Zhang, Sheng Dai, Shizhang Qiao, Polydopamine-inspired, dual heteroatom-doped carbon nanotubes for highly efficient overall water splitting, Advanced Energy Materials, **2016**, 7. DOI:10.1002/aenm.201602068.

[37] Daping He, Yuli Xiong, Jinlong Yang, Xu Chen, Zhaoxiang Deng, Mu Pan, Yadong Li, Shichun Mu, Nanocarbon-intercalated and Fe-N-codoped graphene as a highly active noble-metal-free bifunctional electrocatalyst for oxygen reduction and evolution, Journal of Materials Chemistry A, **2017**, 5, 1930. DOI:10.1039/c5ta09232a.

[38] Pengzuo Chen, Tianpei Zhou, Lili Xing, Kun Xu, Yun Tong, Hui Xie, Lidong Zhang, Wensheng Yan, Wangsheng Chu, Changzheng Wu, Yi Xie, Atomically dispersed iron-nitrogen species as electrocatalysts for bifunctional oxygen evolution and reduction reactions, Angewandte Chemie International Edition, **2016**, 56, 610. DOI:10.1002/anie.201610119.

[39] Dongping Xue, Huicong Xia, Wenfu Yan, Jianan Zhang, Shichun Mu, Defect engineering on carbon-based catalysts for electrocatalytic CO_2 reduction, Nano-Micro Letters, **2020**, 13, 5. DOI:10.1007/s40820-020-00538-7.

[40] Yong Zhao, Jiaojiao Liang, Caiyun Wang, Jianmin Ma, Gordon G. Wallace, Tunable and efficient tin modified nitrogen-doped carbon nanofibers for electrochemical reduction of aqueous carbon dioxide, Advanced Energy Materials, **2018**, 8, 1702524. DOI:10.1002/aenm.201702524.

[41] Hengpan Yang, Yu Wu, Guodong Li, Qing Lin, Qi Hu, Qianling Zhang, Jianhong Liu, Chuanxin He, Scalable production of efficient single-atom copper decorated carbon membranes for CO_2 electroreduction to methanol, Journal of the American Chemical Society, **2019**, 141, 12717. DOI:10.1021/jacs.9b04907.

[42] Guozhu Mao, Haoqiong Hu, Xi Liu, John Crittenden, Ning Huang, A bibliometric analysis of industrial wastewater treatments from 1998 to 2019, Environmental Pollution, **2021**, 275, 115785. DOI:10.1016/j.envpol.2020.115785.

[43] Fenglian Fu, Qi Wang, Removal of heavy metal ions from wastewaters: A review, Journal of Environmental Management, **2011**, 92, 407. DOI:10.1016/j.jenvman.2010.11.011.

[44] Melanie Iwanow, Tobias Gärtner, Volker Sieber, Burkhard König, Activated carbon as catalyst support: Precursors, preparation, modification and characterization, Beilstein Journal of Organic Chemistry, **2020**, 16, 1188. DOI:10.3762/bjoc.16.104.

[45] Tingting Zhang, Yongjun Liu, Lu Yang, Weiping Li, Weida Wang, Pan Liu, Ti-Sn-Ce/bamboo biochar particle electrodes for enhanced electrocatalytic treatment of coking wastewater in a three-dimensional electrochemical reaction system, Journal of Cleaner Production, **2020**, 258, 120273. DOI:10.1016/j.jclepro.2020.120273.

[46] Hongwei Sun, Ting Chen, Lingjun Kong, Quan Cai, Ya Xiong, Shuanghong Tian, Potential of sludge carbon as new granular electrodes for degradation of acid orange 7, Industrial & Engineering Chemistry Research, **2015**, 54, 5468. DOI:10.1021/acs.iecr.5b00780.

[47] Syed M. Husnain, Wooyong Um, Lee Woojin, Yoon-Seok Chang, Magnetite-based adsorbents for sequestration of radionuclides: A review, RSC Advances, **2018**, 8, 2521. DOI:10.1039/C7RA12299C.

[48] Smain Korichi, Aicha Bensmaili, Sorption of uranium (VI) on homoionic sodium smectite experimental study and surface complexation modeling, Journal of Hazardous Materials, **2009**, 169, 780. DOI:10.1016/j.jhazmat.2009.04.014.

[49] Karena W. Chapman, Peter J. Chupas, Tina M. Nenoff, Radioactive iodine capture in silver-containing mordenites through nanoscale silver iodide formation, Journal of the American Chemical Society, **2010**, 132, 8897. DOI:10.1021/ja103110y.

[50] I. Kavanagh, O. Fenton, M. G. Healy, W. Burchill, G. J. Lanigan, D. J. Krol, Mitigating ammonia and greenhouse gas emissions from stored cattle slurry using agricultural waste, commercially available products and a chemical acidifier, Journal of Cleaner Production, **2021**, 294, 126251. DOI:10.1016/j.jclepro.2021.126251.

[51] Yinghui Wang, Yuantao Chen, Meng Zhao, Lili Zhang, Changyou Zhou, Haiyang Wang, Simulated adsorption of iodine by an amino-metal-organic framework modified with covalent bonds, Environmental Science and Pollution Research, **2022**, 29, 88882. DOI:10.1007/s11356-022-21971-8.

[52] Yi Wu, Yang Guo, Rongkui Su, Xiancheng Ma, Qingding Wu, Zheng Zeng, Liqing Li, Xiaolong Yao, Shaobin Wang, Hierarchical porous carbon with an ultrahigh surface area for high-efficient iodine capture: Insights into adsorption mechanisms through experiments, simulations and modeling, Separation and Purification Technology, **2022**, 303, 122237. DOI:10.1016/j.seppur.2022.122237.

[53] Yingfan Chen, Hanxue Sun, Ruixia Yang, Tingting Wang, Chunjuan Pei, Zhentao Xiang, Zhaoqi Zhu, Weidong Liang, An Li, Weiqiao Deng, Synthesis of conjugated microporous polymer nanotubes with large surface areas as absorbents for iodine and CO_2 uptake, Journal of Materials Chemistry A, **2015**, 3, 87. DOI:10.1039/C4TA04235B.

[54] Feng Ren, Zhaoqi Zhu, Xin Qian, Weidong Liang, Peng Mu, Hanxue Sun, Jiehua Liu, An Li, Novel thiophene-bearing conjugated microporous polymer honeycomb-like porous spheres with ultra-high iodine uptake, Chemical Communications, **2016**, 52, 9797. DOI:10.1039/C6CC05188J.

[55] Zhi-Jian Yin, Shun-Qi Xu, Tian-Guang Zhan, Qiao-Yan Qi, Zong-Quan Wu, Xin Zhao, Ultrahigh volatile iodine uptake by hollow microspheres formed from a heteropore covalent organic framework, Chemical Communications, **2017**, 53, 7266. DOI:10.1039/C7CC01045A.

[56] Lin Lin, Heda Guan, Donglei Zou, Zhaojun Dong, Zhi Liu, Feifan Xu, Zhigang Xie, Yangxue Li, A pharmaceutical hydrogen-bonded covalent organic polymer for enrichment of volatile iodine, RSC Advances, **2017**, 7, 54407. DOI:10.1039/C7RA09414K.

[57] Yufei Gao, Ying Huo, Mingyi Chen, Xingdong Su, Jie Zhan, Liang Wang, Feng Liu, Jian Zhu, Yuan Zeng, Jie Fan, Zesheng Li, Rouxi Chen, Hsing-Lin Wang, Phenolic based porous carbon fibers with superior surface area and adsorption efficiency for radioactive protection, Advanced Fiber Materials, **2023**, 5, 1431. DOI:10.1007/s42765-023-00284-6.

[58] Lichao Ge, Can Zhao, Simo Chen, Qian Li, Tianhong Zhou, Han Jiang, Xi Li, Yang Wang, Chang Xu, An analysis of the carbonization process and volatile-release characteristics of coal-based activated carbon, Energy, **2022**, 257, 124779. DOI:10.1016/j.energy.2022.124779.

[59] Lichao Ge, Can Zhao, Tianhong Zhou, Simo Chen, Qian Li, Xuguang Wang, Dong Shen, Yang Wang, Chang Xu, An analysis of the carbonization process of coal-based activated carbon at different heating rates, Energy, **2023**, 267, 126557. DOI:10.1016/j.energy.2022.126557.

[60] Lichao Ge, Can Zhao, Mingjin Zuo, Jie Tang, Wen Ye, Xuguang Wang, Yuli Zhang, Chang Xu, Review on the preparation of high value-added carbon materials from biomass, Journal of Analytical and Applied Pyrolysis, **2022**, 168, 105747. DOI:10.1016/j.jaap.2022.105747.

[61] Can Zhao, Lichao Ge, Xi Li, Mingjin Zuo, Chunyao Xu, Simo Chen, Qian Li, Yang Wang, Chang Xu, Effects of the carbonization temperature and intermediate cooling mode on the properties of coal-based activated carbon, Energy, **2023**, 273, 127177. DOI:10.1016/j.energy.2023.127177.

[62] Yile Zou, Hongfang Wang, Lianfei Xu, Menghao Dong, Boxiong Shen, Xin Wang, Jiancheng Yang, Synergistic effect of CO_2 and H_2O co-activation of Zhundong coal at a low burn-off rate on high performance supercapacitor, Journal of Power Sources, **2023**, 556, 232509. DOI:10.1016/j.jpowsour.2022.232509.

[63] Xinxin Liu, Qingzhao Li, Guiyun Zhang, Yuannan Zheng, Yang Zhao, Preparation of activated carbon from Guhanshan coal and its effect on methane adsorption thermodynamics at different temperatures, Powder Technology, **2022**, 395, 424. DOI:10.1016/j.powtec.2021.09.076.

[64] Jieying Jing, Zemin Zhao, Xuewei Zhang, Jie Feng, Wenying Li, CO_2 capture over activated carbon derived from pulverized semi-coke, Separations, **2022**, 9, 174. DOI:10.3390/separations9070174.

[65] Johanna Muthmann, Christian Bläker, Christoph Pasel, Michael Luckas, Carsten Schledorn, Dieter Bathen, Characterization of structural and chemical modifications during the steam activation of activated carbons, Microporous and Mesoporous Materials, **2020**, 309, 110549. DOI:10.1016/j.micromeso.2020.110549.

[66] Robert F. Service, How far can we push chemical self-assembly? Science, **2005**, 309, 95. DOI:10.1126/science.309.5731.95.

[67] Shuang Zhang, Xiaosheng Zhao, Bo Li, Chiyao Bai, Yang Li, Lei Wang, Rui Wen, Meicheng Zhang, Lijian Ma, Shoujian Li, "Stereoscopic" 2D super-microporous phosphazene-based covalent organic framework: Design, synthesis and selective sorption towards uranium at high acidic condition, Journal of Hazardous Materials, **2016**, 314, 95. DOI:10.1016/j.jhazmat.2016.04.031.

[68] Meicheng Zhang, Yang Li, Chiyao Bai, Xinghua Guo, Jun Han, Sheng Hu, Hongquan Jiang, Wang Tan, Shoujian Li, Lijian Ma, Synthesis of microporous covalent phosphazene-based frameworks for selective separation of uranium in highly acidic media based on size-matching effect, ACS Applied Materials & Interfaces, **2018**, 10, 28936. DOI:10.1021/acsami.8b06842.

[69] Bo Li, Chiyao Bai, Shuang Zhang, Xiaosheng Zhao, Yang Li, Lei Wang, Kuan Ding, Xi Shu, Shoujian Li, Lijian Ma, An adaptive supramolecular organic framework for highly efficient separation of uranium via an in situ induced fit mechanism, Journal of Materials Chemistry A, **2015**, 3, 23788. DOI:10.1039/C5TA07970E.

[70] Hee-Man Yang, Sung-Chan Jang, Sang Bum Hong, Kune-Woo Lee, Changhyun Roh, Yun Suk Huh, Bum-Kyoung Seo, Prussian blue-functionalized magnetic nanoclusters for the removal of radioactive cesium from water, Journal of Alloys and Compounds, **2016**, 657, 387. DOI:10.1016/j.jallcom.2015.10.068.

[71] Xiang Liu, Guan-Ru Chen, Duu-Jong Lee, Tohru Kawamoto, Hisashi Tanaka, Man-Li Chen, Yu-Kuo Luo, Adsorption removal of cesium from drinking waters: A mini review on use of biosorbents and other adsorbents, Bioresource Technology, **2014**, 160, 142. DOI:10.1016/j.biortech.2014.01.012.

[72] Yu-Chen Lai, Yin-Ru Chang, Man-Li Chen, Yu-Kuo Lo, Juin-Yih Lai, Duu-Jong Lee, Poly(vinyl alcohol) and alginate cross-linked matrix with immobilized prussian blue and ion exchange resin for cesium removal from waters, Bioresource Technology, **2016**, 214, 192. DOI:10.1016/j.biortech.2016.04.096.

[73] Jiseon Jang, Dae Sung Lee, Enhanced adsorption of cesium on PVA-alginate encapsulated Prussian blue-graphene oxide hydrogel beads in a fixed-bed column system, Bioresource Technology, **2016**, 218, 294. DOI:10.1016/j.biortech.2016.06.100.

[74] Omar Falyouna, Osama Eljamal, Ibrahim Maamoun, Atsushi Tahara, Yuji Sugihara, Magnetic zeolite synthesis for efficient removal of cesium in a lab-scale continuous treatment system, Journal of Colloid and Interface Science, **2020**, 571, 66. DOI:10.1016/j.jcis.2020.03.028.

[75] Selina Hube, Majid Eskafi, Kolbrún Fríða Hrafnkelsdóttir, Björg Bjarnadóttir, Margrét Ásta Bjarnadóttir, Snærós Axelsdóttir, Bing Wu, Direct membrane filtration for wastewater

treatment and resource recovery: A review, Science of The Total Environment, **2020,** 710, 136375. DOI:10.1016/j.scitotenv.2019.136375.

[76] Qinqin Tao, Xu Zhang, Krishnamoorthy Prabaharan, Ying Dai, Separation of cesium from wastewater with copper hexacyanoferrate film in an electrochemical system driven by microbial fuel cells, Bioresource Technology, **2019,** 278, 456. DOI:10.1016/j.biortech.2019.01.093.

[77] Shengmei Chen, Longtao Ma, Shuilin Wu, Shuyun Wang, Zebiao Li, Adesina Ayotunde Emmanuel, Md Rashedul Huqe, Chunyi Zhi, Juan Antonio Zapien, Uniform virus-like Co-N-Cs electrocatalyst derived from prussian blue analog for stretchable fiber-shaped Zn-air batteries, Advanced Functional Materials, **2020,** 30, 1908945. DOI:10.1002/adfm.201908945.

[78] Aaron Albert Aryee, Farid Mzee Mpatani, Alexander Nti Kani, Evans Dovi, Runping Han, Zhaohui Li, Lingbo Qu, A review on functionalized adsorbents based on peanut husk for the sequestration of pollutants in wastewater: Modification methods and adsorption study, Journal of Cleaner Production, **2021,** 310, 127502. DOI:10.1016/j.jclepro.2021.127502.

[79] Bokseong Kim, Daemin Oh, Sungwon Kang, Youngsug Kim, Sol Kim, Yoonsuhn Chung, Youngkyo Seo, Yuhoon Hwang, Reformation of the surface of Powdered Activated Carbon (PAC) using Covalent Organic Polymers (COPs) and synthesis of a Prussian blue impregnated adsorbent for the decontamination of radioactive cesium, Journal of Alloys and Compounds, **2019,** 785, 46. DOI:10.1016/j.jallcom.2019.01.154.

[80] Xianwen Wang, Liang Cheng, Multifunctional Prussian blue-based nanomaterials: Preparation, modification, and theranostic applications, Coordination Chemistry Reviews, **2020,** 419, 213393. DOI:10.1016/j.ccr.2020.213393.

[81] Avinash A. Kadam, Jiseon Jang, Dae Sung Lee, Facile synthesis of pectin-stabilized magnetic graphene oxide Prussian blue nanocomposites for selective cesium removal from aqueous solution, Bioresource Technology, **2016,** 216, 391. DOI:10.1016/j.biortech.2016.05.103.

[82] Xiaogang You, Libin Yang, Xuefei Zhou, Yalei Zhang, Sustainability and carbon neutrality trends for microalgae-based wastewater treatment: A review, Environmental Research, **2022,** 209, 112860. DOI:10.1016/j.envres.2022.112860.

[83] Kim Yrjälä, Muthusamy Ramakrishnan, Esko Salo, Agricultural waste streams as resource in circular economy for biochar production towards carbon neutrality, Current Opinion in Environmental Science & Health, **2022,** 26, 100339. DOI:10.1016/j.coesh.2022.100339.

[84] Xiaosheng Zhao, Shuang Zhang, Chiyao Bai, Bo Li, Yang Li, Lei Wang, Rui Wen, Meicheng Zhang, Lijian Ma, Shoujian Li, Nano-diamond particles functionalized with single/double-arm amide-thiourea ligands for adsorption of metal ions, Journal of Colloid and Interface Science, **2016,** 469, 109. DOI:10.1016/j.jcis.2016.02.017.

[85] Marek Piechowicz, Carter W. Abney, Xin Zhou, Nathan C. Thacker, Zhong Li, Wenbin Lin, Design, synthesis, and characterization of a bifunctional chelator with ultrahigh capacity for uranium uptake from seawater simulant, Industrial & Engineering Chemistry Research, **2016,** 55, 4170. DOI:10.1021/acs.iecr.5b03304.

[86] Gregg J. Lumetta, Brian M. Rapko, Priscilla A. Garza, Benjamin P. Hay, Robert D. Gilbertson, Timothy J. R. Weakley, James E. Hutchison, Deliberate design of ligand architecture yields dramatic enhancement of metal ion affinity, Journal of the American Chemical Society, **2002,** 124, 5644. DOI:10.1021/ja025854t.

[87] Gregg J. Lumetta, Brian M. Rapko, Benjamin P. Hay, Priscilla A. Garza, James E. Hutchison, Robert D. Gilbertson, A novel bicyclic diamide with high binding affinity for trivalent f-block elements, Solvent Extraction and Ion Exchange, **2003,** 21, 29. DOI:10.1081/sei-120017546.

[88] Lan Xu, Youhoon Chong, Inkyu Hwang, Anthony D'Onofrio, Kristen Amore, G. Peter Beardsley, Chenglong Li, Arthur J. Olson, Dale L. Boger, Ian A. Wilson, Structure-based design, synthesis, evaluation, and crystal structures of transition state analogue inhibitors of inosine monophosphate cyclohydrolase, Journal of Biological Chemistry, **2007,** 282, 13033. DOI:10.1074/jbc.M607293200.

[89] Sinisa Vukovic, Benjamin P. Hay, De novo structure-based design of bis-amidoxime uranophiles, Inorganic Chemistry, **2013,** 52, 7805. DOI:10.1021/ic401089u.

[90] Benjamin P. Hay, De novo structure-based design of anion receptors, Chemical Society Reviews, **2010,** 39, 3700. DOI:10.1039/C0CS00075B.

[91] Lu Zhou, Mike Bosscher, Changsheng Zhang, Salih Özçubukçu, Liang Zhang, Wen Zhang, Charles J. Li, Jianzhao Liu, Mark P. Jensen, Luhua Lai, Chuan He, A protein engineered to bind uranyl selectively and with femtomolar affinity, Nature Chemistry, **2014,** 6, 236. DOI:10.1038/nchem.1856.

[92] Sabine Lattemann, Maria D. Kennedy, Jan C. Schippers, Gary amy: Chapter 2 global desalination situation, Sustainability Science and Engineering, **2010,** 2, 7–39. DOI:10.1016/S1871-2711(09)00202-5.

[93] Yuan Zhou, Richard S. J. Tol, Evaluating the costs of desalination and water transport, Water Resources Research, **2005,** 41. DOI:10.1029/2004WR003749.

[94] M. Busch, W. E. Mickols, Reducing energy consumption in seawater desalination, Desalination, **2004,** 165, 299. DOI:10.1016/j.desal.2004.06.035.

[95] K. Bourouni, M. T. Chaibi, L. Tadrist, Water desalination by humidification and dehumidification of air: State of the art, Desalination, **2001,** 137, 167. DOI:10.1016/S0011-9164(01)00215-6.

[96] G. Hummer, J. C. Rasaiah, J. P. Noworyta, Water conduction through the hydrophobic channel of a carbon nanotube, Nature, **2001,** 414, 188. DOI:10.1038/35102535.

[97] Gamal B. Abdelaziz, Emad M. S. El-Said, Ahmed G. Bedair, Swellam W. Sharshir, A. E. Kabeel, Ashraf Mimi Elsaid, Experimental study of activated carbon as a porous absorber in solar desalination with environmental, exergy, and economic analysis, Process Safety and Environmental Protection, **2021,** 147, 1052. DOI:10.1016/j.psep.2021.01.031.

[98] Emad M. S. El-Said, Mohamed A. Dahab, M. Omara, Gamal B. Abdelaziz, Solar desalination unit coupled with a novel humidifier, Renewable Energy, **2021,** 180, 297. DOI:10.1016/j.renene.2021.08.105.

[99] Gamal B. Abdelaziz, Mohamed A. Dahab, M. A. Omara, Swellam W. Sharshir, Ashraf Mimi Elsaid, Emad M. S. El-Said, Humidification dehumidification saline water desalination system utilizing high frequency ultrasonic humidifier and solar heated air stream, Thermal Science and Engineering Progress, **2022,** 27, 101144. DOI:10.1016/j.tsep.2021.101144.

[100] Emad M. S. El-Said, Gamal B. Abdelaziz, Experimental investigation and economic assessment of a solar still performance using high-frequency ultrasound waves atomizer, Journal of Cleaner Production, **2020,** 256, 120609. DOI:10.1016/j.jclepro.2020.120609.

[101] Guilong Peng, Swellam W. Sharshir, Zhixiang Hu, Rencai Ji, Jianqiang Ma, A. E. Kabeel, Huan Liu, Jianfeng Zang, Nuo Yang, A compact flat solar still with high performance, International Journal of Heat and Mass Transfer, **2021,** 179. DOI:10.1016/j.ijheatmasstransfer.2021.121657.

[102] A. W. Kandeal, Nagi M. El-Shafai, Mohamed R. Abdo, Amrit Kumar Thakur, Ibrahim M. El-Mehasseb, Ibrahem Maher, Maher Rashad, A. E. Kabeel, Nuo Yang, Swellam W. Sharshir, Improved thermo-economic performance of solar desalination via copper chips, nanofluid, and nano-based phase change material, Solar Energy, **2021,** 224, 1313. DOI:10.1016/j.solener.2021.06.085.

[103] Guilong Peng, Swellam W. Sharshir, Yunpeng Wang, Meng An, Dengke Ma, Jianfeng Zang, A. E. Kabeel, Nuo Yang, Potential and challenges of improving solar still by micro/nano-particles and porous materials—A review, Journal of Cleaner Production, **2021,** 311. DOI:10.1016/j.jclepro.2021.127432.

[104] Swellam W. Sharshir, M. Ismail, A. W. Kandeal, Faisal B. Baz, Ayman Eldesoukey, M. M. Younes, Improving thermal, economic, and environmental performance of solar still using floating coal, cotton fabric, and carbon black nanoparticles, Sustainable Energy Technologies and Assessments, **2021,** 48. DOI:10.1016/j.seta.2021.101563.

[105] Swellam W. Sharshir, Mohamed A. Hamada, A. W. Kandeal, Emad M. S. El-Said, Ashraf Mimi Elsaid, Maher Rashad, Gamal B. Abdelaziz, Augmented performance of tubular solar still integrated with cost-effective nano-based mushrooms, Solar Energy, **2021,** 228, 27. DOI:10.1016/j.solener.2021.09.034.

[106] Gamal B. Abdelaziz, Almoataz M. Algazzar, Emad M. S. El-Said, Ashraf Mimi Elsaid, Swellam W. Sharshir, A. E. Kabeel, S. M. El-Behery, Performance enhancement of tubular solar still using nano-enhanced energy storage material integrated with v-corrugated aluminum basin, wick, and nanofluid, Journal of Energy Storage, **2021,** 41. DOI:10.1016/j.est.2021.102933.

[107] Mohammed El Hadi Attia, Abd Elnaby Kabeel, Mohamed Abdelgaied, Gamal B. Abdelaziz, A comparative study of hemispherical solar stills with various modifications to obtain modified and inexpensive still models, Environmental Science and Pollution Research, **2021,** 28, 55667. DOI:10.1007/s11356-021-14862-x.

[108] Gamal B. Abdelaziz, Emad M. S. El-Said, Mohamed A. Dahab, M. A. Omara, Swellam W. Sharshir, Hybrid solar desalination systems review, Energy Sources, Part A: Recovery, Utilization, and Environmental Effects, **2021,** 1. DOI:10.1080/15567036.2021.2005721.

[109] Yang Fu, Gang Wang, Xin Ming, Xinghang Liu, Baofei Hou, Tao Mei, Jinhua Li, Jianying Wang, Xianbao Wang, Oxygen plasma treated graphene aerogel as a solar absorber for rapid and efficient solar steam generation, Carbon, **2018,** 130, 250. DOI:10.1016/j.carbon.2017.12.124.

[110] Wei Hao, Kevin Chiou, Yiming Qiao, Yanming Liu, Chengyi Son, Tao Deng, Jiaxing Huang, Crumpled graphene ball-based broadband solar absorbers, Nanoscale, **2018,** 10, 6306. DOI:10.1039/c7nr09556b.

[111] Xiuqiang Li, Weichao Xu, Mingyao Tang, Lin Zhou, Bin Zhu, Shining Zhu, Jia Zhu, Graphene oxide-based efficient and scalable solar desalination under one sun with a confined 2D water path, Proceedings of the National Academy of Sciences of the United States of America, **2016,** 113, 13953. DOI:10.1073/pnas.1613031113.

[112] Qisheng Jiang, Limei Tian, Kengku Liu, Sirimuvva Tadepalli, Ramesh Raliya, Pratim Biswas, Rajesh R. Naik, Srikanth Singamaneni, Bilayered biofoam for highly efficient solar steam generation, Advanced Materials, **2016,** 28, 9400. DOI:10.1002/adma.201601819.

[113] H. Ghasemi, G. Ni, A. M. Marconnet, J. Loomis, S. Yerci, N. Miljkovic, G. Chen, Solar steam generation by heat localization, Nature Communications, **2014,** 5, 4449. DOI:10.1038/ncomms5449.

[114] A. B. Kuzmenko, E. van Heumen, F. Carbone, D. van der Marel, Universal optical conductance of graphite, Physical Review Letters, **2008,** 100, 117401. DOI:10.1103/PhysRevLett.100.117401.

[115] Huaying Ren, Miao Tang, Baolu Guan, Kexin Wang, Jiawei Yang, Feifan Wang, Mingzhan Wang, Jingyuan Shan, Zhaolong Chen, Di Wei, Hailin Peng, Zhongfan Liu, Hierarchical graphene foam for efficient omnidirectional solar-thermal energy conversion, Advanced Materials, **2017,** 29. DOI:10.1002/adma.201702590.

[116] Liangliang Zhu, Minmin Gao, Connor Kang Nuo Peh, Xiaoqiao Wang, Ghim Wei Ho, Self-contained monolithic carbon sponges for solar-driven interfacial water evaporation distillation and electricity generation, Advanced Energy Materials, **2018,** 8. DOI:10.1002/aenm.201702149.

[117] Yuchao Wang, Lianbin Zhang, Peng Wang, Self-floating carbon nanotube membrane on macroporous silica substrate for highly efficient solar-driven interfacial water evaporation, ACS Sustainable Chemistry & Engineering, **2016,** 4, 1223. DOI:10.1021/acssuschemeng.5b01274.

[118] Sisi Cao, Priya Rathi, Xuanhao Wu, Deoukchen Ghim, Young-Shin Jun, Srikanth Ingamaneni, Cellulose nanomaterials in interfacial evaporators for desalination: A "natural" choice, Advanced Materials, **2021,** 33, e2000922. DOI:10.1002/adma.202000922.

[119] Haoran Li, Zhe Yan, Yan Li, Wenpeng Hong, Latest development in salt removal from solar-driven interfacial saline water evaporators: Advanced strategies and challenges, Water Research, **2020,** 177, 115770. DOI:10.1016/j.watres.2020.115770.

[120] Xiaozhen Hu, Jia Zhu, Tailoring aerogels and related 3D macroporous monoliths for interfacial solar vapor generation, Advanced Functional Materials, **2019,** 30. DOI:10.1002/adfm.201907234.

[121] Xiaona Gong, Liping Zhu, Youyi Xu, Baoku Zhu, Applications of carbon nanotubes (CNTs) in separation membrane materials, Membrane Science and Technology, **2011.** DOI:10.16159/j.cnki.issn1007-8924.2011.05.021.

[122] A. Kalra, S. Garde, G. Hummer, Osmotic water transport through carbon nanotube membranes, Proceedings of the National Academy of Sciences of the United States of America, **2003,** 100, 10175. DOI:10.1073/pnas.1633354100.

[123] Y. Chan, J. M. Hill, Modeling on ion rejection using membranes comprising ultra-small radii carbon nanotubes, The European Physical Journal B, **2012,** 85. DOI:10.1140/epjb/e2012-21029-0.

[124] Arsalan Khalid, Abdulhadi A. Al-Juhani, Othman Charles Al-Hamouz, Tahar Laoui, Zafarullah Khan, Mautaz Ali Atieh, Preparation and properties of nanocomposite polysulfone/multi-walled carbon nanotubes membranes for desalination, Desalination, **2015,** 367, 134. DOI:10.1016/j.desal.2015.04.001.

[125] MaryTheresa M. Pendergast, Eric M. V. Hoek, A review of water treatment membrane nanotechnologies, Energy & Environmental Science, **2011,** 6. DOI:10.1039/C0EE00541J.

[126] Xiaozhen Hu, Weichao Xu, Lin Zhou, Yingling Tan, Yang Wang, Shining Zhu, Jia Zhu, Tailoring graphene oxide-based aerogels for efficient solar steam generation under one Sun, Advanced Materials, **2017,** 29. DOI:10.1002/adma.201604031.

[127] Ken Gethard, Ornthida Sae-Khow, Somenath Mitra, Carbon nanotube enhanced membrane distillation for simultaneous generation of pure water and concentrating pharmaceutical waste, Separation and Purification Technology, **2012,** 90, 239. DOI:10.1016/j.seppur.2012.02.042.

[128] Madhuleena Bhadra, Sagar Roy, Somenath Mitra, Flux enhancement in direct contact membrane distillation by implementing carbon nanotube immobilized PTFE membrane, Separation and Purification Technology, **2016,** 161, 136. DOI:10.1016/j.seppur.2016.01.046.

[129] Bo Zhu, Hui Kou, Zixiao Liu, Zhaojie Wang, Daniel K. Macharia, Meifang Zhu, Binhe Wu, Xiaogang Liu, Zhigang Chen, Flexible and washable CNT-embedded PAN nonwoven fabrics for solar-enabled evaporation and desalination of seawater, ACS Applied Materials & Interfaces, **2019,** 11, 35005. DOI:10.1021/acsami.9b12806.

[130] Zhejun Liu, Haomin Song, Dengxin Ji, Chenyu Li, Alec Cheney, Youhai Liu, Nan Zhang, Xie Zeng, Borui Chen, Jun Gao, Yuesheng Li, Xiang Liu, Diana Aga, Suhua Jiang, Zongfu Yu, Qiaoqiang Gan, Extremely cost-effective and efficient solar vapor generation under nonconcentrated illumination using thermally isolated black paper, Glob Chall, **2017,** 1, 1600003. DOI:10.1002/gch2.201600003.

[131] Lin Zhou, Yingling Tan, Jingyang Wang, Weichao Xu, Ye Yuan, Wenshan Cai, Shining Zhu, Jia Zhu, 3D self-assembly of aluminium nanoparticles for plasmon-enhanced solar desalination, Nature Photonics, **2016,** 10, 393. DOI:10.1038/nphoton.2016.75.

[132] M. F. De Volder, S. H. Tawfick, R. H. Baughman, A. J. Hart, Carbon nanotubes: Present and future commercial applications, Science, **2013,** 339, 535. DOI:10.1126/science.1222453.

[133] K. Koga, G. T. Gao, H. Tanaka, X. C. Zeng, Formation of ordered ice nanotubes inside carbon nanotubes, Nature, **2001,** 412, 802. DOI:10.1038/35090532.

[134] Jian Liu, Guosheng Shi, Pan Guo, Jinrong Yang, Haiping Fang, Blockage of water flow in carbon nanotubes by ions due to interactions between cations and aromatic rings, Physical Review Letters, **2015,** 115, 164502. DOI:10.1103/PhysRevLett.115.164502.

[135] Hubiao Huang, Zhigong Song, Ning Wei, Li Shi, Yiyin Mao, Yulong Ying, Luwei Sun, Zhiping Xu, Xinsheng Peng, Ultrafast viscous water flow through nanostrand-channelled graphene oxide membranes, Nature Communications, **2013,** 4, 2979. DOI:10.1038/ncomms3979.

[136] Kunli Goh, Wenchao Jiang, Huseyin Enis Karahan, Shengli Zhai, Li Wei, Dingshan Yu, Rong Wang, Anthony G. Fane, Yuan Chen, All-carbon nanoarchitectures as high-performance separation membranes with superior stability, Advanced Functional Materials, **2015.** DOI:10.1002/adfm.201502955.

[137] Wei-Song Hung, Chi-Hui Tsou, Manuel De Guzman, Quan-Fu An, Ying-Ling Liu, Ya-Ming Zhang, Chien-Chieh Hu, Kueir-Rarn Lee, Juin-Yih Lai, Cross-linking with diamine monomers to prepare composite graphene oxide-framework membranes with varying d-spacing, Chemistry of Materials, **2014,** 26, 2983. DOI:10.1021/cm5007873.

[138] Y. Su, V. G. Kravets, S. L. Wong, J. Waters, A. K. Geim, R. R. Nair, Impermeable barrier films and protective coatings based on reduced graphene oxide, Nature Communications, **2014,** 5, 4843. DOI:10.1038/ncomms5843.

[139] Pengzhan Sun, Kunlin Wang, Hongwei Zhu, Recent developments in graphene-based membranes: Structure, mass-transport mechanism and potential applications, Advanced Materials, **2016,** 28, 2287. DOI:10.1002/adma.201502595.

[140] Yongil Kim, Jae-Kwang Kim, Christoph Vaalma, Geun Hyeong Bae, Guk-Tae Kim, Stefano Passerini, Youngsik Kim, Optimized hard carbon derived from starch for rechargeable seawater batteries, Carbon, **2018,** 129, 564. DOI:10.1016/j.carbon.2017.12.059.

For Product Safety Concerns and Information please contact our EU
representative GPSR@taylorandfrancis.com
Taylor & Francis Verlag GmbH, Kaufingerstraße 24, 80331 München, Germany